Evolution of Stars and Stellar Populations

Evolution of Stars and Stellar Populations

Maurizio Salaris
Astrophysics Research Institute,
Liverpool John Moores University, UK

Santi Cassisi
INAF-Astronomical Observatory of Collurania,
Teramo, Italy

John Wiley & Sons, Ltd

Copyright © 2005   John Wiley & Sons Ltd, The Atrium, Southern Gate, Chichester,
West Sussex PO19 8SQ, England

Telephone   (+44) 1243 779777

Email (for orders and customer service enquiries): cs-books@wiley.co.uk
Visit our Home Page on www.wiley.com

Reprinted with corrections March 2008

Other Wiley Editorial Offices

John Wiley & Sons Inc., 111 River Street, Hoboken, NJ 07030, USA

Jossey-Bass, 989 Market Street, San Francisco, CA 94103-1741, USA

Wiley-VCH Verlag GmbH, Boschstr. 12, D-69469 Weinheim, Germany

John Wiley & Sons Australia Ltd, 33 Park Road, Milton, Queensland 4064, Australia

John Wiley & Sons (Asia) Pte Ltd, 2 Clementi Loop #02-01, Jin Xing Distripark, Singapore 129809

John Wiley & Sons Canada Ltd, 22 Worcester Road, Etobicoke, Ontario, Canada M9W 1L1

Wiley also publishes its books in a variety of electronic formats. Some content that appears in print
may not be available in electronic books.

Library of Congress Cataloging in Publication Data

Salaris, Maurizio.
    Evolution of stars and stellar populations / Maurizio Salaris, Santi Cassisi.
        p.   cm.
    Includes bibliographical references and index.
    ISBN-13 978-0-470-09219-X (cloth : alk. paper)   ISBN-10 0-470-09219-X (cloth : alk. paper)
    ISBN-13 978-0-470-09220-3 (pbk. : alk. paper)   ISBN-10 0-470-09220-3 (pbk. : alk. paper)
    1. Stars—Evolution. 2. Stars—Populations. 3. Galaxies—Evolution. I. Cassisi, Santi. II. Title.
    QB806.S25 2005
    523.8′8—dc22
                                                                        2005021402

British Library Cataloguing in Publication Data

A catalogue record for this book is available from the British Library

ISBN-13 978-0-470-09219-X (HB)   978-0-470-09220-3 (PB)
ISBN-10 0-470-09219-X (HB)   0-470-09220-3 (PB)

Typeset in 10.5/12.5pt Times by Integra Software Services Pvt. Ltd, Pondicherry, India

To Suzanne and my parents

To all the people who really loved me

Contents

Preface   xi

1 Stars and the Universe   **1**
  1.1 Setting the stage   1
  1.2 Cosmic kinematics   5
    1.2.1 Cosmological redshifts and distances   8
  1.3 Cosmic dynamics   13
    1.3.1 Histories of $R(t)$ 14
  1.4 Particle- and nucleosynthesis   17
  1.5 CMB fluctuations and structure formation   24
  1.6 Cosmological parameters   25
  1.7 The inflationary paradigm   26
  1.8 The role of stellar evolution   28

2 Equation of State of the Stellar Matter   **31**
  2.1 Physical conditions of the stellar matter   31
    2.1.1 Fully ionized perfect gas   35
    2.1.2 Electron degeneracy   38
    2.1.3 Ionization   41
    2.1.4 Additional effects   44

3 Equations of Stellar Structure   **49**
  3.1 Basic assumptions   49
    3.1.1 Continuity of mass   50
    3.1.2 Hydrostatic equilibrium   50
    3.1.3 Conservation of energy   52
    3.1.4 Energy transport   52
    3.1.5 The opacity of stellar matter   66
    3.1.6 Energy generation coefficient   68
    3.1.7 Evolution of chemical element abundances   83
    3.1.8 Virial theorem   86
    3.1.9 Virial theorem and electron degeneracy   89
  3.2 Method of solution of the stellar structure equations   90
    3.2.1 Sensitivity of the solution to the boundary conditions 97
    3.2.2 More complicated cases   98

3.3   Non-standard physical processes   99
    3.3.1   Atomic diffusion and radiative levitation   100
    3.3.2   Rotation and rotational mixings   102

4 Star Formation and Early Evolution   105
4.1   Overall picture of stellar evolution   105
4.2   Star formation   106
4.3   Evolution along the Hayashi track   110
    4.3.1   Basic properties of homogeneous, fully convective stars 110
    4.3.2   Evolution until hydrogen burning ignition   114

5 The Hydrogen Burning Phase   117
5.1   Overview   117
5.2   The nuclear reactions   118
    5.2.1   The p–p chain   118
    5.2.2   The CNO cycle   119
    5.2.3   The secondary elements: the case of ^{2}H and ^{3}He   121
5.3   The central H-burning phase in low main sequence (LMS) stars   123
    5.3.1   The Sun   125
5.4   The central H-burning phase in upper main sequence (UMS) stars   128
5.5   The dependence of MS tracks on chemical composition and
      convection efficiency   133
5.6   Very low-mass stars   136
5.7   The mass–luminosity relation   138
5.8   The Schönberg–Chandrasekhar limit   140
5.9   Post-MS evolution   141
    5.9.1   Intermediate-mass and massive stars   141
    5.9.2   Low-mass stars   142
    5.9.3   The helium flash   148
5.10 Dependence of the main RGB features on physical and chemical
      parameters   149
    5.10.1 The location of the RGB in the H–R diagram   150
    5.10.2 The RGB bump luminosity   151
    5.10.3 The luminosity of the tip of the RGB   152
5.11 Evolutionary properties of very metal-poor stars   155

6 The Helium Burning Phase   161
6.1   Introduction   161
6.2   The nuclear reactions   161
6.3   The zero age horizontal branch (ZAHB)   163
    6.3.1   The dependence of the ZAHB on various physical parameters 165
6.4   The core He-burning phase in low-mass stars   167
    6.4.1   Mixing processes   167

6.5 The central He-burning phase in more massive stars 173
6.5.1 The dependence of the blue loop on various physical parameters 175
6.6 Pulsational properties of core He-burning stars 179
6.6.1 The RR Lyrae variables 181
6.6.2 The classical Cepheid variables 183

7 The Advanced Evolutionary Phases **187**
7.1 Introduction 187
7.2 The asymptotic giant branch (AGB) 187
7.2.1 The thermally pulsing phase 189
7.2.2 On the production of s-elements 194
7.2.3 The termination of the AGB evolutionary phase 195
7.3 The Chandrasekhar limit and the evolution of stars with large CO cores 198
7.4 Carbon–oxygen white dwarfs 199
7.4.1 Crystallization 206
7.4.2 The envelope 210
7.4.3 Detailed WD cooling laws 212
7.4.4 WDs with other chemical stratifications 213
7.5 The advanced evolutionary stages of massive stars 214
7.5.1 The carbon-burning stage 217
7.5.2 The neon-burning stage 219
7.5.3 The oxygen-burning stage 220
7.5.4 The silicon-burning stage 221
7.5.5 The collapse of the core and the final explosion 222
7.6 Type Ia supernovae 224
7.6.1 The Type Ia supernova progenitors 225
7.6.2 The explosion mechanisms 229
7.6.3 The light curves of Type Ia supernovae and their use as distance indicators 230
7.7 Neutron stars 233
7.8 Black holes 236

8 From Theory to Observations **239**
8.1 Spectroscopic notation of the stellar chemical composition 239
8.2 From stellar models to observed spectra and magnitudes 241
8.2.1 Theoretical versus empirical spectra 248
8.3 The effect of interstellar extinction 250
8.4 K-correction for high-redshift objects 253
8.5 Some general comments about colour–magnitude diagrams (CMDs) 254

9 Simple Stellar Populations **259**
9.1 Theoretical isochrones 259
9.2 Old simple stellar populations (SSPs) 264
9.2.1 Properties of isochrones for old ages 264
9.2.2 Age estimates 268

9.2.3   Metallicity and reddening estimates   281
9.2.4   Determination of the initial helium abundance   284
9.2.5   Determination of the initial lithium abundance   287
9.2.6   Distance determination techniques   289
9.2.7   Luminosity functions and estimates of the IMF   301
9.3   Young simple stellar populations   304
9.3.1   Age estimates   304
9.3.2   Metallicity and reddening estimates   309
9.3.3   Distance determination techniques   310

10 Composite Stellar Populations   **315**
10.1 Definition and problems   315
10.2 Determination of the star formation history (SFH)   320
10.3 Distance indicators   327
10.3.1 The planetary nebula luminosity function (PNLF)   329

11 Unresolved Stellar Populations   **331**
11.1 Simple stellar populations   331
11.1.1 Integrated colours   334
11.1.2 Absorption-feature indices   341
11.2 Composite stellar populations   347
11.3 Distance to unresolved stellar populations   347

Appendix I: Constants   **351**

Appendix II: Selected Web Sites   **353**

References   **357**

Index   **369**

Preface

The theory of stellar evolution is by now well established, after more than half a century of continuous development, and its main predictions confirmed by various empirical tests. As a consequence, we can now use its results with some confidence, and obtain vital information about the structure and evolution of the universe from the analysis of the stellar components of local and high redshift galaxies.

A wide range of techniques developed in the last decades make use of stellar evolution models, and are routinely used to estimate distances, ages, star formation histories and the chemical evolution of galaxies; obtaining this kind of information is, in turn, a necessary first step to address fundamental cosmological problems like the dynamical status and structure of the universe, the galaxy formation and evolution mechanisms. Due to their relevance, these methods rooted in stellar evolution should be part of the scientific background of any graduate and undergraduate astronomy and astrophysics student, as well as researchers interested not only in stellar modeling, but also in galaxy and cosmology studies.

In this respect, we believe there is a gap in the existing literature at the level of senior undergraduate and graduate textbooks that needs to be filled. A number of good books devoted to the theory of stellar evolution do exist, and a few discussions about the application of stellar models to cosmological problems are scattered in the literature (especially the methods to determine distances and ages of globular clusters). However, an organic and self-contained presentation of both topics, that is also able to highlight their intimate connections, is still lacking. As an example, the so-called 'stellar population synthesis techniques' – a fundamental tool for studying the properties of galaxies – are hardly discussed in any existing stellar evolution textbook.

The main aim of this book is to fill this gap. It is based on the experience of one of us (MS) in developing and teaching a third year undergraduate course in advanced stellar astrophysics and on our joint scientific research of the last 15 years. We present, in a homogeneous and self-contained way, first the theory of stellar evolution, and then the related techniques that are widely applied by researchers to estimate cosmological parameters and study the evolution of galaxies.

The first chapter introduces the standard Big Bang cosmology and highlights the role played by stars within the framework of our currently accepted cosmology. The two following chapters introduce the basic physics needed to understand how stars work, the set of differential equations that describes the structure and evolution of stars, and the numerical techniques to solve them.

Chapters 4 to 7 present both a qualitative and quantitative picture of the life cycle of single stars (although we give some basic information about the evolution of interacting binary systems when dealing with Type Ia supernovae progenitors) from their formation to the final stages. The emphasis in our presentation is placed on those properties that are needed to understand and apply the methods discussed in the rest of the book, that is, the evolution with time of the photometric and chemical properties (i.e. evolution of effective temperatures, luminosities, surface chemical abundances) of stars, as a function of their initial mass and chemical composition.

The next chapter describes the steps (often missing in stellar evolution books) necessary to transform the results from theoretical models into observable properties. Finally, Chapters 9 to 11 present an extended range of methods that can be applied to different types of stellar populations – both resolved and unresolved – to estimate their distances, ages, star formation histories and chemical evolution with time, building on the theory of stellar evolution we have presented in the previous chapters.

We have included a number of references which are not meant to be a totally comprehensive list, but should be intended only as a first guide through the vast array of publications on the subject.

This book has greatly benefited from the help of a large number of friends and colleagues. First of all, special thanks go to David Hyder (Liverpool John Moores University) and Lucio Primo Pacinelli (Astronomical Observatory of Collurania) for their invaluable help in producing many of the figures for this book. Katrina Exter is warmly thanked for her careful editing of many chapters; Antonio Aparicio, Giuseppe Bono, Daniel Brown, Vittorio Castellani, Carme Gallart, Alan Irwin, Marco Limongi, Marcella Marconi and Adriano Pietrinferni are acknowledged for many discussions, for having read and commented on various chapters of this book and helped with some of the figures. We are also indebted to Leo Girardi, Phil James, Kevin Krisciunas, Bruno Leibundgut, Luciano Piersanti and Oscar Straniero for additional figures included in the book.

Sue Percival and Phil James are warmly acknowledged for their encouragement during the preparation of the manuscript, Suzanne Amin and Anna Piersimoni for their endless patience during all these months. We are also deeply indebted to Achim Weiss and the Max Planck Institut für Astrophysik (MS), Antonio Aparicio and the Instituto de Astrofisica de Canarias (SC) for their invitation and hospitality. During our stays at those institutes a substantial part of the manuscript was prepared. Finally, we wish to send a heartfelt thank-you to all colleagues with whom we have worked in the course of these wonderful years of fruitful scientific research.

Liverpool   **Maurizio Salaris**
Teramo   **Santi Cassisi**
March 2005

1 Stars and the Universe

1.1 Setting the stage

Stars are not distributed randomly in the universe, but are assembled through gravitational interactions into galaxies. Typical distances between stars in a given galaxy are of the order of 1 parsec (pc) whereas distances between galaxies are typically of the order of 100 kpc–1 Mpc (1 pc is the distance at which the semi-major axis of Earth orbit subtends an angle of 1 arcsecond; this corresponds to $\sim$3.26 light years, where one light year is the distance travelled by light in one year, i.e. 9.4607×10^{17} cm).

There are three basic types of galaxies: spirals, ellipticals and irregulars (see Figure 1.1). Spiral galaxies (our galaxy, the Milky Way, is a spiral galaxy) constitute more than half of the bright galaxies that we observe within $\sim$100 Mpc of the Sun. They generally comprise a faint spherical halo, a bright nucleus (or bulge) and a disk that contains luminous spiral arms; spirals have typical masses of the order of $10^{11} M_{\odot}$ ($1 M_{\odot}$ denotes one solar mass, i.e. 1.989×10^{33} g). Spirals are divided into normal and barred spirals, depending on whether the spiral arms emerge from the nucleus or start at the end of a bar springing symmetrically from the nucleus. Dust and young stars are contained in the disk whereas the nucleus and halo are populated by older stars. Elliptical galaxies account for $\sim$10 per cent of the bright galaxies, have an elliptical shape, no sign of a spiral structure nor of dust and young stars, a mass range between $\sim$10^5 and $\sim$10^{12} $M_{\odot}$, and in general resemble the nuclei of spirals. There is no sign of significant rotational motions of the stars within ellipticals, whereas stars in the disks of spirals show ordered rotational motion.

These two broad types of galaxies are bridged morphologically by the so-called lenticular galaxies, which make up about 20 per cent of the galaxies, and look like elongated ellipticals without bars and spiral structure. The third broad group of galaxies are the irregulars, that show no regular structure, no rotational symmetry and are relatively rare and faint.

Evolution of Stars and Stellar Populations   Maurizio Salaris and Santi Cassisi
© 2005 John Wiley & Sons, Ltd

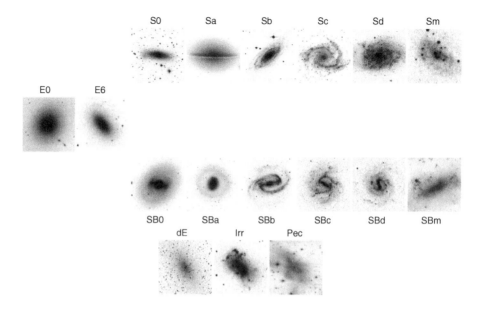

Figure 1.1 The so-called *tuning fork diagram*, i.e. the galaxy morphological classification. Elliptical galaxies are denoted by E (the various subclasses are denoted by the approximate value of the ellipticity) spirals by S, barred spirals by SB; examples of Dwarf elliptical (dE), Irregular (Irr) and Peculiar (Pec) galaxies are also displayed (courtesy of P. James)

Many galaxies show various types of non-thermal emission over a large wavelength range, from radio to X-ray, and are called active galaxies. These active galaxies display a large range of properties that can probably be explained invoking one single mechanism (possibly related to accretion of matter onto a black hole); the difference in their properties is most likely due to the fact that we are observing the same kind of object at different angles, and therefore we see radiation from different regions within the galaxy. Examples of active galaxies are the Seyfert galaxies, radio galaxies, BL Lac objects and quasars. There are also so-called starburst galaxies, e.g. galaxies displaying a mild form of activity, and showing a strong burst of star formation.

For many years it was believed that galaxies extend as far as they are visible. However, starting from the 1970s, the orbits of neutral hydrogen clouds circling around individual spiral galaxies provided rotation curves (e.g. rotational velocity as a function of the distance d from the galactic centre) that, instead of dropping as $\sqrt{d}$ beyond the edge of the visible matter distribution (as expected from Keplerian orbits after the limit of the mass distribution is reached) show a flat profile over large distances well beyond this limit. This can be explained only by a steady increase with distance of the galaxy mass, beyond the edge of the visible mass distribution. This dark matter reveals its presence only through its gravitational pull, since it does not produce any kind of detectable radiation.

Even galaxies are not distributed randomly in the universe, but are aggregated in pairs or groups, which in turn are often gathered into larger clusters of galaxies. Our galaxy (often referred to as the Galaxy) belongs to the so-called Local Group of galaxies, that includes about 20 objects (mainly small) among them the Large Magellanic Cloud (LMC) the Small Magellanic Cloud (SMC) and Andromeda (M31). The nearest cluster of galaxies is the Virgo cluster (at a distance of about 20 Mpc). Further away are other galaxy clusters, among them the Coma cluster, located at a distance of about 100 Mpc, that contains thousands of objects. Deep galaxy surveys (e.g. the APM, COSMOS, 2dF and SDSS surveys) have studied and are still probing the distribution of galaxies in the universe, and have revealed even more complex structures, like filaments, sheets and superclusters, that are groupings of clusters of galaxies.

Dark matter is also found within clusters of galaxies. This can be inferred studying the X-ray emission of the hot ionized intracluster gas that is accelerated by the gravitational field of the cluster. A rough comparison of visible and dark matter contribution to the total matter density of the universe tells us that about 90 per cent of the matter contained in the universe is dark.

It is evident from this brief description that overall the universe appears to be clumpy, but the averaged properties in volumes of space of the order of 100 Mpc are smoother, and the local inhomogeneities can be treated as perturbations to the general homogeneity of the universe.

The dynamical status of the universe is revealed by spectroscopic observations of galaxies. The observed redshift of their spectral lines shows that overall galaxies are receding from us (in the generally accepted assumption that the observed redshift is due to the Doppler effect) with a velocity v that increases linearly with their distance D, so

$$v = H_0 \times D,$$

as first discovered by Hubble and Humason during the 1920s (hence the name of Hubble law for this relationship). The constant H_0 is called the Hubble constant. Taken at face value this relationship seems to locate us in a privileged point, from where all galaxies are escaping. However, if one considers the overall homogeneity of the universe, the same Hubble law has to apply to any other location and the phenomenon of the recession of the galaxies might be looked upon as an expansion of the universe as a whole; a useful and widely used analogy is that of the two-dimensional surface of a balloon that is being inflated. If the galaxies are points drawn on the surface of the balloon, they will appear to be receding from each other in the same way as the Hubble law, irrespective of their location.

Superimposed on the general recession of galaxies are local peculiar velocities due to the gravitational pull generated by the local clumpiness of the universe. For example, the Milky Way and M31 are moving towards each other at a speed of about 120 km s^{-1}, and the Local Group, is approaching the Virgo Cluster at a speed of $\sim$170 km s^{-1}. On a larger scale, the Local Group, Virgo Cluster and thousands of other galaxies are streaming at a speed of about $600\,\mathrm{km\,s^{-1}}$ towards the so called Great Attractor, a concentration of mass in the Centaurus constellation, located at a distance of the order of $70\,\mathrm{Mpc}$. These peculiar velocities become negligible with respect to the general recession of the galaxies (Hubble flow) when considering increasingly distant objects, for which the recession velocity predicted by the Hubble law is increasingly high.

Another discovery of fundamental importance for our understanding of the universe was made serendipitously in 1965 by Penzias and Wilson. Observations of electromagnetic radiation in a generic frequency interval reveal peaks associated with discrete sources – i.e. stars or galaxies – located at specific directions; when these peaks are eliminated there remains a dominant residual radiation in the microwave frequency range. The spectrum of this cosmic microwave background (CMB) radiation is extremely well approximated by that of a black body with a temperature of 2.725 K. After removing the effect of the local motion of the Sun and of our galaxy, the CMB temperature is to a first approximation constant when looking at different points in the sky, suggesting a remarkable isotropy which is hard to explain in terms of residual emission by discrete sources. From the CMB temperature one easily obtains the energy density associated to the CMB, ϵ_{CMB}, given that $\rho_{\mathrm{CMB}} = \epsilon_{\mathrm{CMB}}/c^2 \sim 4.64 \times 10^{-34}\,\mathrm{g\,cm^3}$

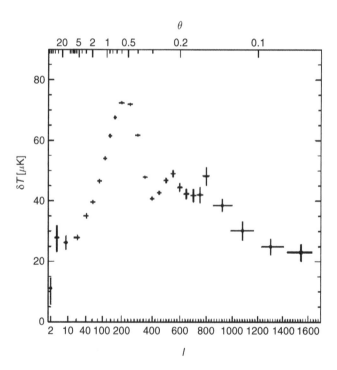

Figure 1.2 Plot of the CMB temperature fluctuations (in units of 10^{-6} K) as a function of the angular scale in degrees (upper horizontal axis) and the so-called wave number $l \sim \pi/\theta$ (lower horizontal axis); this is also called the power spectrum of the CMB fluctuations

(c denotes the speed of light, $2.998 \times 10^{10}\,\mathrm{cm\,s^{-1}}$). This CMB photon density is the dominant component of the present radiation density in the universe; a rough comparison of ρ_{CMB} with the present matter density ρ shows that at the present time the density associated with the photons is about three orders of magnitude lower than the matter density, including the dark matter contribution.

In 1992 the COBE satellite first discovered tiny variations δT of the CMB temperature, of the order of $\delta T/T \sim 10^{-5}$ (where T is the global mean of the CMB temperature) when looking at different points in the sky. By computing the average over the sky of the ratio $\delta T/T$ (temperature fluctuation) measured from any two points separated by an angle θ, one obtains what is called the angular power spectrum of the CMB temperature anisotropies, displayed in Figure 1.2. This power spectrum shows the existence of a series of peaks located at specific angular scales.

A comprehensive theory for the structure and evolution of the universe must be able to explain the basic observations outlined above in terms of evolutionary processes rooted in accepted physics theories. The following sections introduce briefly the Hot Big Bang theory, which is the presently widely accepted cosmological theory. Detailed presentations of cosmology at various levels of complexity can be found in [11], [57], [118] and [142].

1.2 Cosmic kinematics

A cornerstone of the Big Bang theory is the so-called cosmological principle: it states that the large-scale structure of the universe is homogeneous and isotropic. Homogeneity means that the physical properties of the universe are invariant by translation; isotropy means that they are also rotationally invariant. Both these properties can be applied only considering average properties of large volumes of space, where the local structures (galaxies, clusters of galaxies) are smeared out over the averaging volumes.

As discussed before, the adequacy of the cosmological principle can be empirically verified by studying the distribution of clusters of galaxies on scales of the order of 100 Mpc and by the isotropy of the CMB. Locally the universe is clumpy, but this clumpiness disappears when averaging the matter density over large enough volumes. In this way the local clumpiness is treated as a perturbation to the general smoothness of the universe. The universe is then treated as a fluid whose particles are galaxies, moving according to the Hubble law; within this picture of a cosmic fluid the cosmological principle implies that every co-moving observer (i.e. moving with the Hubble flow) in the cosmic fluid has the same history.

A first step when discussing events happening in the universe is to set up an appropriate coordinate system. For the time coordinate a natural choice is to use standard clocks co-moving with the cosmic fluid, that will define a cosmic time t; an operational way to synchronize t for co-moving observers at different locations is to set t to the same value when each observer sees that a property of the cosmic fluid, i.e. the average local density of matter ρ, has reached a certain agreed value. After synchronization, by virtue of the cosmological principle, the two observers must be able to measure exactly the same value (possibly different from the one at the time of synchronization) of that property whenever their clocks show the same time.

As for the three spatial coordinates, the cosmological principle greatly restricts the possible geometries. The assumption of homogeneity and isotropy requires that the tridimensional space has a single curvature, i.e. it must have the same value at all positions, but can in principle depend on time. The space–time interval ds between two events in an homogeneous and isotropic static space can be written as follows

$$ds^2 = c^2 dt^2 - \left(\frac{dr^2}{1 - Kr^2} + r^2 d\theta^2 + r^2 \sin^2 \theta d\phi^2 \right)$$

where K is the spatial curvature, dt the cosmic time separation, r the radial coordinate and θ and ϕ the polar and azimuthal angles in spherical coordinates, respectively. The expansion (or contraction) of the universe can be accounted for by redefining the radial coordinate r as $r \equiv R(t)\chi - \chi$ being dimensionless – and the curvature K as $K(t) \equiv k/R(t)^2$. The constant k and coordinate χ are defined in a way that $k = +1$ for a positive spatial curvature, $k = 0$ for a flat space and $k = -1$ for a negative curvature. $R(t)$ is the so-called cosmic scale factor, that has the dimension of a distance and is dependent on the cosmic time t. With these substitutions one obtains the so-called Friedmann–Robertson–Walker (FRW) metric:

$$ds^2 = c^2 dt^2 - R(t)^2 \left(\frac{d\chi^2}{1 - k\chi^2} + \chi^2 d\theta^2 + \chi^2 \sin^2 \theta d\phi^2 \right) \tag{1.1}$$

The values of the three spatial coordinates χ, θ and ϕ are constant for an observer at rest with respect to the expansion of the cosmic fluid. One can easily see that the factor $R(t)$ in Equation (1.1) allows a scaling of the spatial surfaces that depends only on time, thus preserving the homogeneity and isotropy dictated by the cosmological principle. It is important to stress that it is only by virtue of the cosmological principle that we can uniquely define a four-dimensional coordinate system co-moving with the cosmic flow. As an example, the definition of cosmic time would be impossible in a universe without homogeneity and isotropy, because we could not synchronize the various clocks using mean properties (that would not be the same everywhere at a given time t) of the cosmic flow.

The geometrical properties of the three-dimensional space determined by the value of k can be briefly illustrated as follows. Let us consider at a cosmic instant t a sphere with centre at an arbitrary origin where $\chi = 0$, and surface located at a fixed value χ; the difference between the coordinates of the centre of the sphere and the surface is equal to $r = R(t)\chi$. The area A of the spherical surface of coordinate radius $r = R(t)\chi$ is, by definition, $A = 4\pi r^2 = 4\pi R(t)^2 \chi^2$. The physical radius R_p of the spherical surface is the distance between the centre and surface of the sphere measured with a standard rod at the same cosmic time t. This means that one has to determine the interval Δs^2 between the two events assuming $dt = 0$, so that $R_p = \sqrt{-\Delta s^2}$. From the FRW metric one obtains

$$R_p = R(t) \int_0^\chi \frac{d\chi}{\sqrt{1 - k\chi^2}} \tag{1.2}$$

R_p is equal to $R(t)\sin^{-1}\chi$, $R(t)\chi$ and $R(t)\sinh^{-1}\chi$ when $k = 1, 0$ and -1, respectively. When $k = 0$ one has $\chi = R_p/R(t)$, and $A = 4\pi R_p^2$, i.e. r is equal to R_p and the area A increases as R_p^2, as in Euclidean geometry. When $k = +1$ one has $r = R(t)\sin(R_p/R(t))$ and $A = 4\pi R(t)^2 \sin^2(R_p/R(t))$, which reaches a maximum value $A = 4\pi R(t)^2$ when $R_p = (\pi/2)R(t)$, then becomes zero when $R_p = \pi R(t)$ and has in general a periodic behaviour. This means that in the case of $k = +1$ space is closed and the periodicity corresponds to different circumnavigations. In the case of $k = -1$ then $A = 4\pi R(t)^2 \sinh^2(R_p/R(t))$, which increases with R_p faster than in the case of a Euclidean space.

It is easy to see how simply $R(t)$ describes the observed expansion of the universe. Let us set $\chi = 0$ at the location of our own galaxy, that is approximately co-moving with the local cosmic fluid (hence its spatial coordinate does not change with time) and consider another galaxy – also at rest with respect to the expansion of the universe – whose position is specified by a value χ of the radial coordinate (the angles θ and ϕ are assumed to be equal to zero for both galaxies). Its proper distance (defined in the same way as for the proper radius R_p discussed before) D at a given cosmic time t is given by:

$$D = R(t) \int_0^\chi \frac{d\chi}{\sqrt{1 - k\chi^2}}$$

As in the case of Equation (1.2) D is equal to $R(t)\sin^{-1}\chi$, $R(t)\chi$ and $R(t)\sinh^{-1}\chi$ when $k = 1, 0$ and -1, respectively. The velocity v of the recession of the galaxy due to the expansion of the universe is

$$v = \frac{dD}{dt} = \frac{dR(t)}{dt} \int_0^\chi \frac{d\chi}{\sqrt{1 - k\chi^2}} = \frac{dR(t)}{dt} \frac{1}{R(t)} D$$

This looks exactly like the Hubble law; in fact, by writing

$$H(t) = \frac{dR(t)}{dt} \frac{1}{R(t)} \tag{1.3}$$

we obtain

$$v = H(t) \times D$$

$H(t)$ corresponds to the Hubble constant and one can notice that its value can change with cosmic time. The value of $H(t)$ determined at the present time is denoted as H_0.

This result is clearly independent of the location of the origin for the radial coordinate χ, since any position in the universe is equivalent according to the cosmological principle. It is important to notice that locally, e.g. within the solar system or within a given galaxy, one cannot see any effect of the cosmic expansion, since the local gravitational effects dominate. For distances large enough $(D > c/H(t))$ the last equation predicts recession velocities larger than the speed of light, an occurrence that seems to go against special relativity. The contradiction is, however, only apparent, given that galaxies recede from us faster than the speed of light (superluminal recession) because of the expansion of space; locally, they are at rest or moving in their local inertial reference frame with peculiar velocities $\ll c$.

In the following section we will briefly describe the observational counterpart of $v = H(t) \times D$ and show how it probes the evolution of the kinematic status of the universe.

1.2.1 Cosmological redshifts and distances

What we measure to estimate the recession velocity of galaxies is a redshift z, that can be related to the change of $R(t)$ with time. Consider light reaching us (located at $\chi = 0$) from a galaxy at a radial coordinate χ. Two consecutive maxima of the electromagnetic wave are emitted at times t_e and $t_e + \delta t_e$ and received at times t_0 and $t_0 + \delta t_0$; if $\delta t_e = \delta t_0$ we would not observe any redshift since the wavelength of the electromagnetic wave is given by the spatial distance between the two consecutive maxima, i.e. the observed wavelength is $\lambda_0 = c\delta t_0$, and the emitted one is $\lambda_e = c\delta t_e$. We will now find the relationship between δt_e and δt_0. Since $ds = 0$ for light, we have

$$\int_{t_e}^{t_0} \frac{dt}{R(t)} = \frac{1}{c} \int_0^{\chi} \frac{d\chi}{\sqrt{1 - k\chi^2}}$$

$$\int_{t_e + \delta t_e}^{t_0 + \delta t_0} \frac{dt}{R(t)} = \frac{1}{c} \int_0^{\chi} \frac{d\chi}{\sqrt{1 - k\chi^2}}$$

for the first and second maximum, respectively. The right-hand side of both equations is the same, therefore we can write

$$\int_{t_e + \delta t_e}^{t_0 + \delta t_0} \frac{dt}{R(t)} - \int_{t_e}^{t_0} \frac{dt}{R(t)} = 0$$

The first term on the left-hand side of the previous equation can be rewritten as

$$\int_{t_e + \delta t_e}^{t_0 + \delta t_0} \frac{dt}{R(t)} = \int_{t_e}^{t_0} \frac{dt}{R(t)} + \int_{t_0}^{t_0 + \delta t_0} \frac{dt}{R(t)} - \int_{t_e}^{t_e + \delta t_e} \frac{dt}{R(t)}$$

and therefore

$$\int_{t_0}^{t_0 + \delta t_0} \frac{dt}{R(t)} - \int_{t_e}^{t_e + \delta t_e} \frac{dt}{R(t)} = 0$$

The intervals δt_e and δt_0 are negligible compared with the expansion timescale of the universe, and therefore $R(t)$ is to a good approximation constant during these two time intervals; inserting this condition into the previous equation provides

$$\frac{\delta t_0}{R(t_0)} = \frac{\delta t_e}{R(t_e)}$$

The redshift $z = (\lambda_0 - \lambda_e)/\lambda_e$ is therefore given by

$$z = \frac{\delta t_0}{\delta t_e} - 1 = \frac{R(t_0)}{R(t_e)} - 1 \tag{1.4}$$

In an expanding universe $z > 0$ (since $R(t_0) > R(t_e)$) as observed. If the redshift is small enough, i.e. t_e is close to t_0 in cosmological terms, we can expand $R(t_e)$ about t_0 using the Taylor formula, and retain only the terms up to the second order:

$$R(t_e) = R(t_0) + (t_e - t_0)\frac{dR(t_0)}{dt} + \frac{1}{2}(t_e - t_0)^2\frac{d^2R(t_0)}{dt^2}$$

We can now define H_0 as

$$H_0 \equiv H(t_0) = \frac{dR(t_0)}{dt}\frac{1}{R(t_0)}$$

i.e. the present value of the Hubble constant, and the so-called deceleration parameter

$$q_0 \equiv -\frac{d^2R(t_0)}{dt^2}\frac{1}{R(t_0)H_0^2} \tag{1.5}$$

Both H_0 and q_0 are related to the present rate of expansion of the universe. H_0 measures the actual expansion rate, whilst q_0 is positive if the expansion is slowing down (hence the name deceleration parameter) or negative if the opposite is true. With these definitions the second-order expansion of $R(t_e)$ can be rewritten as

$$R(t_e) = R(t_0)\left[1 + H_0(t_e - t_0) - \frac{1}{2}q_0H_0^2(t_e - t_0)^2\right]$$

and after additional manipulations one obtains the following useful results:

$$z = H_0(t_0 - t_e) + H_0^2(t_0 - t_e)^2\left(1 + \frac{1}{2}q_0\right) \tag{1.6}$$

$$t_0 - t_e = \frac{1}{H_0}\left[z - \left(1 + \frac{1}{2}q_0\right)z^2\right] \tag{1.7}$$

$$\chi = \frac{c}{R(t_0)H_0}\left[z - \frac{1}{2}(1 + q_0)z^2\right] \tag{1.8}$$

These relationships between z, H_0 and q_0 hold in the case of a redshift due to the expansion of the universe. Superimposed on the expansion of the universe are local peculiar velocities (e.g. blue- and redshifts) due to the motions caused by local anisotropies in the matter distribution; an example is local motions in clusters and groups of galaxies due to the gravitational potential of the cluster itself. These effects are minimized by observing suitably distant objects, where the velocities corresponding to the expansion of the universe become so large that they make local peculiar motions negligible.

From an observational point of view, the Hubble law needs, in addition to the measurements of the redshift z, an estimate of galaxy distances. This is usually done by comparing the observed flux l received from certain standard candles (i.e. objects of known intrinsic luminosity L) with their intrinsic luminosities. Traditionally one uses the inverse square law to determine the distance:

$$d = \left(\frac{L}{4\pi l} \right)^{1/2} \tag{1.9}$$

This result is based on the conservation of energy and assumes a flat static space. In cosmology, the distance obtained through Equation (1.9) is called the luminosity distance, and is denoted by d_L.

Consider a light source located at a radial co-moving coordinate χ; at a given cosmic time t_e the source emits photons that reach the observer located at $\chi = 0$ at time t_0. By the time the light reaches the observer it is distributed uniformly across a sphere of coordinate radius $R(t_0)\chi$. The area of the spherical surface at the observer location centred at the source is therefore given by $4\pi R(t_0)^2 \chi^2$. The photons emitted by the source are redshifted by the expansion of the universe, and their energy is therefore reduced by a factor $(1 + z)$ when measured by the observer; this is because the wavelength is increased by a factor $(1 + z)$ and the photon energy is proportional to the inverse of its wavelength. There is also an additional reduction by a factor $(1 + z)$ due to the so-called time dilation effect, i.e. the observer receives less photons per unit time than emitted at the source. This can easily be understood by means of the same arguments as were applied in the case of the wave maxima, that led to the notion of redshift. We found before that the time between two consecutive maxima at emission is different from that at reception; the same holds for the time interval between photons emitted by the source, and implies that the rate of reception of photons is different from the rate of emission. Taking into account these two effects, conservation of energy dictates that:

$$l = \frac{1}{(1+z)^2} \frac{L}{4\pi R(t_0)^2 \chi^2}$$

We now define the luminosity distance d_L of the observed source, according to Equation (1.9); one obtains $d_L = R(t_0)\chi(1+z)$, which can be rewritten using Equation (1.8) as (retaining the terms up to the second order in z):

$$d_L = \frac{cz}{H_0}\left[1 + \frac{1}{2}(1 - q_0)z\right] \tag{1.10}$$

The first term is the empirical Hubble law, with the recession velocity given by the product cz. The higher-order correction term is proportional to the deceleration parameter q_0 and starts to play a role when $z > 0.1$.

Another way to determine cosmological distances is to consider objects (e.g. galaxies) with known diameter D_p, and compare the measured angular diameters Θ with the intrinsic ones. One can define a diameter distance d_{D_p} as

$$d_{D_p} = \frac{D_p}{\Theta} \tag{1.11}$$

which is equal to d_L for a flat static space. Consider an object located at the radial co-moving coordinate χ, that emits light at time t_e; if the observer is located at $\chi = 0$ and receives the light from the object at t_0, the relationship between D_p and Θ can easily be obtained by determining $\sqrt{\Delta s^2}$ where Δs is obtained integrating the FRW metric with $dt = d\chi = d\phi = 0$. This provides $d_{D_p} = R(t_e)\chi$. By comparing the latter equation with $d_L = R(t_0)\chi(1 + z)$ obtained before and using the definition of z we obtain

$$d_{D_p} = \frac{d_L}{(1 + z)^2}$$

In principle d_{D_p} is different from d_L, but the two distances converge to the same value when $z \to 0$.

It should be clear from this brief discussion that the empirical study of the trends of d_L and d_{D_p} with redshift z provides an estimate of the kinematical parameters H_0 and q_0. A third possible method to determine the kinematical status of the universe involves number counts of galaxies with a flux greater than some specified value l ($N(l)$). Assuming there are n galaxies per unit volume, in a static flat universe (with uniform distribution of galaxies) one expects

$$N(l) = \frac{4}{3}\pi n \left(\frac{L}{4\pi l}\right)^{3/2}$$

where L is the intrinsic galaxy luminosity, supposed constant. For an expanding universe it can be shown that (as a second-order approximation in z)

$$N(l) = \frac{4\pi n(t_0)}{3}\left(\frac{L}{4\pi l}\right)^{3/2}\left[1 - \frac{3H_0}{c}\left(\frac{L}{4\pi l}\right)^{1/2}\right]$$

where $n(t_0)$ is the number density at the present time (e.g. in the low redshift universe); notice that by a fortuitous cancellation of terms this relationship does not depend on q_0. The correction term to the static flat case is always negative, so that in principle one should always observe fewer sources than predicted by the simple $l^{-3/2}$ formula.

There are many practical difficulties in implementing these three tests; the reason is that we are assuming the existence of perfect standard candles and the absence of evolutionary effects on the size, and brightness of galaxies. Evolutionary effects are particularly important since a high redshift means a time far in the past, when galaxies had a very different age from the present one. A detailed discussion of these classical cosmological tests and the related observational problems can be found in [187]. In recent years the class of stellar objects called *Type Ia supernovae* (see Section 7.6) has been used as an effective standard candle and applied with great success to study the d_L–z relationship (see [146]).

We conclude this section by discussing briefly the concept of particle horizon in an FRW expanding universe. In general, as the universe expands and ages, a generic observer is able to see increasingly distant objects as the light they emitted has time to arrive at the observer's location. This implies that as time increases, increasingly larger regions of space come into causal contact with the observer, who will therefore be able to 'see' increasingly larger portions of the universe. We can ask ourselves what is the co-moving coordinate χ_H of the most distant galaxy we can see at a given cosmic time t. Increasing values of χ_H with time mean that we are actually seeing more and more distant galaxies (supposed to be at rest with respect to the cosmic expansion) as the time increases. Consider a radially travelling photon, for which $ds = 0$. From the FRW metric we obtain

$$\int_0^t \frac{dt'}{R(t')} = \frac{1}{c} \int_0^{\chi_H} \frac{d\chi}{\sqrt{1 - k\chi^2}}$$

and therefore

$$\chi_H = \sin\left(c \int_0^t \frac{dt'}{R(t')} \right) \quad k = +1$$

$$\chi_H = c \int_0^t \frac{dt'}{R(t')} \quad k = 0$$

$$\chi_H = \sinh\left(c \int_0^t \frac{dt'}{R(t')} \right) \quad k = -1 \tag{1.12}$$

If the space has $k = 0$ or $k = -1$ it is in principle possible, for specific forms of $R(t)$, to have an infinite χ_H; this means that all galaxies in the universe might eventually be visible at a certain time t for particular forms of the function $R(t)$. If $k = +1$ the behaviour of χ_H is periodic, and if the argument of the sine function is equal to or larger than π, one can sweep the entire universe.

1.3 Cosmic dynamics

The previous discussion about the kinematics of the cosmic fluid was based exclusively on the properties of the FRW metrics which, in turn, depend only on the hypothesis of homogeneous and isotropic cosmic fluid. To determine the behaviour of $R(t)$ with cosmic time t and the value of k we need to apply a theory for the physical force(s) governing the evolution of the cosmic fluid. The only fundamental interaction able to bridge the relevant cosmological scale is the gravitational force, therefore we need to use a theory of gravity – the general relativity theory – to describe the evolution of FRW universes.

The case of a space with the FRW metrics provides the equation

$$\left(\frac{dR(t)}{dt}\right)^2 = -kc^2 + \frac{8\pi G\rho(t)R(t)^2}{3}$$
(1.13)

where G is the gravitational constant $(6.6742 \times 10^{-8}\,\text{dyn cm}^2\,\text{g}^{-2})$ and ρ is the matter density. Equation (1.13) was obtained in 1922 by Friedmann, who solved Einstein's field equations for an isotropic and homogeneous universe. As we will see in a moment, these equations predict an expanding universe. A more general form of the field equations contains the constant Λ – called the cosmological constant – introduced by Einstein in 1917 in order to obtain static universes (the expansion of the universe had not been discovered yet). It is important to notice that the value of Λ must be small in absolute terms, since the planetary motions in the solar system are well described by the Einstein field equations with $\Lambda = 0$. Including Λ in the gravitational field equations provides

$$\left(\frac{dR(t)}{dt}\right)^2 = -kc^2 + \frac{(8\pi G\rho(t) + \Lambda)R(t)^2}{3}$$
(1.14)

It is clear that the evolution of $R(t)$ is controlled by the density (ρ), the geometry (k) and the cosmological constant (Λ). By using the definition of $H(t)$ one can rewrite Equation (1.14) as

$$H(t)^2 = -\frac{kc^2}{R(t)^2} + \frac{8\pi G\rho(t)}{3} + \frac{\Lambda}{3}$$
(1.15)

It is customary to introduce the critical density $\rho_c \equiv 3H(t)^2/(8\pi G)$ and define the density parameter $\Omega_\rho = \rho/\rho_c$, an equivalent for the cosmological constant $\Omega_\Lambda = \Lambda/(3H(t)^2)$, and the sum $\Omega = \Omega_\rho + \Omega_\lambda$. With these definitions Equation (1.15) becomes

$$(1 - \Omega)H(t)^2 R(t)^2 = -kc^2$$
(1.16)

We can immediately see from this form of the Friedmann equation that there is an intimate connection between the density of matter plus the cosmological constant, and the geometry of space. $\Omega = 1$ gives a flat space, $\Omega > 1$ a positive curvature, and $\Omega < 1$ a negative curvature. It is also important to notice the obvious fact that Ω changes with time, since $H(t)$ and $R(t)$ both change with t, but the product kc^2 is a constant.

1.3.1 Histories of $R(t)$

Equation (1.14) enables us to perform a simple analysis of the behaviour of $R(t)$ for various model universes, once an additional equation for the density is obtained; this equation can be determined by applying the first principle of thermodynamics to the cosmic fluid. In an isolated system the first law of thermodynamics states that $dU = -PdV$ where U is the internal energy of the system, V its volume and P the pressure. The internal energy is ρc^2 times the volume V (i.e. the energy associated with the rest mass of the matter) so that the time evolution of the system according to the first law is

$$\frac{d(\rho(t)c^2 V(t))}{dt} = -P\frac{dV(t)}{dt}$$

which can also be rewritten as

$$\frac{d(\rho(t)c^2 R(t)^3)}{dt} = -P\frac{dR(t)^3}{dt} \tag{1.17}$$

using the fact that the volume V scales as $R(t)^3$. Let us now assume that the density is dominated by matter and not by radiation; this is a very good assumption since observationally – as discussed before – one finds that at the present time the matter density is about three orders of magnitude larger than the density associated with radiation ($\rho_r = \epsilon_r/c^2$, where ϵ_r is the photon energy density). If the matter is non-relativistic (a correct assumption for almost the whole evolution of the universe) its pressure is negligible with respect to ρc^2 and Equation (1.17) provides

$$\frac{d\rho(t)}{dt}\frac{1}{\rho(t)} = -3\frac{dR(t)}{dt}\frac{1}{R(t)} \tag{1.18}$$

which implies

$$\rho(t)R(t)^3 = \rho(t_0)R(t_0)^3$$

where t_0 is the present cosmic time and t a generic value. This reflects the simple fact that the density of non-relativistic matter is decreasing because of dilution as space is expanding. If photons were to be the dominant contributor to the total density, the previous relationship would be different. In fact, for photons (and more generally for relativistic particles) $P = (\rho_r c^2)/3$ and P is no longer negligible with respect to U. Therefore Equation (1.17) would provide

$$\frac{d\rho_r(t)}{dt}\frac{1}{\rho_r(t)} = -4\frac{dR(t)}{dt}\frac{1}{R(t)} \tag{1.19}$$

and

$$\rho_r(t)R(t)^4 = \rho_r(t_0)R(t_0)^4$$

The scaling of $\rho(t)$ with $R(t)^4$ is firstly due to the decrease of the number density of photons as $R(t)^{-3}$ when the universe expands (since the volume increases as $R(t)^3$). In addition, the energy of individual photons decreases as $R(t)^{-1}$ because of the cosmological redshift and therefore both ϵ_r and ρ_r decrease with time as $R(t)^{-4}$, faster than the matter density.

By considering a matter dominated universe one now can rewrite Equation (1.14) as

$$\left(\frac{dR(t)}{dt}\right)^2 = -kc^2 + \frac{8\pi G\rho(t_0)R(t_0)^3}{3R(t)} + \frac{\Lambda R(t)^2}{3} \tag{1.20}$$

Differentiation of this equation with respect to t provides:

$$\frac{d^2R(t)}{dt^2} = -\frac{4\pi G\rho(t_0)R(t_0)^3}{3R(t)^2} + \frac{\Lambda R(t)}{3} \tag{1.21}$$

This equation shows clearly how the self gravitation of matter (represented by ρ) acts to slow down the expansion of the universe, because it appears as a negative contribution to the acceleration of $R(t)$. On the other hand, a positive Λ acts like a negative density and tends to accelerate the expansion of the universe; a particular choice of Λ makes the universe static (although in a situation of unstable equilibrium). The term $(\Lambda R(t))/3$ is often called the cosmic repulsion term.

It is now easy to determine some general properties of $R(t)$ in a matter dominated universe. If Λ is zero or negative the acceleration of $R(t)$ is always negative; at some time in the past $R(t)$ must have reached zero and therefore ρ was infinite (i.e. a singularity is attained). It is natural to set the zero point of the cosmic time at this instant, which can also be considered the origin of the universe. As for the future evolution, if Λ is negative, $R(t)$ will also intersect the t axis some time in the future (hence a final implosion) since the expansion will slow down, eventually stop and then reverse to a contraction. If Λ is zero the acceleration can become zero in the future if $R(t)$ becomes infinite, and therefore the expansion can slow down without ever being followed by a contraction. The precise behaviour depends in this case on the value of k. If $k = -1$ or 0 the future collapse is avoided, but not if $k = +1$.

If Λ is positive then $R(t)$ is not always decelerating and there is the possibility of avoiding a singularity in the past. In fact, if $k = +1$ one can obtain from Equations (1.20) and (1.21) that in the past there has been a minimum of $R(t)$ different from zero, given by $R_{min}^3 = (4\pi G\rho(t_0)R(t_0)^3)/\Lambda$ if the cosmological constant satisfies the following relation: $\Lambda < (c^6)/(4\pi G\rho(t_0)R(t_0)^3)^2$. As for the future evolution, if $k = 0$

or -1 the expansion continues forever, whereas if $k=+1$ the expansion may vanish and then be followed by a contraction, depending upon the value of Λ.

For historical interest we show briefly how it is possible to obtain a static universe by tuning the value of Λ. In a static universe both $R(t)$ and $\rho(t)$ are constant, and both velocity and acceleration of $R(t)$ are equal to zero. With these constraints Equations (1.20) and (1.21) provide $\Lambda = 4\pi\rho(t_0)G$, $k/R^2 = (4\pi\rho(t_0)G)/c^2$, where R denotes the constant value of $R(t)$. Since R has to be positive and k can be only equal to 0, $+1$, -1, we have that a static universe will have $k=+1$ and $R=c/\sqrt{4\pi\rho(t_0)G}$.

We conclude by providing analytical relationships between $R(t)$ and t for the case of flat geometry, i.e. $\Omega = 1$ and $k=0$, and arbitrary values of Λ, which are relevant to the presently favoured cosmological model. With this choice of parameters the universe began from a singular state ($R=0$ and $\rho=\infty$ at $t=0$) and Equation (1.20) gives (see also Figure 1.3):

$$R(t) = R(t_0)\left(\frac{8\pi G\rho(t_0)}{\Lambda}\right)^{1/3}\sinh^{2/3}\left(\frac{1}{2}t\sqrt{3\Lambda}\right) \quad \Lambda > 0$$

$$R(t) = R(t_0)(6\pi G\rho(t_0))^{1/3}t^{2/3} \quad \Lambda = 0$$

$$R(t) = R(t_0)\left(\frac{8\pi G\rho(t_0)}{|\Lambda|}\right)^{1/3}\sin^{2/3}\left(\frac{1}{2}t\sqrt{3|\Lambda|}\right) \quad \Lambda < 0 \qquad (1.22)$$

For $\Lambda = 0$ one obtains the very simple result $q_0 = 1/2$, $H(t) = 2/(3t)$ and therefore the age of the universe is $t_0 = 2/(3H_0)$. The quantity $1/H_0$ is often called Hubble time.

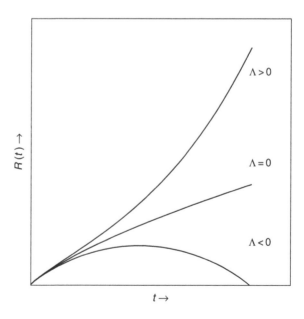

Figure 1.3 Qualitative behaviour of the scale factor $R(t)$ with respect to the cosmic time t for models with $\Omega = 1$ and $k=0$

1.4 Particle- and nucleosynthesis

We have already noticed that the density of matter in an expanding universe decreases with time slower than the density of photons. This means that as we go backwards in time the radiation density increases faster than the density of matter. Therefore, there must be a point in time when the two densities were equal and before that the universe was radiation dominated. If the actual densities of matter and radiation are $\rho(t_0)$ and $\rho_r(t_0)$, respectively, the equality is attained at

$$\frac{R(t_0)}{R(t_E)} = \frac{\rho(t_0)}{\rho_r(t_0)} = 1 + z_E$$

where t_E and z_E are, respectively, the cosmic time and redshift of matter–radiation equality; their values are of the order of 10^4–10^5 years and 10^3, respectively. It is worth noticing at this stage that the results about the trend of $R(t)$ with t we gave in the previous section were obtained assuming a matter dominated universe (negligible radiation density) at all time. The onset of a radiation dominated universe at the beginning of the evolution does not, however, alter the general results regarding the occurrence of an initial singularity, and also the quantitative relationship between $R(t)$ and t is not substantially changed, since – as we will soon see – the radiation dominated era lasts only a short time compared with the timescale of cosmological evolution.

In case of radiation $\epsilon_r = aT_r^4$ (where T_r is the radiation temperature and $a = 7.566 \times 10^{-15} \text{erg cm}^{-3} \text{K}^{-4}$) $T_r(t) \propto R(t)^{-1}$ and therefore the radiation temperature was steadily increasing in the past. Calculations of the interaction cross section between photons and matter and the expansion rate of the universe show that during the radiation dominated epoch the interaction rate was high enough to ensure that at each instant there was (to a good approximation) thermodynamical equilibrium (see Chapter 2) i.e. photons followed a black-body distribution of energies characterized by the same temperature $T = T_r$ for both radiation and matter. During the radiation era $\rho_r(t)$ becomes so large that the contribution of the terms containing k and Λ in Equation (1.14) are negligible and we can write

$$\left(\frac{dR(t)}{dt} \right)^2 = \frac{8\pi G \rho_r(t) R(t)^2}{3}$$

This equation, in conjunction with $\epsilon_r = aT^4$ and $T_r(t)R(t) = T_r(t_0)R(t_0)$ does provide

$$R(t) = R(t_0) T_r(t_0) \left(\frac{32\pi Ga}{3c^2} \right)^{1/4} t^{1/2}$$

$$T(t) = \left(\frac{3c^2}{32\pi Ga} \right)^{1/4} t^{-1/2}$$

$$\rho(t) = \frac{3}{32\pi G} t^{-2} \tag{1.23}$$

These formulae for the early evolution of ρ and T do not contain adjustable constants; however, they do not strictly apply when approaching t_E, since in this case the matter contribution to ρ is not negligible, and also the contributions of the curvature and the cosmological constant may play a role.

There is also a minimum time t_P (Planck time) below which we cannot describe the evolution of the universe with Equation (1.23) due to quantum uncertainty. This stems from the uncertainty principle applied to the pair of physical variables energy E and time t, i.e. $\Delta E \Delta t > h/(2\pi)$ where h is the Planck constant ($h = 6.626 \times 10^{-27}$ erg s). Consider a length (Planck length) $l_P = ct_P$ that defines a region in causal contact at time t_P. A mass $m_P \sim \rho_P l_P^3$ is associated with this length scale (ρ_P is the density of matter at $t = t_P$) hence an energy $m_P c^2 = \rho_P (ct_P)^3 c^2$. The uncertainty relationship can therefore be rewritten as $\rho_P (ct_P)^3 c^2 t_P = \rho_P c^5 t_P^4 > (h/2\pi)$. From Equation (1.23) we have $\rho_P \approx 1/(Gt_P^2)$, and consequently $\rho_P c^5 t_P^4 \approx (c^5 t_P^4)/(Gt_P^2) > (h/2\pi)$, that provides $t_P > ((hG)/(2\pi c^5))^{1/2} \sim 10^{-43}$ s. Due to this quantum uncertainty we cannot be completely sure that there has really been a singularity at $t = 0$.

When $t \sim 10^{-43}$ s the universe was extremely hot, the temperature being of the order of 10^{32} K, that corresponds to an energy of $\sim 10^{19}$ GeV (an energy of 1 eV corresponds to a temperature of 1.1605×10^4 K from the relationship $E = K_B T$ where K_B is the Boltzmann constant equal to 1.3807×10^{-16} erg K). According to the currently accepted cosmology and particle physics theories, it is during the first epochs after the singularity that today's stable particles – the proton–neutron pair, electrons in a number that compensates for the electric charge of the protons, neutrinos – were produced. A description of what happened during those first moments of the evolution of the universe – the so-called Big Bang – has to be based on the knowledge of the four fundamental interactions (gravitational, strong, weak, electromagnetic) which we briefly summarize below.

According to the standard model of particle physics, the fundamental interacting particles – quarks and leptons – are all fermions (particles with spin 1/2). Leptons are the negatively charged particles electron, μ and τ, and the associated neutrinos ν_e, ν_μ, ν_τ. The rest masses are ~ 0.0005 GeV for the electron (1 eV $= 1.7827 \times 10^{-33}$ g using the relationship $E = mc^2$), 0.106 GeV for μ and 1.178 GeV for τ. Neutrinos are supposed to be massless although recent experiments suggest a mass different from zero, but not yet well determined. There are six quark species (positively and negatively charged) called 'down', 'up', 'strange', 'charm', 'bottom' and 'top'. Their mass increases from ~ 0.31 GeV for the 'down' quark up to ~ 177 GeV for the 'top' quark. In addition, there are antiparticles for each lepton and quark.

The gravitational interaction involves all particles, it is described by general relativity, and is supposed to be mediated by a boson (particle with integer spin) called graviton. At the moment there is no established quantum theory of gravity, which is the reason why we cannot try to described what happened at $t < t_P$. The strong interaction is a short-range interaction mediated by gluons, a family of eight massless bosons, and involves the so-called hadrons (respectively baryons, like protons and neutrons, and mesons, the most relevant of them being the pions π^0, π^+, π^-) which are made of combinations of quarks. Baryons are made of triplets of quarks, mesons of pairs quark–antiquark (e.g. the proton is made by two 'up' and one 'down' quark, the neutron by two 'down' and one 'up' quark). The weak interaction involves all particles, has short range and is mediated by the W^+, W^- and Z^0 bosons, with masses of the order of ~ 90 GeV. The electromagnetic interaction is a long-range interaction acting among charged particles, and is mediated by the photon (a massless boson).

According to the so-called Grand Unified Theories (GUT) the strong, weak and electromagnetic interactions were all unified into a single force mediated by superheavy bosons with masses of the order of 10^{15} GeV. This idea stems from the successful unification of weak and electromagnetic interactions into the electroweak force that separates into the two components at sufficiently low energies. If it is possible to unify the strong with the electroweak force at even higher energies has yet to be seen; there are various theories that are, however, difficult to test experimentally. Interestingly a GUT prediction is that the proton should decay with a timescale of $\sim 10^{32}$–10^{33} yr (this hypothetical decay has not been observed yet). A further goal of physics is to unify gravity with the other three forces (a unification which should happen at energies higher than GUT). Whether this is possible – in spite of various attempts – remains to be seen.

If GUT are a viable proposition (at the moment there is no experimental confirmation of their predictions) the physical conditions during the first moments right after the singularity were adequate to attain the unification of the four fundamental interactions. Following this line of thought one expects that the steady temperature decrease caused by the expansion of the universe has caused a number of spontaneous symmetry breaking, that have generated the separate interactions we see today. At $t = 10^{-43}$ s the gravitational force has separated from the other three interactions which are still unified into a single force. At energies between 10^{16} and 10^{14} GeV (between 10^{-38} and 10^{-35} seconds after the singularity) the strong force separated from the electroweak one. The superheavy bosons disappear rapidly due to annihilation or decay processes. At this stage the universe is made of leptons, antileptons, quarks, antiquarks, gluons and four bosons that mediate the electroweak interaction (and probably gravitons). At energies of the order of 10^2 GeV (about 10^{-11} seconds after the singularity) the electroweak interaction separates into the electromagnetic and weak one. The leptons (massless until this moment) acquire mass through the Higgs mechanism (probably also the neutrinos) and the bosons that mediated the electroweak interaction give rise to the massive W^+, W^-, Z^0 bosons and photons. Below ~ 90 GeV the massive bosons disappear through annihilation or decay. At this stage the universe was made of photons (and probably gravitons) quark–antiquark and lepton–antilepton pairs. By about 10^{-6} seconds after the singularity, quarks combine into hadrons.

Photons and matter are in equilibrium through absorption and creation–annihilation processes. Particles and antiparticles continually annihilate each other but more pairs are produced from the high-energy photon field as long as $K_B T > 2mc^2$ where m is the rest mass of the particle and antiparticle pair, and T is the temperature. This means that the number of a given particle species with mass m and the photon number are about the same as long as the previous inequality is satisfied. When the temperature goes below $2mc^2/K_B$, the particle–antiparticle pairs of mass m annihilate, without being replaced by newly produced pairs. If there is an asymmetry between the number of particles and antiparticles, after the annihilation only the residual number of surviving particles or antiparticles will be left. We do not have, to date, any empirical evidence for the existence of antimatter in the universe. The antiprotons observed in the cosmic rays are consistent with the hypothesis of production by interaction of cosmic rays with the interstellar medium; therefore, at least for the Galaxy, there is no evidence of antimatter. Absence of γ rays from a cluster of galaxies due to nucleon–antinucleon annihilations is further evidence against the existence of antimatter in the universe. In order to have a universe populated only by matter, it is necessary to postulate an asymmetry between matter and antimatter so that the annihilation processes destroyed all antimatter leaving the excess of matter that we see today. To explain the observed ratio between photons and matter an initial matter–antimatter asymmetry of only $\sim(1/10^8)$ particles is needed. Proposed mechanisms to explain this asymmetry involve processes acting when the temperature drops below the threshold for the separation of the strong force from the electroweak one, but no definitive solution to this problem (unless one invokes an ad hoc initial condition) has been found yet.

At energies of the order of 1 GeV (about 10^{-5} s after the singularity) nucleons and antinucleons annihilate, leaving the small excess of nucleons arising from the asymmetry discussed before. When the energy is down to about 130 MeV pairs π^+–π^- annihilate and π^0 particles decay into photons

The μ leptons annihilate with the corresponding antiparticles at about 100 MeV (the more massive τ leptons annihilated at higher energies). As for the nucleons, protons (p) and neutrons (n) were constantly being transformed into each other via the following reactions:

$$n \leftrightarrow p + e^- + \bar{\nu}_e$$

$$n + e^+ \leftrightarrow p + \bar{\nu}_e$$

$$n + \bar{\nu}_e \leftrightarrow p + e^-$$

involving electrons e^-, positrons e^+, electron neutrinos ν_e and electron antineutrinos $\bar{\nu}_e$. The conversions from one particle to the other were easily accomplished as long as the energy was above 1.293 MeV, e.g. the energy corresponding to the mass difference between proton and neutron. The direct and inverse reactions were so frequent that an equilibrium was established between the number densities of protons and neutrons, given by

$$\frac{n_n}{n_p} = e^{(-1.293\text{MeV})/K_B T}$$

At 100 MeV $n_n/n_p \sim 0.99$, decreasing down to ~ 0.22 when the energy is ~ 1 MeV. At about 3 MeV the neutrinos decouple, i.e. they do not interact any longer with the rest of the matter[1]; their decoupling happens after the annihilation of muons and

---

[1] Neutrinos have a very small interaction cross section. However, right after the singularity the universe was so dense that neutrinos were also tightly coupled to the other components of the cosmic fluid.

before the annihilation of electrons and positrons. From this moment on neutrinos travel essentially undisturbed by the other particles, their temperature still decreasing (they are relativistic particles) as $1/R(t)$. When the energy goes below $\sim 1\,\mathrm{MeV}$ electron–positron pairs annihilate, leaving a small remainder excess of electrons. Neutrons cannot any longer be replenished fast enough, due to the electron–positron annihilation and the fast expansion of the universe (it is now about 20 seconds after the singularity) and the number ratio $(n_n/n_p) \sim 0.224$ attained at electron–positron annihilation decreases due to the neutron decay (half-life of the order of 10 minutes). Between 100 and 200 seconds after the singularity the energy has fallen to about $0.1\,\mathrm{MeV}$ $(T \sim 10^9\,\mathrm{K})$ and the nuclear fusion reaction

$$p + n \rightarrow {}^2D + \gamma$$

becomes an efficient producer of deuterium (at higher energies the deuterium produced was photodissociated by the energetic photons). Helium production reactions become efficient when ^2D is sufficiently abundant:

$$^2D + {}^2D \rightarrow {}^3H + p$$
$$^3H + {}^2D \rightarrow {}^4He + n$$
$$^2D + p \rightarrow {}^3He + \gamma$$
$$^3He + n \rightarrow {}^4He + \gamma$$

Since there are no stable nuclei with atomic mass 5 to 8, and because of the fast expansion of the universe that lowers the energies of the particles involved on very short timescales – the energy drops below $\sim 30\,\mathrm{keV}$ $(1\,\mathrm{keV} = 10^3\,\mathrm{eV})$ e.g. $\sim 3 \times 10^8\,\mathrm{K}$ about 20 minutes after the singularity – the nucleosynthesis leaves about a fraction of 0.75 (by mass) of protons (hydrogen), ~ 0.25 of helium, $\sim 10^{-4}$ of deuterium, $\sim 10^{-5}$ of ^3He and $\sim 10^{-10}$ of lithium. The precise values of these abundances (see Figure 1.4) are determined by the competition between the expansion rate of the universe and the nucleon density. The formation of He is limited only by the availability of neutrons; to a good approximation the helium abundance is therefore set by the neutron abundance at the beginning of the nucleosynthesis. It therefore depends, albeit only mildly, on the value of the matter density, increasing for increasing density.

More precisely, the primordial element abundances depend on the density of baryonic matter that we denote with ρ_b and $\Omega_b = \rho_b/\rho_c$. It is important to notice that ρ_b is approximately the same as the density of baryonic plus leptonic matter, since the number of electrons equals the proton numbers to achieve charge neutrality, but electrons are about 10^3 times lighter than baryons. From now on we will denote with baryonic density the density of baryonic plus leptonic matter.

It is generally assumed that the density of dark matter does not play any role in this cosmological nucleosynthesis, since it is supposed to affect the cosmic fluid only via its contribution to the gravitational interaction after the end of the radiation dominated epoch.

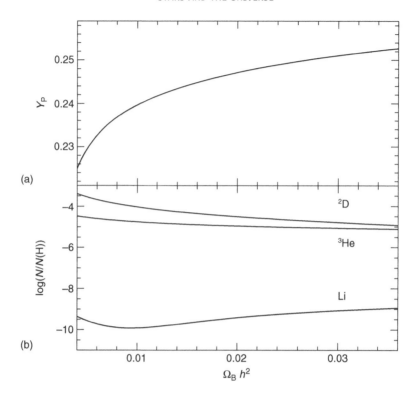

Figure 1.4 Abundances of ^{4}He ((a) in mass fraction), deuterium, lithium and ^{3}He ((b) number ratios with respect to hydrogen) produced during the primordial nucleosynthesis, as a function of the product $\Omega_B h^2$. Ω_B denotes the present (at $t = t_0$) value of the baryon density in units of the critical density, $h = H_0/(100 \, \text{km} \, \text{Mpc} \, \text{s}^{-1})$

The primordial abundances of deuterium and ^{3}He are decreasing functions of Ω_b; this is explained by the fact that the higher the matter density, the higher the temperature at which these elements reach an abundance high enough to start the production of helium, and consequently the higher their destruction rate. The behaviour of lithium is more complicated; for present values of Ω_b below ≈ 0.002[2] lithium is produced by direct fusion of helium with ^{3}H, whereas at higher densities (for present values of Ω_b larger than ≈ 0.02) it is produced by fusion of helium and ^{3}H producing ^{7}Be that transforms into lithium by electron captures. In both cases the final abundance of lithium increases with Ω_b. However, there is an intermediate region showing a dip in the abundance, due to the efficiency of a destruction reaction that involves a proton capture and a consequent decay into two helium nuclei.

---

[2] Once the present value of the density parameter Ω_b or Ω_r is given, its value at any earlier (or future) epochs can be obtained from Equations (1.18) and (1.19).

As the universe expands and cools, one reaches the time when the energy densities of matter and radiation are equal (at redshift ~ 3000) and immediately after that the matter density starts to dominate. Even so, matter and radiation are still tightly coupled through electron scattering processes. The universe is ionized, and matter is made mostly of protons (hydrogen nuclei) and free electrons. When the temperature drops below the ionization energy of hydrogen (13.6 eV) the ionization fraction stays close to one, due to the large excess number of photons over baryons (photons dominate by number, although matter dominates energetically and therefore gravitationally) so that the number of photons in the high-energy tails of the black-body spectrum is high enough to keep the matter fully ionized. Eventually the temperature, and therefore the number density, of sufficiently energetic photons drops so low that recombination prevails. It is at this time, $\sim 10^{5.5}$ years after the singularity (i.e. at redshift ~ 1000 and $T \sim 4000$ K) that the first atoms form. The resulting dearth of free electrons has the immediate consequence of reducing the efficiency of electron scattering, so that matter and radiation decouple. From this moment on the temperatures of radiation and matter become different and start to evolve separately; radiation no longer interacts with matter and can travel undisturbed through space, since the number of particles of matter is too low to produce significant interactions. The radiation temperature T_r is reduced according to $T_r \propto R(t)^{-1}$, and the black-body spectrum it had at decoupling is preserved. This last point can be demonstrated as follows. For a black-body spectrum the number of photons with frequencies between ν and $\nu + d\nu$ contained in a volume of space $V(t)$ at cosmic time t is given by

$$dN(t) = \frac{8\pi\nu^2 V(t)d\nu}{c^3(e^{(h\nu/K_B T_r(t))} - 1)} \tag{1.24}$$

At a later time t' the frequency will be redshifted to $\nu' = \nu R(t)/R(t')$ and therefore $d\nu' = d\nu R(t)/R(t')$. The volume will have expanded to $V(t') = V(t)R(t')^3/R(t)^3$, but the number of photons within $V(t')$ will be the same as the number within $V(t)$ because of conservation (no appreciable interactions with the matter happen); the temperature $T_r(t)$ will have also changed according to $T(t') = T(t)R(t)/R(t')$. By imposing $dN(t') = dN(t)$ and rewriting Equation (1.24) expressing ν, $V(t)$ and $T_r(t)$ in terms of ν', $V(t')$ and $T_r(t')$ according to the relationships given before, one obtains that

$$dN(t') = \frac{8\pi\nu'^2 V(t')d\nu'}{c^3(e^{(h\nu'/K_B T_r(t'))} - 1)}$$

i.e. of the same form as Equation (1.24).

This black-body radiation, homogeneous and isotropic (because of the cosmological principle) with a temperature T_r nowadays of the order of ~ 3 K (as obtained from $T_r \propto R(t)^{-1}$) is the theoretical counterpart of the observed CMB.

1.5 CMB fluctuations and structure formation

According to the scenario presented above, the CMB is the relic of the hot phase before decoupling, and provides us with information about the state of the universe when its age was only about a few 10^5 yr. The wealth of structures populating the universe nowadays suggests the existence of some density inhomogeneities in the cosmic fluid that have grown with time; if the universe was perfectly isotropic and homogeneous no structures would have formed with time, whereas in case of inhomogeneities, regions denser than the background tend to contract and get denser still, inducing a growth of the initial perturbation.

In 1970 Peebles, Yu, Sunyaev and Zel'dovich predicted that these inhomogeneities had to be imprinted in the CMB as the tiny temperature fluctuations that have recently been detected. In very simple terms, fluctuations of the local density of matter would have behaved as sound waves (with their fundamental mode plus overtones) in the cosmic fluid before recombination, with the photons providing the restoring force. The matter we are considering here is the baryonic matter, to which the photons are tightly coupled, whereas the dark matter did not have any interaction with photons. At recombination the photons started to travel unimpeded through space for the first time; photons released from denser, hotter regions were more energetic than photons released from more rarefied regions. These temperature differences were thus frozen into the CMB at recombination and are detected today. The shape of the observed CMB power spectrum is explained when one assumes that the phases of all the sound waves were synchronized at birth – i.e. that they were all triggered at the same time – and that the initial disturbances were approximately equal on all scales, e.g. the fluctuations on small scales had approximately the same magnitude as those affecting larger regions.

The first and highest peak in the CMB power spectrum (see Figure 1.2) corresponds to the fundamental wave of this acoustic oscillation; subsequent peaks represent the overtones. The power spectrum shows a strong drop off after the third peak (an effect known as Silk damping) due to the process of recombination. Since the recombination is not instantaneous, during its development the photon mean free path starts to progressively increase, producing a flow of photons from regions of high densities to lower density zones, that smooths out the small-scale temperature fluctuations. Most importantly, amplitude and location of the peaks are closely related to a number of cosmological parameters (for more details see, for example, the discussion in [112]); in particular, the location of the first peak is mainly related to the geometry of the three-dimensional space, whereas the ratio of the heights of the first to second peak is strongly dependent on Ω_b. Also the values of the Hubble constant and of the cosmological constant affect both the location and the amplitudes of the peaks albeit with different sensitivities.

Up to decoupling the density perturbations of the baryonic matter did not grow substantially (in fact the fluctuations of the CMB temperature are extremely small) due to the damping effect of the tightly coupled photons, and their evolution can be followed analytically with linear approximations. However, the perturbations involving dark matter are supposed to be able to grow more substantially (no interaction with the photons and no imprint left on the CMB power spectrum) and after baryonic matter decoupled from the radiation field, it could fall into and enhance the potential well of the dark matter condensations, thus starting to build up the structures we see today. The end point of the evolution of the primordial density fluctuations is the present statistical distribution of matter. This is generally very complicated, varying from point to point with objects of different sizes and masses (alternatively, fluctuations of various wavelengths and amplitudes) and its study is fundamental in order to determine the mechanisms of structure formation. A particular sticking point is the so-called biasing, i.e. the fact that the light distribution may not faithfully trace the mass distribution (we can only detect the luminous matter directly, not the dark matter). Studies not only of the light but also of the gravitational field in the observable structures can overcome this problem, i.e. by determining the peculiar velocities induced by the mass distribution through the gravitational interaction.

Numerical simulations of the evolution of perturbation and structure formation (see, for example [110]) have shown that the best assumption about the dark matter dynamical status is that it is 'cold', i.e. it has a negligible velocity dispersion; a 'hot' dark matter, i.e. matter with a large velocity dispersion, has been excluded, since it does not allow the formation of galaxies. As for the nature of this mysterious dark matter, various hypothetical particles predicted by GUT have been proposed as viable candidates; at the moment the question is still wide open, since – as mentioned before – there is no experimental confirmation for any of the proposed GUT and therefore for their predictions.

1.6 Cosmological parameters

Combining the location and amplitude of the peaks in Figure 1.2 to a number of constraints obtained from the spatial distribution of galaxies, the d_L–z empirical relationship using Type Ia supernovae as standard candles shown in Figure 1.5, and the empirical determination of the Hubble law at low redshifts, one can obtain a consistent picture for the fundamental cosmological parameters, within the framework of the Big Bang cosmology.

According to these estimates (see Table 1.1) the three-dimensional geometry of the universe is flat to a high degree of accuracy and the total density is dominated by the cosmological constant, whereas matter makes only ~25 per cent of the total. In addition, the matter density appears dominated by the elusive dark matter, while the familiar baryonic matter makes only a negligible fraction of the total Ω. The estimated (small) value of the present baryonic matter density also agrees with some determinations of the ^{2}D abundance in high redshift gas clouds (supposed to be of primordial origin) and recent estimates of the initial He content in Galactic globular clusters (made of stars formed close to the Big Bang epoch). Since we do not know what dark matter is and what physical energy is represented by the cosmological constant, we are in the situation of ignoring the origin of more than 90 per cent of

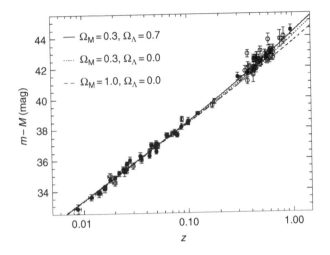

Figure 1.5 Empirical d_L–z relationship using Type Ia supernovae as distance indicators (filled and open circles; data from [146] and [164]) compared with the theoretical results for three different choices of the matter (baryonic plus dark – Ω_M) and cosmological constant (Ω_Λ) density parameters (but the same value of H_0). The luminosity distance d_L is related to the displayed $(m - M)$ – called distance modulus – by $(m - M) = 5\log(d_L) - 5$, where d_L is in parsec and $(m - M)$ is in magnitudes (see Chapter 8)

Table 1.1 Basic cosmological parameters (from [10])

Temperature CMB	T_{CMB} (K)	2.725 ± 0.002
Hubble constant	H_0 (km Mpc s^{-1})	71^{+4}_{-3}
Total density	Ω	1.02 ± 0.02
Cosmological constant density	Ω_Λ	0.73 ± 0.04
Baryon density	Ω_b	0.044 ± 0.004
Dark matter density	Ω_{DM}	0.22 ± 0.04
Photon density	$\Omega_\gamma (10^{-5})$	(4.800 ± 0.014)
Age of the universe	t_0 (Gyr)	13.7 ± 0.2
Redshift of matter–energy equality	z_{eq}	3233^{+194}_{-210}
Redshift of decoupling	z_{dec}	1089 ± 1
Age at decoupling	t_{dec} (10^3 yr)	379^{+8}_{-7}

the matter/energy content of the universe, although we can 'feel' its presence from its gravitational influence.

1.7 The inflationary paradigm

According to the results reported in Table 1.1 the universe is flat, with $\Omega = 1$ to a very good approximation. The flatness is potentially a problem because Equation (1.16) tells us that is an unstable condition. Consider a matter dominated universe. In this case one can neglect Λ in Equation (1.15), multiply both terms by $3/(8\pi G\rho(t))$ and obtain

$$\frac{3H(t)^2}{8\pi G\rho(t)} - 1 = -\frac{3kc^2}{8\pi G\rho(t)R(t)^2}$$

which can be rewritten as

$$(\Omega^{-1} - 1)\rho(t)R(t)^2 = \text{constant}$$

The right-hand side of this equation is a constant, and therefore we can write, for two different values of the cosmic time t,

$$(\Omega^{-1} - 1)\rho(t)R(t)^2 = (\Omega_0^{-1} - 1)\rho(t_0)R(t_0)^2$$

where the right-hand side contains the present values and the left-hand side the corresponding values at a given earlier time t. This latter equation can also be rewritten as

$$(\Omega^{-1} - 1) = (\Omega_0^{-1} - 1)\frac{\rho(t_0)}{\rho(t)}\left(\frac{R(t_0)}{R(t)}\right)^2$$

and since $R(t) = R(t_0)/(1+z)$, $\rho(t) = \rho(t_0)(1+z)^3$, we obtain

$$(\Omega^{-1} - 1) = \frac{(\Omega_0^{-1} - 1)}{(1+z)}$$

From this equation we can easily see that if Ω was at the beginning only slightly different from unity, then it could not possibly be equal to unity nowadays. For example, about 1 second after the Big Bang ($z \approx 10^{11}$) Ω had to be different from unity by less than $\approx 2 \times 10^{-13}$, in order to have Ω within 0.02 of unity today. This is the so called flatness problem: why was Ω so finely tuned?

The second problem faced by our understanding of the universe is the so-called horizon problem. The CMB across the sky is to a very good approximation isotropic, thus confirming one of the assumptions behind the cosmological principle. However, the size of the region in causal contact with a given observer increases with time for a flat universe, and its size at decoupling was much smaller than at the present time, corresponding to about only one degree in the sky today. Why are two points at the opposite sides of the sky at the same temperature (apart from the small primordial fluctuations) even though no information was able to travel from one to the other at decoupling?

A third question is: what is the origin of the primordial fluctuations and why were they triggered all at the same time?

One can, in principle, consider these three occurrencies as the initial conditions of our universe; however, to avoid such a finely tuned choice of initial conditions, the so-called inflationary paradigm was proposed in the 1980s by Guth, Linde, Sato, Albrecht and Steinhardt. The central idea is that there is a period in the early universe where a term Λ_{inf} – originated by some hypothetical quantum field – analogous to the cosmological constant dominates Equation (1.15), that can therefore be rewritten as

$$H(t)^2 = \frac{\Lambda_{inf}}{3}$$

The solution of this equation, after recalling the definition of $H(t)$, and assuming a constant Λ_{inf}, is

$$R(t) = R(t_i)e^{\sqrt{\Lambda_{inf}/3}\,t} = R_i e^{H(t)t}$$

if t is much larger than the cosmic time $t = t_i$ of the beginning of the Λ_{inf} dominated epoch.

   If this exponential expansion (inflation) is long enough, it will drive Ω towards 1, irrespective of its initial value; this happens because $R(t)$ increases exponentially, $H(t)$ is constant (its value set by the value of Λ_{inf}) and therefore, following Equation (1.16), $\Omega \to 1$ if $R(t)$ has increased enough during this phase. Moreover, during inflation, a very small patch of the universe can grow to enormous dimensions, so that the isotropy of the CMB temperature, we see today, arose from a very small causally connected region that underwent an inflationary growth. An expansion by a factor of $\approx 10^{30}$ solves both the flatness and horizon problem without invoking ad hoc initial conditions. The quantum field that originated Λ_{inf} is expected to experience quantum fluctuations that were stretched by the inflation to the scales we see imprinted in the CMB. Therefore the simultaneous triggering of the primordial fluctuations is due to the onset of inflation. The general belief is that the inflation occurred when the strong force separated from the electroweak one, at about $t = 10^{-35}$ s, and lasted until about $t = 10^{-32}$ s.

1.8 The role of stellar evolution

The theory of stellar evolution that we will present in the following chapters is devoted to unveiling the physical and chemical properties of the stars populating the universe, and their development with time. The role played by stars in our understanding of the mechanisms driving the evolution of the matter created during the Big Bang is paramount. Take the human body as an example. Its chemical composition comprises ~50 per cent of carbon, ~20 per cent of oxygen, ~8.5 per cent of nitrogen plus ~10 per cent of heavier elements, the remaining fraction being hydrogen. Apart from hydrogen, no other elements were produced during the primordial nucleosynthesis.

   After the groundbreaking study by Burbidge, Burbidge, Fowler and Hoyle ([26]) we know that all elements heavier than helium are produced in stars and injected into the interstellar medium by mass loss processes and explosions. Out of this chemically enriched matter new generations of stars are formed and the cycle is perpetuated until interstellar gas is available to form new stars. Stellar evolution theory is therefore a fundamental tool to understand the chemical composition of the present universe and how it evolved from the Big Bang epoch. It also provides powerful tools to study the timescale and mechanisms for the formation of the stellar populations and galaxies we see in the universe, as well as cosmic yardsticks to estimate distances and investigate the kinematical status of the universe.

Starting from the basics of stellar physics, in the following chapters we will build up a comprehensive picture of how stars form and evolve, how their physical and chemical properties change with time, and how one can take advantage of this knowledge to address broader questions related to galaxy formation and evolution and the estimate of the main cosmological parameters.

2   Equation of State of the Stellar Matter

2.1 Physical conditions of the stellar matter

Stars are made of a mixture of gas (molecules, atoms, ions and electrons) plus radiation. At temperatures higher than 10^4–10^5 K the atoms are partly or completely ionized, i.e. they have lost part or all of their electrons; below a few thousand K molecules are formed. Free electrons come mainly from the ionization of atoms and, in the case of very high temperatures and low densities, also from electron–positron pair production at the expenses of the radiation field. They are distributed among the ions, and all the gas particles are embedded into a sea of photons.

The thermodynamical properties of the stellar matter are described by the equation of state (EOS) which uniquely determines the fractions of free electrons, neutral and ionized atoms (and molecules), their ionization states, the pressure P and all other thermodynamic quantities as functions of density ρ, temperature T and chemical composition of the gas. This chemical composition is parametrized by the so-called mean molecular weight μ, which is the ratio between the average mass of a gas particle ($<m>$) and the atomic mass unit ($m_H = 1.6605 \times 10^{-24}$ g). Changing the chemical composition of the stellar gas does change the average mass of the gas particles, with a consequent change of μ. The average gas particle mass can also be written as the ratio between the density ρ and the number of particles per unit volume n. From this relationship and $\mu = <m>/m_H$, we obtain

$$\mu = \frac{\rho}{n m_H}$$

We consider, as an example, a fully ionized gas; by denoting with X_i, A_i and Z_i the mass fraction (normalized to unity), atomic weight and atomic number, respectively,

Evolution of Stars and Stellar Populations   Maurizio Salaris and Santi Cassisi
© 2005 John Wiley & Sons, Ltd

of element i, the total number of particles per unit volume is given by

$$n = \sum_i \frac{X_i}{A_i m_H} \rho + \sum_i \frac{X_i Z_i}{A_i m_H} \rho$$

where the first term represents the contribution of the ions, and the second one the contribution of the electrons. The ion contribution provides the ion mean molecular weight

$$\mu_i = \frac{1}{\sum_i \frac{X_i}{A_i}}$$

whereas the electron contribution provides the mean electron molecular weight

$$\mu_e = \frac{1}{\sum_i \frac{X_i Z_i}{A_r}}$$

It is evident from the previous definitions that the molecular weight of the gas is the harmonic mean of the ion and electron molecular weights

$$\frac{1}{\mu} = \frac{1}{\mu_i} + \frac{1}{\mu_e}$$

In stellar astrophysics it is common practice to use the symbols X, Y and Z to denote the mass fractions of hydrogen, helium and all other elements heavier than helium (called metals) respectively; these three parameters are related through the normalization $X + Y + Z = 1$. For the metals, the distribution of the individual fractional abundances has to be specified too. The solar heavy element distribution is considered to be the standard metal mixture, and is made up of ~48 per cent (in mass fraction) of oxygen, ~5 per cent of nitrogen, ~17 per cent of carbon and much smaller amounts of other metals (see Table 2.1). Representative values of Y, Z are $Y \sim 0.25$, $Z \sim 0$ for the chemical composition of the matter produced during the primordial nucleosynthesis, $Y \sim 0.25$, $Z \sim 10^{-5}$–10^{-2} for the initial chemical composition of oldest stellar populations in our galaxy (hereafter referred to as Galaxy) – the so-called Population II; $Y \sim 0.27$, $Z \sim 0.02$ for the Sun, $Y \sim 0.30$, $Z \sim 0.03$–0.04 for the more metal-rich stellar populations in the Galaxy – the so-called Population I.

With the above definition of X, Y and Z, we obtain

$$n = \left(2X + \frac{3}{4}Y\right) \frac{\rho}{m_H} + \sum_i \frac{X_i}{A_i m_H} \rho + \sum_i \frac{X_i Z_i}{A_i m_H} \rho$$

where the index i runs over all elements heavier than H and He. Since $\sum_i X_i/A_i$ is much smaller than 1 and $Z_i/A_i \sim 1/2$, we have that

$$n \sim \left(2X + \frac{3}{4}Y + \frac{Z}{2}\right) \frac{\rho}{m_H}$$

Table 2.1 Solar distribution of the most abundant metals (abundances in mass fraction normalized to unity, from [88])

Element	Mass fraction
C	$1.73285E{-}01$
N	$5.31520E{-}02$
O	$4.82273E{-}01$
Ne	$9.86680E{-}02$
Na	$1.99900E{-}03$
Mg	$3.75730E{-}02$
Al	$3.23800E{-}03$
Si	$4.05200E{-}02$
P	$3.55000E{-}04$
S	$2.11420E{-}02$
Cl	$4.56000E{-}04$
Ar	$5.37900E{-}03$
K	$2.10000E{-}04$
Ca	$3.73400E{-}03$
Ti	$2.11000E{-}04$
Cr	$1.00500E{-}03$
Mn	$5.48000E{-}04$
Fe	$7.17940E{-}02$
Ni	$4.45900E{-}03$

and

$$\mu = \frac{1}{2X + \frac{3}{4}Y + \frac{Z}{2}}$$

When the gas chemical composition has a higher fraction of heavier elements, the denominator of this formula decreases and μ increases.

The EOS of the stellar matter can be directly obtained from the computation of the Helmholtz free energy F, in the framework of what is usually called the 'chemical picture'. An alternative way to determine the EOS – that we do not discuss here – is to follow the so-called 'physical picture' described, for example, in [165].

From the point of view of statistical mechanics the free energy of a gas is given by $F = -K_B T \ln(\Xi)$ where K_B is the Boltzmann constant, and Ξ is the so-called partition function of the system, defined as

$$\Xi = \int_E g(E') e^{-E'/K_B T} dE'$$

i.e. an integral over all the available energies E, with $g(E')dE'$ giving the number of states available between E' and $E' + dE'$. When dealing with quantum systems with discrete energy levels

$$\Xi = \sum_E g(E') e^{-E'/K_B T}$$

where $g(E')$ is the number of states with energy E' available to the system.

Given a chemical composition of the stellar matter, one has first to determine the equilibrium values for the numbers N_s of the various possible particle species s, e.g. free electrons, neutral atoms, atoms in the various possible ionization states, fully ionized atoms, molecules, etc.. A very useful property of the free energy F is that it attains a minimum for the equilibrium values of N_s, once the temperature T and density ρ are fixed. To be more specific, we denote with F' a generic expression for the free energy of our system made of a set of arbitrary N_s numbers for the various particle species s, at a given temperature and density. The equilibrium values of N_s and the corresponding free energy F are obtained by minimizing F', with the constraints (called stoichiometric constraints)

$$\frac{dF'}{dN_j} - \frac{dF'}{dN_{j+1}} = \frac{dF'}{dN_e}$$

for an ionization process of the kind $j \leftrightarrow (j+1) + e$, and

$$\frac{dF'}{dN_{AB}} - \frac{dF'}{dN_A} = \frac{dF'}{dN_B}$$

for a molecular dissociation process like $AB \leftrightarrow A + B$, plus the total number and charge conservations. In principle, a term of radiation free energy should be added to F'; however, since it does not depend on the particle concentration, it is not required in the minimization of F', and can be added to F after the minimization procedure.

Once the equilibrium values for N_s are determined, one can compute straightforwardly the appropriate value of F for the system, at the given T and ρ. By expressing F as the free energy per gram, one obtains all the relevant thermodynamical quantities from the well-known relationships that follow (as usual, subscripts denote the quantities that are kept fixed in the differentiation; a subscript μ means that the chemical composition is kept fixed). Internal energy per gram E:

$$E = F - T \left(\frac{dF}{dT} \right)_{\rho,\mu} \tag{2.1}$$

pressure:

$$P = \rho^2 \left(\frac{dF}{d\rho} \right)_{T,\mu} \tag{2.2}$$

specific heats at constant volume, c_V and constant pressure, c_P

$$c_V = \left(\frac{dE}{dT}\right)_{\rho,\mu} \qquad c_P = c_V + P\frac{\chi_T^2}{\rho T \chi_\rho} \qquad (2.3)$$

where $\chi_\rho = (d\ln(P)/d\ln(\rho))_{T,\mu}$, $\chi_T = (d\ln(P)/d\ln(T))_{\rho,\mu}$. By introducing the ratio $\gamma = c_P/c_V$, we can define the adiabatic gradient

$$\nabla_{ad} \equiv \frac{d\ln(T)}{d\ln(P)} = \frac{\gamma - 1}{\gamma \chi_T} \qquad (2.4)$$

As we shall see a little later, the knowledge of P, ρ, c_P, χ_T, χ_ρ and ∇_{ad} is sufficient for the computation of stellar models.

2.1.1 Fully ionized perfect gas

A very simple EOS is that of a perfect (or ideal) monoatomic gas. We will often use this EOS to exemplify the thermodynamical behaviour of the matter in stars. Two conditions define a perfect gas. The first one is that the potential energy of interaction between the gas particles is negligible with respect to their kinetic energy; the second one is that de Broglie wavelength ($\lambda_{dB} = h/p$, where h is the Planck constant and p the momentum of the particle) associated to the gas particles is much smaller than their mean separation. This implies that the momenta p and kinetic energies E_{kin} of the various particle species follow a Maxwell distribution, with

$$n(p)dp = \sqrt{2}\pi N \left(\frac{1}{\pi m K_B T}\right)^{3/2} p^2 e^{-(p^2/2mK_B T)} dp$$

$$n(E_{kin})dE_{kin} = 2\pi N \left(\frac{1}{\pi K_B T}\right)^{3/2} \sqrt{E_{kin}} e^{-(E_{kin}/K_B T)} dE_{kin}$$

being the number of particles with, respectively, momenta between p and $p + dp$ and kinetic energy between E_{kin} and $E_{kin} + dE_{kin}$ (we assumed nonrelativistic energies as appropriate for stellar conditions, but see also the discussion later on about quantum effects). N denotes the total number of particles in the system and m their mass.

The most probable value of E_{kin} for a particle is equal to $K_B T$, whereas the mean value is equal to $(3/2)K_B T$. Using these results, the first condition for a perfect gas implies $K_B T > (Z^2 e^2)/d$ where Z is the charge of the dominant ionic species, and d the average distance between ions. If n is the number of particles per unit volume, one has that

$$\frac{4\pi}{3}nd^3 = \frac{4\pi\rho}{3\mu m_H}d^3 = 1$$

from which one obtains (neglecting a numerical factor close to unity)

$$d \sim \left(\frac{\mu m_{\mathrm{H}}}{\rho} \right)^{1/3}$$

The condition involving the de Broglie wavelength can be written as

$$\frac{h}{p} = \frac{h}{(2mK_{\mathrm{B}}T)^{1/2}} \ll d$$

where m is the particle mass.

One can immediately see that for a given value of d the electron de Broglie wavelength becomes comparable to the electron mean distances well before the same occurs for massive particles like protons and neutrons. The reason is that the electron mass is three orders of magnitude smaller than the proton and neutron masses. The free energy F per unit mass of a perfect gas is given by

$$F = -K_{\mathrm{B}}T \sum_{s} N_{s} \left[\frac{3}{2} \ln(T) - \ln(N_{s}\rho) + \ln(G_{s}) + 1 + \ln(g_{s}) \right] + F_{\mathrm{int}} \qquad (2.5)$$

where N_{s} is the number of particle species s in the unit mass (e.g. electrons, neutral molecules and atoms, fully ionized atoms, partially ionized ions), g_{s} is the statistical weight (the number of states with the same energy available to an individual particle, i.e. $g_{\mathrm{s}} = 2$ for electrons, because of their spin 1/2) and

$$G_{s} = (2\pi K_{B} m_{s}/h^{2})^{3/2}$$

with m_{s} being equal to the mass of the particle species s. F_{int} is the free energy of only the particle species with bound states (they are also included in the sum over the index s) due to their internal degrees of freedom. These particle species – which we label with the index j – can be in various possible ionization states k, and for a given ionization state can be in many possible excitation states i. F_{int} is given by

$$F_{\mathrm{int}} = -K_{\mathrm{B}}T \sum_{j,k} N_{j,k} \ln \left(\frac{e^{-E_{j,k,0}/K_{\mathrm{B}}T}}{g_{k}} \Xi_{j,k}^{\mathrm{int}} \right)$$

This sum involves only particle species j with bound states (e.g. not fully ionized atoms) and runs over all their ionization states k; the energy $E_{j,k,0}$ is the energy of the ground state of the species j in the ionization state k. Ξ^{int} is the internal partition function of species j in the ionization state k, that accounts for all possible excited states i

$$\Xi_{j,k}^{\mathrm{int}} = \sum_{i} g_{i} e^{(E_{j,k,i} - E_{j,k,0})/K_{\mathrm{B}}T}$$

To Equation (2.5) for the free energy of a perfect gas one has to add the contribution of the radiation, supposed to be in thermodynamical equilibrium with matter, i.e.

$$F_{rad} = -\frac{4}{3}\frac{\sigma T^4}{c\rho}$$

where $\sigma = 5.67051 \times 10^{-5}\,erg\,cm^{-2}\,s^{-1}\,K^{-4}$ is the Stefan–Boltzmann constant that is related to the black-body constant a by the relationship $a = 4\sigma/c$.

We consider the case when matter is fully ionized; in this case F_{int} is equal to zero because there are no species with bound states. We also neglect F_{int} when atoms and molecules are neutral and $K_B T$ is much smaller than the smallest possible energy difference between the internal states of the bound particle species. In this case the electronic degrees of freedom (and the vibrational and rotational modes of molecules) are not excited; therefore F_{int} contributes as a constant to the internal energy E, but gives zero contribution to all other relevant thermodynamical quantities obtained from differentiating either F or E.

We can now use Equation (2.1) and write the total number of particles in a unit mass as $1/(\mu m_H)$ to obtain the following result for the internal energy E per unit mass of a perfect monatomic gas of matter plus radiation in thermodynamical equilibrium:

$$E = \frac{3}{2}\frac{1}{\mu m_H}K_B T + \frac{aT^4}{\rho} \tag{2.6}$$

From Equation (2.2), we obtain the well-know relationship between P and ρ:

$$P = \frac{K_B}{\mu m_H}\rho T + \frac{aT^4}{3} \tag{2.7}$$

where the second term represents the radiation pressure P_{rad}. If the radiation free energy is negligible,

$$\chi_T = \chi_\rho = 1, \quad c_P = \frac{5}{2}\frac{K_B}{\mu m_H}, \quad c_V = \frac{3}{2}\frac{K_B}{\mu m_H}, \quad \nabla_{ad} = 0.4$$

In the case of non-negligible contribution from radiation, we define $\zeta \equiv P_{gas}/P$, where P is the total pressure due to the gas plus radiation; with this definition of ζ one has that $1 - \zeta = P_{rad}/P$. The general expression for E will be

$$E = \frac{K_B T}{\mu m_H}\left[\frac{3}{2} + \frac{3(1-\zeta)}{\zeta}\right]$$

from which one obtains

$$c_P = \frac{K_B}{\mu m_H}\left[\frac{3}{2} + \frac{3(4+\zeta)(1-\zeta)}{\zeta^2} + \frac{4-3\zeta}{\zeta^2}\right]$$

and

$$\nabla_{ad} = \frac{1 + \frac{(1-\zeta)(4+\zeta)}{\zeta^2}}{\frac{5}{2} + \frac{4(1-\zeta)(4+\zeta)}{\zeta^2}}$$

When $\zeta \to 1$ then $P_{gas} \to P$ and $c_P \to \frac{5}{2}\frac{K_B}{\mu m_H}$, $\nabla_{ad} \to 0.4$. In the case where $\zeta \to 0$ then $P \to P_{rad}$, $c_P \to \infty$, and $\nabla_{ad} \to 0.25$.

This simple EOS is very useful for heuristic purposes and can be applied to fully ionized matter as well as to matter where atoms and eventually molecules are neutral and do not interact with each other.

To describe the stellar matter comprehensively under all possible ranges of physical conditions encountered during stellar evolution, three additional effects have to be included in the EOS, namely non-ideal interactions among the charged particles, the ionization process of atoms (and eventually molecules), and quantum effects on the momentum and energy distribution of the gas particles.

2.1.2 Electron degeneracy

Quantum effects become important when, as mentioned before, the interparticle distance is comparable to the de Broglie wavelength, i.e. when the density is high enough at a given temperature or, conversely, the temperature is low enough at a given density. It has been shown already that electrons are the first particle species to be affected, and we will discuss this case in more detail.

When quantum effects are important, the distribution of the individual electron E_{kin} values cannot follow the Maxwell distribution any more, because – due to the Pauli exclusion principle – each individual quantum state can be occupied at most by one particle. This occurrence modifies the expression that gives the number of particles per unit volume with momenta between p and $p + dp$ ($n(p)dp$) which is given by the Fermi–Dirac distribution

$$n(p)dp = \frac{8\pi p^2}{h^3}\left(\frac{1}{1 + e^{(-\eta + E_{kin}/K_B T)}}\right)dp$$

where E_{kin} is the energy associated with the momentum p and η denotes the so-called degeneracy parameter (electrons are said to undergo 'degeneracy' when they obey this law for the momentum distribution). The degeneracy parameter η is related to the electron number density n_e (which can also be expressed in terms of ρ and μ_e as described above) by the constraint

$$n_e = \frac{8\pi}{h^3}\int_0^\infty \frac{p^2 dp}{1 + e^{(-\eta + E_{kin}/KT)}}$$

In the case of non-relativistic energies the number of electrons per unit volume with energy between E_{kin} and $E_{kin} + dE_{kin}$ is given by

$$n(E_{kin})dE_{kin} = \frac{8\sqrt{2}\pi m_e^{3/2}}{h^3} \left(\frac{E_{kin}^{1/2}}{1 + e^{(-\eta + E_{kin}/K_B T)}} \right) dE_{kin}$$

It can be demonstrated that in the limit of low density and high temperature (non-relativistic energies), i.e. when the condition for the onset of quantum effects is not satisfied, η becomes large and negative and the momentum distribution of the degenerate electrons becomes the Maxwell distribution of a perfect classical gas.

In case of low temperatures and high densities, η is large and positive; if we define the so-called Fermi energy E_F as $\eta = E_F/K_B T$, the term $1/[1+e^{(-\eta + E_{kin}/K_B T)}]$ is equal to one when $E_{kin} < E_F$, whereas it is equal to zero when $E_{kin} > E_F$. The transition from one to zero when $E_{kin} \sim E_F$ becomes steeper, the larger the value of η. When η is equal to ∞ it shows as a discontinuity at $E_{kin} = E_F$ (see Figure 2.1). The case of η approaching ∞ is called complete degeneracy. It is particularly instructive to consider the idealized case of $T = 0$. If $T = 0$ then $\eta = \infty$, and only the quantum states with $E_{kin} \leq E_F$ are populated since $1/[1 + e^{(-\eta + E_{kin}/K_B T)}]$ is equal to zero when E_{kin} is larger than E_F. From the relationship between n_e and η we obtain

$$n_e = \frac{\rho}{\mu_e m_H} = \frac{8\pi p_F^3}{3h^3}$$

where p_F is the momentum corresponding to the Fermi energy E_F. An increase of the density increases p_F and E_F, up to a point when relativistic effects have to be accounted for.

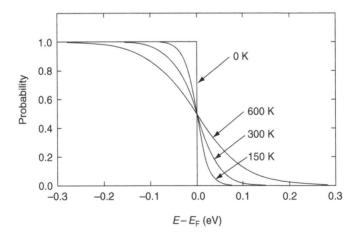

Figure 2.1   Value of the occupation probability $1/[1 + e^{(E_{kin} - E_F)/K_B T}]$ as a function of the electron energy in regime of degeneracy, for four different temperatures

We now give some more detail for the case where $T = 0$. If $T = 0$ the internal energy of the electrons (equal to their kinetic energy if no other particle interactions are considered) is equal to the free energy F. For non-relativistic energies $E_F = p_F^2/2m_e$, and from the previous relationship between ρ and p_F one obtains that $\rho \propto E_F^{3/2}$. The internal energy (hence F) of the electrons can be determined by integrating $n(E_{kin})E_{kin}dE_{kin}$ between 0 and E_F:

$$E = \frac{1}{\rho} \int_0^{E_F} \frac{8\sqrt{2}\pi m_e^{3/2}}{h^3} E_{kin}^{3/2} dE_{kin}$$

(we need the density ρ in the denominator if we want to compute the energy per unit mass) which gives $E = F \propto (1/\rho)E_F^{5/2}$. Recalling that $\rho \propto E_F^{3/2}$ and employing Equation (2.2) we obtain

$$P \propto \rho^{5/3}$$

Inclusion of the neglected constants provides

$$P = \frac{1}{20}\left(\frac{3}{\pi}\right)^{2/3}\frac{h^2}{m_e m_H^{5/3}}\left(\frac{\rho}{\mu_e}\right)^{5/3} = 1.0036 \times 10^{13}\left(\frac{\rho}{\mu_e}\right)^{5/3}$$

where the numerical constant is given in cgs units. It is very important to realize that in the case of electron degeneracy the electron pressure does not depend on the temperature. This has important consequences for stellar evolution, as we will see in the following chapters. Consider now a mixture of classical ions following the perfect gas law and non-relativistic degenerate electrons of arbitrary degeneracy parameter; one can estimate the ratio

$$\frac{P_i}{P_i + P_e} \approx 2.5\frac{\mu_e}{\mu_i}\frac{1}{\eta}$$

where P_i and P_e are the ion and electron pressure, respectively. The higher the degeneracy parameter, the lower the contribution of the ions to the total gas pressure (but the ions provide the main contribution to the density ρ). As a general rule the electron pressure always dominates when electrons are degenerate. The electron pressure is always strong enough to prevent further substantial contractions (hence a substantial increase in density) of the stellar layers involved (for more detail see Section 3.1.8) thus averting the onset of quantum effects for the ions, with the exception of the neutron stars. For this same mixture of ions and electrons the internal energy is equal to

$$E = \frac{3}{2}\frac{P}{\rho}$$

where $P = P_i + P_e$,

$$c_P \sim \frac{5}{2}\frac{P}{\rho T}\frac{3\mu_e}{2\mu_i\eta}$$

and

$$\nabla_{ad} = 0.4$$

Let us step back to the degenerate electrons and assume now that they have energies in the extreme relativistic limit, i.e. $E_{kin} = pc$; in this case $\rho \propto E_F^3$. By using $E_{kin} = pc$ it is easy to see that E can be obtained from

$$E = \frac{1}{\rho} \int_0^{p_F} \frac{8\pi p^2 pc}{h^3} dp$$

This provides $E = F \propto (1/\rho)E_F^4$. By using $\rho \propto E_F^3$ and Equation (2.2) we finally obtain that $P \propto \rho^{4/3}$. Inclusion of the neglected constants provides

$$P = \left(\frac{3}{\pi}\right)^{1/3} \frac{hc}{8m_H^{4/3}} \left(\frac{\rho}{\mu_e}\right)^{4/3} = 1.2435 \times 10^{15} \left(\frac{\rho}{\mu_e}\right)^{4/3}$$

where the numerical constant is again given in cgs units. For a mixture of classical perfect ions and extremely relativistic degenerate electrons of arbitrary degeneracy parameter,

$$\frac{P_i}{P_i + P_e} \approx 4 \frac{\mu_e}{\mu_i} \frac{1}{\eta}.$$

Moreover,

$$E = \frac{3P}{\rho} - \frac{3K_B T}{2\mu_i m_H}$$

$$c_P = \frac{4P}{\rho T} \frac{3\mu_e}{\mu_i} \frac{1}{\eta} - \frac{3}{2} \frac{K_B}{\mu_i m_H}$$

$$\nabla_{ad} = 0.5$$

The most general case of a system made of a perfect fully ionized classical ion gas, radiation, and electrons that can be relativistically or non-relativistically (partially or fully) degenerate is more elaborate, and the reader is referred to [62], [81] and [207] for a detailed discussion on this subject.

2.1.3 Ionization

Up to now we have discussed the EOS of a fully ionized perfect gas including the effect of electron degeneracy. Another physical process that has to be accounted for is the ionization of the atoms, since the degree of ionization of the stellar matter becomes smaller when approaching the cooler stellar surface. In the atmosphere of

the Sun, for instance, hydrogen and helium are neutral. When the gas is partially
ionized the thermodynamic properties of the stellar matter do depend on the degree of
ionization. As an example, we mention the molecular weight μ, ∇_{ad}, c_P; in particular,
c_P largely increases in ionization regions, with respect to the value determined before
for a fully ionized perfect gas, whereas ∇_{ad} decreases below 0.4.

To obtain an equation providing the number of atoms in a given ionization state, we
start by considering a generic free energy computed for a fixed chemical composition

$$F' = -K_B T \sum_s N_s \left[\frac{3}{2} \ln T - \ln(N_s \rho) + \ln(G_s) + 1 + \ln(g_s) \right] + F_{int}$$

with

$$F_{int} = -K_B T \sum_{j,k} N_{j,k} \ln \left(\frac{e^{-E_{j,k,0}/K_B T}}{g_k} \Xi_{j,k}^{int} \right)$$

and

$$\Xi_{j,k}^{int} = \sum_i g_i e^{(E_{j,k,i} - E_{j,k,0})/K_B T}$$

We consider temperatures so high that the electronic degrees of freedom are
excited, and assume that the free electrons are not degenerate. We now impose the
condition

$$\frac{dF'}{dN_{j,k}} - \frac{dF'}{dN_{j,k+1}} = \frac{dF'}{dN_e}$$

for an ionization process of the kind $k \leftrightarrow (k + 1) + e$, and differentiate appropriately
the free energy with respect to the number of electrons, and of the ions in ionization
states k and k + 1. The contribution to dF'/dN_e comes only from the first term in F',
whereas $dF'/dN_{j,k}$ and $dF'/dN_{j,k+1}$ have contributions from both the first and second
term (F_{int}) of the free energy, since both ionic species have internal bound states. By
noticing that $m_{j,k} \sim m_{j,k+1}$ – i.e. that the masses of atoms of species j ionized k and
k + 1 times are the same – we obtain for the number density of particles of a given
species j in ionization states k and k + 1

$$\frac{n_{j,k+1} n_e}{n_{j,k}} = \frac{\Xi_{j,k+1}^{int}}{\Xi_{j,k}^{int}} \frac{g_e (2 \pi m_e K_B T)^{3/2}}{h^3} e^{-\chi_{j,k}/K_B T} \tag{2.8}$$

where n_e is the number density of electrons, $g_e = 2$, $\Xi_{j,k}^{int}$ the internal partition function
of the ion j in the ionization state k, and

$$\chi_{j,k} = E_{j,k+1,0} - E_{j,k,0}$$

i.e. the energy (always positive) required to remove an electron from the ground state of a k-times ionized particle of species j. Equation (2.8) is the so-called Saha equation; together with the charge and mass conservation the Saha equation determines all the individual number densities of the various particle species, and therefore the value of F and other thermodynamical quantities in the ionization regions. In the case where the free electrons are partially degenerate, the Saha equation can be written as

$$\frac{n_{j,k+1}}{n_{j,k}}e^{\eta} = \frac{\Xi^{int}_{j,k+1}}{\Xi^{int}_{j,k}}e^{-\chi_{j,k}/K_B T} \tag{2.9}$$

where η is the degeneracy parameter for the free electron gas.

Even a cursory look at Equation (2.8) discloses at least one problem. The partition function of isolated atoms formally diverges, because of the infinite number of bound states. Finite values are obtained only when the sum is truncated. A second problem stems from the fact that Equation (2.8) predicts that an ideal ionizing gas at any temperature recombines at sufficiently high densities; this yields the unphysical result that at high densities and high temperatures the gas is predominantly neutral. Both problems are, however, just artifacts of the ideal gas model, that assumes negligible interparticle interactions. In reality the atoms in the stellar gas are not completely isolated, and the Coulomb interaction with neighbouring particles strongly perturbs the higher quantum states of the ions, which are also the less tightly bound. This is easy to envisage given that, in order to determine the electrostatic potential acting on electrons belonging to a given ion, one has to superimpose the electrostatic potential of all the neighbouring particles. Obviously the higher quantum states are the most affected, and as the density raises they are broadened into distributions resembling conduction bands and ultimately destroyed. This phenomenon is often called depression of the continuum.

As far as the ionization process is concerned, the net effect of Coulomb interactions with neighbouring particles is to allow only a finite number of bound states. This, in turn, truncates the partition function avoiding its divergence – since one has to sum over a finite number of excited states only – and also lowers the value of $\chi_{j,k}$, hence increasing the efficiency of the ionization process (pressure ionization effect). This continuum depression is often crudely described by the following approximation

$$\chi'_{j,k} = \chi_{j,k} - \frac{(k+1)e^2}{R_D}$$

when the density exceeds a prescribed value. In this equation R_D is the Debye radius (see our discussion about nuclear reactions for an explanation of the physical meaning of the Debye radius)

$$R_D = \left(\frac{K_B T m_H}{4\pi e^2 \rho \delta^2}\right)^{1/2}$$

and

$$\delta^2 = \sum_i (Z_i^2 + Z_i) \frac{X_i}{A_i}$$

where the sum runs over all atomic species i in the stellar matter, and Z_i, A_i are the individual ionic charges and atomic weights.

A more thermodynamically consistent way to approach this problem is to introduce in the expression for F some additional factors modelling the pressure ionization mechanism, and derive consistently (as shown before) the degree of ionization and thermodynamical quantities. Unfortunately we still lack a definitive theory for the pressure ionization, although various methods can be found in the literature and have been applied to the computation of the stellar EOS ([74, 100, 132, 192]).

2.1.4 Additional effects

The previous discussion about the ionization process has led us to address the problem of non-ideal interactions among the charged particles in the stellar matter. One of the effects of these interactions is to pressure ionize the atoms, but this is not the full story. In general, the EOS deviates from that of an ideal gas because of the effect of Coulomb interactions between the charged particles. Due to the tendency of electrons (negative charge) to cluster around ions (positive charge) additional contributions to the free energy of the system have to be accounted for, whose net effect is to add a negative term to the pressure and internal energy of the stellar matter, for a given pair of density and temperature values. The larger the density at a given temperature, the smaller the mean interparticle distance and the larger the interactions.

The computation of the non-ideal corrections to the EOS is non-trivial, and there is no general analytical solution covering the entire range of physical conditions experienced by the stellar matter. These corrections are usually parametrized as a function of $\Gamma \equiv (Ze)^2/(dK_BT)$, where Z is the charge of the ions, and d is the mean distance between ions. The parameter Γ measures the ratio between the electrostatic interactions and the thermal energy of the ions; when the Coulomb interactions are negligible, i.e. when the density is sufficiently low at a given temperature, $\Gamma \sim 0$. In the case of a fully ionized mixture of various chemical species Γ is defined as

$$\Gamma = \frac{<Z>^{1/3} e^2}{d' K_B T} < Z^{5/3} >$$

with

$$d' = \left(\frac{4\pi\rho}{3\mu_i m_H} \right)^{-1/3}$$

where μ_i is the ionic molecular weight and

$$< Z^\nu > = \frac{\sum_1 n_1 Z_1^\nu}{\sum_1 n_1}$$

n_1 and Z_1 being the number and charge of ions of species 1.

As mentioned before, corrections to the free energy F are usually expressed as a function of Γ; from the value of F corrected for non-ideal effects one then obtains, as usual, all other thermodynamical quantities. When $\Gamma < 0.05$ – a range typical for stars like the present Sun – the correction to the free energy is proportional to $\Gamma^{3/2}$ (Debye–Hückel correction) whereas at higher densities the dependence on Γ is more complicated.

We will discuss with some more detail the properties of the EOS when Γ is large in the sections devoted to white dwarfs and neutron stars. Here we just mention that when Γ increases the kinetic energy distribution of the ions increasingly differs from the Maxwell distribution, due to the electrostatic interactions. When $\Gamma = 1$ the stellar matter behaves like a liquid, and at increasingly larger values the ions tend to form a rigid lattice. The transition to this state is called crystallization, and is associated with the release of latent heat, which has relevant implications for stellar evolution, as we will see later on.

We conclude this section with a very brief mention of two other effects to be included in the computation of the stellar EOS. The first one is the formation of molecules, a process efficient only in the external layers of cool stars. When k_i neutral atoms of type i and k_j neutral atoms of type j combine to form a molecule x, the dissociation equilibrium of the molecule x is usually written as

$$\frac{n_{i,0}^{k_i} n_{j,0}^{k_j}}{n_x} = K_{i,j,x}$$

where $n_{i,0}$ and $n_{j,0}$ are the number densities of the neutral atoms i and j, and n_x is the number density of the formed molecule x. $K_{i,j,x}$ is called the dissociation rate and it is a function of the temperature T. Analytical formulae for the computation of the dissociation rate as a function of T for various molecules can be found in the literature ([139]). In most cases it is sufficient to consider only the formation of the H_2 molecule, for which

$$\log(K_{i,j,1}) = 12.5335 - 4.9252\,\theta + 0.0562\,\theta^2 - 0.0033\,\theta^3$$

where $\theta = 5040/T$ and T is in K.

The equation for the molecular dissociation has then to be coupled to the Saha equation in order to compute all the ion and molecule abundances.

The second effect is the production of electron–positron pairs; this happens at temperatures higher than $\sim 10^9$ K, when the photons of the radiation field have energies larger than $2m_e c^2$. Under these conditions, electron–positron pairs are created spontaneously, and since the produced pair annihilates rapidly, an equilibrium is reached between photons and electron–positron pairs. This reaction adds particles to the gas, and therefore has to be taken into account in the computation of the EOS. The inclusion of pair production in the EOS is discussed in, for example, [62] and [207].

We close this chapter by showing in Figure 2.2 a map of the temperature–density diagram, according to a modern EOS used in stellar evolution computation. The temperature–density stratification of selected stellar models is also displayed. The short-dashed line in Figure 2.2 corresponds to the boundary between the radiation dominated and matter dominated EOS. The long-dashed line in Figure 2.2 corresponds to the boundary where the Coulomb effect is just beginning to be important at this metallicity (solar). The Coulomb effect goes as the cube of the nuclear charge so cores of stars where helium has been substantially processed by nuclear burning (see Chapter 6) have a much larger Coulomb corrections than implied by this figure. The alternating short- and long-dashed line corresponds to the boundary where the pressure ionization is just beginning to be important. The diagonal jagged line continued by the $\log(T) = 6$ isotherm corresponds to the high-density, low-temperature calculational limit of FreeEOS. The medium-weight lines labelled H_2, H, He, and He^+ correspond to the midpoints of the hydrogen dissociation and ionization zones, and the mid-points of the helium first and second ionization zones. The thick solid

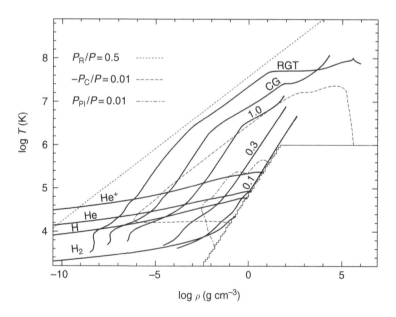

Figure 2.2 Mapping of the temperature–density diagram according to the EOS of the stellar matter, using the publicly available FreeEOS (http://freeeos.sourceforge.net) (courtesy of A. Irwin)

lines labelled by 0.1, 0.3, 1.0, RGT, and CG correspond to the internal temperature and density profiles of Zero Age Main Sequence models with 0.1, 0.3, and 1.0 solar masses, and Red Giant tip (see Section 5.2.2) and clump giant (solar metallicity analog of Zero Age Horizontal Branch star – see Chapter 6) models with $1M_\odot$. These stellar interior results were calculated using the FreeEOS.

3 Equations of Stellar Structure

3.1 Basic assumptions

The standard theory of stellar evolution is based on the following assumptions.

- Stars are spherically symmetric systems made of matter plus radiation. The effects of rotation and magnetic fields are negligible.

- The evolution of the physical and chemical quantities describing a star is slow, i.e. the temporal evolution of the stellar structure can be described by a sequence of models in hydrostatic equilibrium. The assumption of hydrostatic equilibrium (as we will see below) implies that the pressure has to increase toward the centre. In order to increase the pressure, the equation of state dictates that density and temperature have to increase too.

- The matter in each stellar layer is very close to local thermodynamic equilibrium. This hypothesis implies that the average distance travelled by particles between collisions – the mean free path – is much smaller than the dimension of the system, i.e. the radius of the star, and that the time elapsed between collisions is much smaller than the timescale for the change of the microscopic properties of the gas. The consequence of this hypothesis is that, at each point within the star, radiation can be well described by the Planck function corresponding to the unique temperature in common with the matter. This also means that each stellar layer can be assumed to behave like a black body, with (almost) no net energy flux absorbed or emitted. In reality there must be a small outgoing flux, otherwise stars would not shine. This asymmetry is, however, extremely small; in the case of the Sun, the flux at the surface is only $\approx 10^{-13}$ times the flux emitted by 1 cm^2 of a black body at the temperature typical of the Sun centre ($\sim 10^7$ K).

- The only mechanism of chemical element transport within stars is convection, i.e. the effect of rotational mixing and atomic diffusion is negligible.

Evolution of Stars and Stellar Populations   Maurizio Salaris and Santi Cassisi
© 2005 John Wiley & Sons, Ltd

This set of assumptions suffices to provide a system of differential equations able to describe the stellar structure and its temporal evolution. The equations of stellar structure we are going to describe will first be derived – for heuristic purposes – employing the distance r from the stellar centre as an independent variable, and then transformed to equations that use the local mass value m_r as an independent variable. This choice is particularly convenient because stars are subject to large changes of radius during their evolution, so that the local value of r is not always associated to the same mass layer. The whole solution and analysis is simplified and clearer when using the local mass as an independent variable. It is important to note that when the equations are written in terms of m_r, the index r in m_r is simply an index running over the stellar mass layers.

The equations of stellar structure are five differential equations which describe the run of pressure, temperature, luminosity, radius and chemical element abundances as a function of m_r at a given time t, and their evolution with t. The solution of these equations requires the knowledge of auxiliary functions like the equation of state and opacity of the stellar matter, and the rate of energy generation.

3.1.1 Continuity of mass

According to the assumption of spherical symmetry all parameters describing the star depend only on one quantity, i.e. the distance r from the centre. By denoting with ρ the value of the matter density at a generic point r within the star, the mass contained within a sphere of radius r centred on the centre of the star, is given by

$$m_r = \int_0^r 4\pi r'^2 \rho \, dr'$$

Differentiating this equation with respect to the distance from the centre provides

$$\frac{dm_r}{dr} = 4\pi r^2 \rho \qquad (3.1)$$

the 'continuity of mass' equation. One can straightforwardly rewrite Equation (3.1) using m_r as an independent variable:

$$\frac{dr}{dm_r} = \frac{1}{4\pi r^2 \rho} \qquad (3.2)$$

3.1.2 Hydrostatic equilibrium

We now determine the equation of motion of a generic infinitesimal cylindrical volume element with axis along the radial direction, located between radii r and $r + dr$. By denoting with dA the area of its base (perpendicular to the radial direction) and considering the density ρ constant within this volume, we can obtain the mass dm contained in this element according to:

$$dm = \rho \, dr \, dA$$

We neglect rotation and consider self-gravity and internal pressure as the only forces in action. The mass enclosed within the radius r acts as a gravitational mass located at the centre of the star; this generates an inward gravitational acceleration

$$g(r) = \frac{Gm_r}{r^2}$$

Due to spherical symmetry, the pressure forces acting on both sides perpendicular to the radial direction are balanced, and only the pressure acting along the radial direction has to be determined. The force acting on the top of the cylinder is $P(r + dr)dA$, whereas the force acting on the base of the element is $P(r)dA$. By writing

$$P(r + dr) = P(r) + \frac{dP}{dr} dr$$

and remembering that $dr \, dA = dm/\rho$, the equation of motion for the volume element can be written as

$$\frac{d^2 r}{dt^2} dm = -g(r)dm - \frac{dP}{dr} \frac{dm}{\rho}$$

We may now divide both sides of this equation by dm and use the definition of $g(r)$ to obtain

$$\frac{dP}{dr} = -\frac{Gm_r \rho}{r^2} - \rho \frac{d^2 r}{dt^2} \tag{3.3}$$

The condition of hydrostatic equilibrium means that $d^2 r/dt^2 = 0$, hence

$$\frac{dP}{dr} = -\frac{Gm_r \rho}{r^2} \tag{3.4}$$

the 'equation of hydrostatic equilibrium'. This equation implies that the pressure decreases outwards, since the right-hand side is always negative.

We can now try to estimate to what degree the condition of hydrostatic equilibrium is satisfied in real stars. Let us consider the Sun and suppose that a fraction δ of the local gravitational acceleration $g = Gm_r/r^2$ is not compensated by the pressure forces. If a mass element at the surface starts at rest with an acceleration δg, its inward displacement s will be given by $s \sim (1/2)\delta g t^2$. A change by just 5 per cent of the solar radius $R_\odot$ would happen in a time $t \approx 10^3/\delta^{1/2}$ s. Paleontology tells us that the solar radius has not changed by that much for at least 10^6 years ($\sim 3 \times 10^{13}$ s) therefore δ must be less than at least $\approx 10^{-20}$; this means that the hydrostatic equilibrium condition is verified to a very high degree of accuracy.

We can easily rewrite Equation (3.4) using m_r as an independent variable by noting that $dm_r = 4\pi r^2 \rho dr$, obtaining

$$\frac{dP}{dm_r} = -\frac{Gm_r}{4\pi r^4} \tag{3.5}$$

3.1.3 Conservation of energy

If some energy is produced in a spherical shell of thickness dr, located at distance r from the centre of the star, the local luminosity is equal to

$$dL_r = 4\pi r^2 \rho \epsilon dr$$

where ϵ denotes the coefficient of energy generation per unit time and unit mass. This gives

$$\frac{dL_r}{dr} = 4\pi r^2 \rho \epsilon \tag{3.6}$$

the 'equation of conservation of energy'. In terms of m_r we obtain

$$\frac{dL_r}{dm_r} = \epsilon \tag{3.7}$$

3.1.4 Energy transport

If there is energy production and this energy is transported through the star to be released from the surface, we need an equation that describes this process. Inside a star energy can be transported either by random motions of the constituent particles, or by organized large-scale motions of the matter.

In the case of random motions the particles move due to the kinetic energy associated with their temperature, and interact with surrounding particles having covered a given mean free path, thereby transferring energy from hot to cold regions (diffusion approximation).

Radiative transport

Consider a net flux of photons crossing a volume element of unit area and depth dr located at a distance r from the centre of the star. While crossing this volume of stellar matter, due to interactions with the neighbouring particles, some photons will be extracted from the net outgoing flux and redistributed isotropically within the stellar-structure. By denoting with F_{rad} the flux of energy (energy per unit time and unit surface) associated with the outgoing photons, the momentum dp transferred from the photons to the volume element is equal to

$$dp = \frac{dF_{\text{rad}}}{c} = \frac{F_{\text{rad}}}{c} \frac{dr}{l}$$

where l is the photon mean free path. On the other hand dp is also equal to the opposite of the change dP_{rad} of the pressure exerted by the photons over the length dr; hence we obtain

$$dP_{\text{rad}} = -\frac{F_{\text{rad}}}{c} \frac{dr}{l}$$

A basic assumption in this derivation is that the properties of the photons do not change along their mean free path. We now introduce the opacity coefficient κ_{rad}, defined as $\kappa_{\text{rad}}\rho \equiv 1/l$, which is a measure of the probability that the photons experience one interaction per unit length. Using the definition of κ we obtain

$$\frac{dP_{\text{rad}}}{dr} = -\frac{\kappa_{\text{rad}}\rho}{c} F_{\text{rad}} \tag{3.8}$$

The assumption of local thermodynamic equilibrium provides $P_{\text{rad}} = aT^4/3$. Therefore $dP_{\text{rad}}/dr = (4/3)aT^3(dT/dr)$, and Equation (3.8) becomes

$$\frac{dT}{dr} = -\frac{3\kappa_{\text{rad}}\rho}{4acT^3} F_{\text{rad}} \tag{3.9}$$

This is the equation of radiative transport in stellar interiors, when energy is carried by photons; κ_{rad} is the radiative opacity due to the interactions of photons with the surrounding particles. Equation (3.9) shows that whenever there is a temperature gradient there will always be a radiative flux, although the latter may not be the total outgoing flux. If the total energy flux is carried by photons, Equation (3.9) becomes

$$\frac{dT}{dr} = -\frac{3\kappa_{\text{rad}}\rho}{4acT^3} \frac{L_r}{4\pi r^2} \tag{3.10}$$

This equation accounts for the processes of absorption and re-emission of the photons in the stellar interior through the mean opacity coefficient κ_{rad}.

To compute κ_{rad}, one needs to take into account the wavelength dependence of the processes of interaction between stellar matter and photons (see Section 3.1.5) described by monochromatic opacities κ_ν. The mean radiative opacity coefficient κ_{rad} can be computed from κ_ν as follows.

Equation (3.8) also holds when considering separately photons of various frequencies, i.e.

$$\frac{dP_{r,\nu}}{dr} = -\frac{\kappa_\nu \rho}{c} F_\nu \tag{3.11}$$

where F_ν is the monochromatic flux of frequency ν, κ_ν the associated monochromatic opacity, and $P_{r,\nu}$ the corresponding radiation pressure. This equation can be integrated over the frequency spectrum, providing

$$\frac{dP_{rad}}{dr} = -\frac{\rho}{c} \int_0^\infty \kappa_\nu F_\nu d\nu$$

This relationship has to be equivalent to Equation (3.8), therefore

$$\frac{1}{\kappa_{rad}} = \frac{\int_0^\infty F_\nu d\nu}{\int_0^\infty \kappa_\nu F_\nu d\nu} \tag{3.12}$$

This equation shows that κ_{rad} depends not only on the monochromatic opacities κ_ν, but also on the monochromatic fluxes F_ν. For radiation in thermodynamic equilibrium $P_{r,\nu}$ is given by

$$P_{r,\nu} = \frac{4\pi}{3c} B_\nu(T)$$

where $B_\nu(T)$ is the Planck function

$$B_\nu(T) = \frac{2h\nu^3}{c^2} \frac{1}{e^{h\nu/K_B T} - 1}$$

Equation (3.11) can then be used to obtain

$$F_\nu = -\frac{4\pi}{3\kappa_\nu \rho} \frac{dB_\nu(T)}{dr}$$

When this new expression for F_ν is inserted in Equation (3.12), and recalling that $dB_\nu(T)/dr = (dB_\nu(T)/dT)(dT/dr)$, one obtains:

$$\frac{1}{\kappa_{rad}} = \frac{\int_0^\infty \frac{1}{k_\nu} \frac{dB_\nu(T)}{dT} d\nu}{\int_0^\infty \frac{dB_\nu(T)}{dT} d\nu} \tag{3.13}$$

The radiative opacity determined by means of this equation is called Rosseland mean opacity; it is easy to notice that its value is dominated by the frequency intervals where the monochromatic opacity κ_ν is small.

This treatment of radiative transport is valid only in the case of the stellar interiors, where collisions are frequent, the mean free path of the various particles is much smaller than the dimension of the star, and the matter properties do not vary much along a photon mean free path. To be more specific, we can define a new quantity, the optical depth τ, given by

$$\tau = \int_r^\infty \kappa_{rad} \rho \, dr$$

This is a measure of the probability that photons interact with the stellar matter before being radiated away. The diffusion approximation for the radiative transport breaks down when τ is lower than ≈ 1–10. This means that one has to integrate the equations of stellar structure from the centre up to a point where $\tau \sim 1$, that will constitute a sort of 'surface' for the stellar model. The layers where τ is lower than ~ 1 are called 'stellar atmosphere', and are crucial to both predict the spectrum of the radiation emitted by the star, and provide the outer boundary condition necessary to solve the equations of stellar structure.

The mean free path of photons in the solar atmosphere is typically of the order of 10^7 cm, to be compared with a typical value of the order of 1 cm in the solar interior (the solar radius $R_\odot$ is equal to 6.9599×10^{10} cm). The treatment of the radiative energy transport in the stellar atmosphere is much more complicated and we refer the reader to [138] for a comprehensive presentation of the topic. We will come back later to this issue.

Conductive transport

Let us now consider the case of energy transport due to the constituents of the stellar matter other than photons (conductive transport), i.e. free non-degenerate electrons. The energy flux transferred to a volume element of unit area and depth dr by an outgoing flux of electrons is given approximately by

$$F_e \sim -N_e v l \frac{dE}{dr}$$

where N_e is the number of electrons per unit volume, v their average velocity, l the mean free path and E their average kinetic energy. Since $E \propto K_B T$,

$$F_e \sim -K_B N_e v l \frac{dT}{dr}$$

which has the same form as Equation (3.9); the quantity that multiplies dT/dr can therefore also be written as in Equation (3.9), introducing the electron opacity κ_e. We will discuss the computation of κ_e later in this chapter.

This relationship also approximates to the equation for the energy transport due to heavier constituents of the stellar gas, i.e. ions. A detailed study of the processes of interaction among the gas constituents shows that electron transport is very inefficient with respect to radiation, and energy transport due to heavier particles is even less efficient due mainly to the fact that ions are outnumbered by electrons, and they move more slowly at a given temperature. However, in the case where electrons are degenerate, they are able to transport energy much more efficiently (i.e. they have a longer mean free path than in the case of non-degenerate electrons). Therefore the total energy flux will be $F = F_{rad} + F_e$. In its more general formulation the equation of the radiative plus conductive energy transport for the stellar interiors becomes:

$$\frac{dT}{dr} = -\frac{3\kappa\rho}{4acT^3} \frac{L_r}{4\pi r^2} \tag{3.14}$$

where κ is the total opacity of the stellar matter, given by the harmonic mean

$$\frac{1}{\kappa} = \frac{1}{\kappa_{rad}} + \frac{1}{\kappa_e}$$

If electron conduction is effective, $\kappa_e \ll \kappa_{rad}$ and therefore $\kappa \sim \kappa_e$. We can now rewrite this equation for the radiative plus conductive transport in terms of m_r as an independent variable:

$$\frac{dT}{dm_r} = -\frac{3\kappa}{64\pi^2 ac} \frac{L_r}{r^4 T^3} \qquad (3.15)$$

It is useful to transform Equation (3.15) further into a more general expression that can also be used in the case of convective transport. We consider first a generic logarithmic gradient $\nabla \equiv d\ln(T)/d\ln(P)$ that can also be written as $\nabla = (dT/dP)(P/T)$. This implies that $dT/dm_r = \nabla(T/P)dP/dm_r$, and therefore

$$\frac{dT}{dm_r} = -\frac{T}{P}\nabla\frac{Gm_r}{4\pi r^4} \qquad (3.16)$$

having used the right-hand side of the equation of hydrostatic equilibrium (Equation (3.5)) in place of dP/dm_r. In the case of radiative plus conductive transport we denote ∇ with ∇_{rad} and from Equation (3.15) we obtain

$$\nabla_{rad} = \frac{3}{16\pi acG}\frac{\kappa L_r P}{m_r T^4} \qquad (3.17)$$

The same Equation (3.16) can be employed to describe the convective energy transport as well, provided that the appropriate value of ∇ is used.

Convective transport

The third form of energy transport efficient in stars, besides radiation and electron conduction, is convection. Convection is a mechanism of energy and chemical element transport that involves organized large-scale motions of matter in the stellar interior.

Matter inside the stars is never at rest, but usually the movements of the gas elements are small, random perturbations around their equilibrium positions. Under certain conditions, these small random perturbations can trigger large-scale motions that involve sizable fractions of the total stellar mass. These large-scale motions are called convection, and are equivalent to the motion of water elements in a kettle heated from below. Hot gas elements may rise to the top – thereby transporting energy from the hottest to the coolest regions – where they cool and then fall down as cold material.

In order to include convective transport in the stellar model computation one needs to find a criterion for the onset of convection, and the expression of the temperature gradient in a convection zone. It is important to be aware that the treatment of convection in the stellar interior is extremely complicated and needs the introduction of various approximations. This stems from the fact that the flow of gas in a stellar convective region is turbulent, in the sense that the velocity and all other properties of the flow vary in a random and chaotic way. The random nature of the turbulence precludes computations based on a complete description of the motion of all the fluid particles, based on the solution of the Navier–Stokes equations for the stellar fluid (an introduction to the computation of turbulent flows is given in [229]). Instead, we have to adopt a model for the convective transport, that can provide only mean approximate values for the properties of the flow of gas in the stellar convective regions. The model adopted in stellar evolution studies is extremely simple; the gas flow is made of gas elements with a certain characteristic size, the same in all dimensions, that move by a certain mean free path before dissolving. All gas elements have the same physical properties at a given distance r from the star centre. Columns of upward and downward moving gas bubbles are envisaged; upward moving elements start from a given layer, cover a mean free path and then dissolve, releasing their excess heat into the surrounding gas, and are replaced at their starting point by the downward moving elements, that thermalize with the surrounding matter, thus perpetuating the cycle.

The mean free path and characteristic size of the convective elements are assumed to be same (they will be denoted by Λ, the so called 'mixing length') and equal to a multiple of the local pressure scale height H_p, defined as

$$\frac{1}{H_p} = -\frac{1}{P}\frac{dP}{dr} = -\frac{d\ln P}{dr}$$

Usually Λ is written as $\Lambda = \alpha_{ml}H_p$, α_{ml} being a constant to be empirically calibrated. By using the equation of hydrostatic equilibrium, and denoting with g the local acceleration of gravity ($g = (Gm_r)/r^2$) one obtains that $H_p = P/(g\rho)$.

The need for a mean free path in the framework of the mixing length theory can be easily explained as follows. Consider the cross sections of the rising and falling gas columns; if originally in a given layer of a stellar convective region the cross sections were the same, the rising gas (always in pressure equilibrium with the surroundings) will expand by a factor e after a distance equal to H_p; this means that at this point within the star there is much less space available for the falling gas. On the other hand, the amount of falling material must be the same as the rising one, otherwise the star would either dissolve or concentrate all the mass in the interior, thus violating the hydrostatic equilibrium condition. The only solution is that after a distance Λ of the order of H_p part of the material stops and inverts its motion.

This model for convection (which we will denote as 'mixing length theory', and was developed by [14]) is very different from the properties of convective flows in laboratory conditions. In this latter case the gas particles have a spectrum of length scales, some with length comparable to that of the flow boundaries, as well as elements of intermediate and small size; it is at the scale of the smallest elements that the energy of the turbulent flow is dissipated. In spite of this difference from laboratory experiments (that cannot, however, reproduce the conditions typical of stellar interiors) the mixing length theory provides a reasonable qualitative picture of the convective energy transport in stars. By using this simple model we can now determine a local criterion for the onset of convection following the derivation in [113]. 'Local' means that the criterion can be applied on a layer-by-layer basis to check its stability, and involves only physical and chemical quantities evaluated at the layer itself. This is very practical, because in this way there is no need to account for the behaviour of other parts of the star. In realistic cases convective motions are not only dependent on the local conditions but, in principle, have to be coupled to the neighbouring layers. We will see later in this section that the lack of 'non-locality' in the convection treatment of stellar interiors causes some relevant uncertainties.

Let us consider a bubble of gas inside a star, at rest at a distance r from the centre. The bubble will have a pressure P_0, temperature T_0, density ρ_0 and molecular weight μ_0 equal to those of the environment, supposed to be in radiative equilibrium (in this section 'radiative' actually means 'radiative plus conductive') as depicted in Figure 3.1. If the random motions displace the bubble by a small amount Δr away from the equilibrium position, the equation of motion for an element of unit volume can be written as (assuming the viscosity is negligible)

$$\rho \frac{d^2 \Delta r}{dt^2} = -g\Delta\rho$$

where $\Delta\rho$ is the density difference $\rho_{\text{bubble}} - \rho_{\text{surr}}$ between the bubble (supposed to have constant density) and the surroundings, and g is the local acceleration of gravity. One reasonable assumption made in this derivation is that the motion of the bubble is fast enough so that all time derivatives of the mean stellar properties are equal to zero.

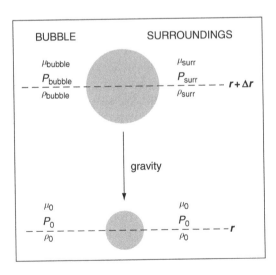

Figure 3.1   Illustration of the physical scenario for the onset of convection

We now assume that along the displacement Δr the bubble is always in pressure equilibrium with the surroundings, i.e. $\Delta P = (P_{\text{bubble}} - P_{\text{surr}}) = 0$, and that μ_{bubble} is always equal to its initial value μ_0 (there is no matter exchange with the surroundings).

The assumption of pressure equilibrium means that the motion of the bubble has to happen with a speed lower than the local sound speed. In the presence of a molecular weight gradient $d\mu/dp$ throughout the region, the difference $\Delta\mu = (\mu_{\text{bubble}} - \mu_{\text{surr}})$ will be equal to $\Delta\mu = \mu_0 - [\mu_0 + (d\mu/dr)\Delta r]$, that provides

$$\Delta\mu = -\frac{d\mu}{dr}\Delta r$$

Using the relationship $d\ln(\mu) = (1/\mu)d\mu$ one gets $\Delta\mu = -\mu(d\ln(\mu)/d\ln(P))$ $(d\ln(P)/dr)\Delta r$. Differentiating with respect to time one obtains

$$\frac{d\Delta\mu}{dt} = -\mu\frac{d\ln(\mu)}{d\ln(P)}\frac{d\ln(P)}{dr}\frac{d\Delta r}{dt}$$

The temperature difference $\Delta T = (T_{\text{bubble}} - T_{\text{surr}})$ depends on the difference between the temperature gradients in the bubble and in the surroundings. It is useful to write ΔT in the following form

$$\Delta T = \left[\left(\frac{dT}{dr}\right)_{\text{ad}} - \left(\frac{dT}{dr}\right)_{\text{rad}}\right]\Delta r - \beta\Delta T dt$$

where $\beta\Delta T$ denotes the rate of temperature change due to energy losses from the bubble, whose efficiency is governed by the parameter β. If $\beta = 0$ the displacement of the bubble is adiabatic, hence the presence of $(dT/dr)_{\text{ad}}$ (adiabatic temperature gradient – $(dT/dr)_{\text{rad}}$ is the temperature gradient for radiative transport assumed to be efficient in the surroundings). By introducing the logarithmic gradient ∇ and differentiating with respect to t, one obtains

$$\frac{d\Delta T}{dt} = T\frac{d\ln(P)}{dr}(\nabla_{\text{ad}} - \nabla_{\text{rad}})\frac{d\Delta r}{dt} - \beta\Delta T$$

If $\Delta P = 0$, and on the assumption that the differences ΔT, $\Delta\rho$ and $\Delta\mu$ are small, we obtain from the EOS

$$\chi_\rho\frac{\Delta\rho}{\rho} + \chi_T\frac{\Delta T}{T} + \chi_\mu\frac{\Delta\mu}{\mu} = 0$$

where

$$\chi_\rho = (d\ln(P)/d\ln(\rho))_{T,\mu}, \; \chi_T = (d\ln(P)/d\ln(T))_{\rho,\mu}, \; \chi_\mu = (d\ln(P)/d\ln(\mu))_{\rho,T}$$

For a perfect gas with negligible radiation $\chi_\rho = \chi_T = 1$, and $\chi_\mu = -1$.

We have derived in this way the following set of four homogeneous equations for the four unknowns ΔT, $\Delta \rho$, $\Delta \mu$ and Δr:

$$\rho \frac{d^2 \Delta r}{dt^2} + g \Delta \rho = 0 \tag{3.18}$$

$$\frac{d \Delta \mu}{dt} + \mu \frac{d \ln(\mu)}{d \ln(P)} \frac{d \ln(P)}{dr} \frac{d \Delta r}{dt} = 0 \tag{3.19}$$

$$\frac{d \Delta T}{dt} + T \frac{d \ln(P)}{dr} (\nabla_{\text{rad}} - \nabla_{\text{ad}}) \frac{d \Delta r}{dt} + \beta \Delta T = 0 \tag{3.20}$$

$$\chi_\rho \frac{\Delta \rho}{\rho} + \chi_T \frac{\Delta T}{T} + \chi_\mu \frac{\Delta \mu}{\mu} = 0 \tag{3.21}$$

One can search for solutions of the form $\Delta x = A e^{nt}$; by inserting into the respective equations this functional dependence for ΔT, $\Delta \rho$, $\Delta \mu$ and Δr, one obtains a non-trivial solution when the determinant derived from the coefficients of A_t, A_ρ, A_μ and A_r is equal to zero, i.e.

$$n^3 + n^2 \beta + n \left[g \frac{\chi_T}{\chi_\rho} \frac{d \ln(P)}{dr} \left(\nabla_{\text{rad}} - \nabla_{\text{ad}} + \frac{\chi_\mu}{\chi_T} \frac{d \ln(\mu)}{d \ln(P)} \right) \right] + \left(\beta g \frac{\chi_\mu}{\chi_\rho} \frac{d \ln(P)}{dr} \frac{d \ln(\mu)}{d \ln(P)} \right) = 0 \tag{3.22}$$

In the case of adiabatic motion of the bubble, $\beta = 0$ and the previous equation reduces to

$$n^2 = -g \frac{\chi_T}{\chi_\rho} \frac{d \ln(P)}{dr} \left(\nabla_{\text{rad}} - \nabla_{\text{ad}} + \frac{\chi_\mu}{\chi_T} \frac{d \ln(\mu)}{d \ln(P)} \right)$$

The quantity n in the case of adiabatic displacement is called the Brunt–Väisälä frequency. If the right-hand side of this equation is positive, n is real and positive, and the amplitude of Δr (and ΔT, $\Delta \rho$, $\Delta \mu$) will increase exponentially, therefore starting a large-scale convective motion. If the right-hand side is negative, n is a complex number with real part equal to zero; this results in an oscillatory motion whose frequency would be equal to n, but with amplitude (given by the real part of n) equal to zero. Therefore, if $\beta = 0$, the condition for the onset of convection is

$$\nabla_{\text{rad}} > \nabla_{\text{ad}} - \frac{\chi_\mu}{\chi_T} \frac{d \ln(\mu)}{d \ln(P)} \tag{3.23}$$

This is the so-called Ledoux criterion. In a region of uniform chemical composition this condition becomes $\nabla_{\text{rad}} > \nabla_{\text{ad}}$, the so-called Schwarzschild criterion.

If $d\ln(\mu)/d\ln(P) > 0$, as is generally true inside stars (the molecular weight decreases towards the surface because heavier elements are produced in the centre) the molecular weight gradient has the effect of stabilizing the matter against convection, because χ_μ is negative, χ_T is positive, and the term added to ∇_{ad} in Equation (3.23) is overall positive. In the case where the environment shows a gradient in molecular weight due to the presence of ionization regions, the convection criterion is again the Schwarzschild one, because the matter inside the convective bubble will also undergo ionization, and therefore $\Delta\mu \sim 0$.

If $\beta > 0$, one must find solutions with the real part positive; if we call α_i the coefficient of the ith power of n in Equation (3.22), we obtain solutions with a positive real part if at least one of the following conditions is satisfied: $\alpha_0 < 0$, $\alpha_1 < 0$, $\alpha_1\alpha_2 - \alpha_0\alpha_3 < 0$. This implies that the layer is convective if at least one of the following conditions is satisfied

$$\frac{d\ln(\mu)}{dr} > 0 \tag{3.24}$$

$$\nabla_{rad} > \nabla_{ad} - \frac{\chi_\mu}{\chi_T}\frac{d\ln(\mu)}{d\ln(P)} \tag{3.25}$$

$$\nabla_{rad} > \nabla_{ad} \tag{3.26}$$

The first condition is the so called Rayleigh–Taylor instability. We will briefly see how this case will be relevant for white dwarf stars (see Section 7.4). In the case of $d\ln(\mu)/dr < 0$ (equivalent to $d\ln(\mu)/d\ln(P) > 0$) Equation (3.25) implies Equation (3.26), as discussed before. This means that the criterion for the onset of convection can be summarized as follows:

$$\frac{d\ln\mu}{dr} > 0$$

or

$$\frac{d\ln\mu}{dr} \le 0 \quad \text{and} \quad \nabla_{rad} > \nabla_{ad}$$

These criteria are valid irrespective of the value of β, i.e. irrespective of the efficiency of the energy losses, as long as β is not zero.

The physical meaning of the criteria discussed above is easy to grasp. Consider a uniform chemical composition, $\beta = 0$ and temperature increasing towards the centre. When $\nabla_{rad} > \nabla_{ad}$, the temperature of the bubble after a displacement Δr towards the stellar surface will be higher than the surrounding; since the pressure within the bubble is the same as that of the surroundings, its density is lower and consequently will receive an acceleration towards the surface. If the displacement is towards the centre, the temperature within the bubble will be lower than the surroundings, implying higher density and an acceleration towards the centre. In both cases the displacement will be amplified, thus inducing large-scale motions. If $d\ln(\mu)/dr < 0$

the density of the environment, at fixed pressure and temperature, will be lower than the case of uniform μ for displacements towards the surface, and higher for displacements towards the centre. This means that the density of the bubble will be lower (higher) than the density of the surroundings if the radiative gradient is larger than the adiabatic one minus an additional amount proportional to the value of the molecular weight gradient.

We now address the case of efficient energy losses and $d\ln\mu/dr \leq 0$. The temperature of a rising bubble is higher than the surroundings, but the mean molecular weight is larger than that of the surroundings and this will prevent upward movements larger than a given length if ∇_{rad} is just larger than ∇_{ad}. The bubble is then pushed downwards, but because of the energy losses the temperature of the bubble is lower than the surroundings when it is moving towards the starting position, so that the element returns to the original position with a larger velocity than in the case of the initial random oscillations in the equilibrium state. On the downward path the same effect operates but with the opposite sign, and cycle after cycle the oscillations increase steadily thus inducing large-scale motions. In the presence of these oscillational motions a less efficient mixing is expected than in the previous cases. We will come back to this point in the chapter about the hydrogen burning phases.

As already mentioned, these criteria for the onset of convection are local. The boundary of the convective regions are fixed by the layer where ∇_{rad} gets lower than ∇_{ad}, that means, at the layer where the random motions of the gas are not accelerated and amplified. However, bubbles coming from inside the convective region are still able to cross (overshoot) this formal boundary because they may have a non-negligible velocity there, and only after crossing the boundary are they slowed down and their motion is halted. The question is: how far do these convective elements travel inside the convectively stable region? A definitive answer to this question requires the adoption of a well tested and established non-local theory of stellar convection that is, however, still lacking.

One could think of applying the concepts of the mixing length theory but, given its local nature, this does not provide a satisfactory answer. In fact, at a given point in the overshoot region the velocity and the temperature excess of a convective element depends on both the local quantities and the amount of braking the element has experienced during the previous path. Rough estimates based on the mixing length theory provide an overshoot distance λ_{OV} (expressed in terms of the pressure scale height H_p at the Schwarzschild convective boundary) either negligible, e.g. $\lambda_{OV} < 0.01H_P$, or much more extended, e.g. $\lambda_{OV} \sim 0.5H_P$. What is done in practice, as for the calibration of the mixing length parameter, is to use empirical constraints to calibrate λ_{OV}.

Having established that the convective transport is efficient when $\nabla_{rad} > \nabla_{ad}$, we have to find an equation for the value of the temperature gradient in a convective region. As a general rule the actual gradient in a convective region, ∇_{conv}, has to satisfy the condition $\nabla_{conv} > \nabla_{ad}$, i.e. it has to be superadiabatic. This is easily understood by imagining a rising gas bubble that dissolves after a length Λ, and releases an amount of heat per unit volume $\delta Q = \rho c_P \delta T$, where c_P is the specific heat per unit mass at constant pressure (surrounding matter has the same pressure of the gas within the bubble) and δT the temperature difference between bubble and surrounding. The difference δT – neglecting energy losses along the path – is proportional to the difference between the adiabatic gradient associated with the motion of the bubble and the temperature gradient in the environment. If the two gradients were to be the same, no heat would be released and no energy transport is possible. Heat is exchanged only if the environment is cooler than the bubble, hence its temperature gradient has to be larger than the adiabatic one.

In the case of convective regions in the stellar interiors, a good approximation is to use the value of the local adiabatic gradient, because the density in the stellar core is so high that the actual convective gradient has to be only negligibly superadiabatic in order to transport the flux of energy. Unfortunately, this simplification is not possible when convection involves layers close to the stellar surface. In this case one needs to find a more complex expression for ∇_{conv}; the equations for ∇_{conv} generally used in stellar evolution computations are based on the so-called mixing length theory (MLT)[1]. The reader has to be warned that various approximations and somewhat arbitrary constant factors enter the MLT; different choices for the values of these factors are possible, but the resulting superadiabatic gradients can essentially be made equivalent by simply rescaling the value of the parameter $\alpha_{ml} = \Lambda/H_P$. An important assumption, as in the derivation for the convection criteria, is that the difference of temperature and density between the convective elements and their surroundings is always small. The goal of the following analysis is to obtain two equations for the unknown gradient and convective velocity, in the framework of the MLT.

We start by enforcing the condition that at a given layer within the convective region the total energy flux is the sum of the flux carried by radiation and convection

$$F_r = \frac{L_r}{4\pi r^2} = F_{rad} + F_{conv}$$

where the radiative flux is given by (see Equation (3.9))

$$F_{rad} = -\frac{4acT^3}{3\kappa\rho}\frac{dT}{dr}$$

dT/dr being the actual temperature gradient in the convective region. By neglecting the energy losses during the displacement of the gas bubbles, the convective flux transported by the matter elements that move upwards with a velocity v is given by

$$F_{conv} = \frac{1}{2}\rho v c_P \left[\left(\frac{dT}{dr}\right)_{ad} - \left(\frac{dT}{dr}\right)\right]\Lambda$$

---

[1] A more sophisticated treatment of surface convection used in some stellar evolution models is the so-called Full Spectrum of Turbulence theory, described in [34].

where the term $(1/2)\, \rho v$ provides the flux of mass per square centimeter per second. The factor $(1/2)$ takes into account the fact that at each layer approximately half of the matter is rising and half is moving downwards. One approximates the density of the bubble with the one of the environment, and all the relevant quantities are evaluated at the layer r, instead of being averages over the mean free path Λ. One needs now an expression for the velocity of the convective elements, that can be derived from the equation of motion. The net force per unit volume f_r acting on the convective element at the layer r is given by $f_r = -g\Delta\rho(r)$. If $\Delta\rho(r)$ (the difference in density between bubble and environment) is approximately zero at r, where the bubble starts its movement, and increases linearly with r (e.g. $\Delta\rho(r) \propto \Delta r$) at a point $r + \Delta r$ we will have $f_{\Delta r} = -g\Delta\rho(\Delta r)$; in this equation we have assumed g is unchanged along the path and denote with $\Delta\rho(\Delta r)$ the density difference evaluated at $r + \Delta r$. The work done per unit volume in moving the bubble through the distance Δr is given by

$$W(\Delta r) = -g \int_0^{\Delta r} \Delta\rho((\Delta r)')d(\Delta r)' = -\frac{1}{2}g\Delta\rho(\Delta r)\Delta r$$

since we have assumed that $\Delta\rho \propto \Delta r$. It is now necessary to average $W(\Delta r)$ over all possible values of Δr, and the average value is usually set to $(1/4)W(\Lambda)$. This choice takes into account the fact that gas bubbles tend to dissolve on average after a length Λ, and that part of the work goes into energy losses from the bubble, and transfer of kinetic energy to the surroundings. Therefore the average value for the work done is

$$< W(\Delta r) >= \frac{1}{4}W(\Lambda) = -\frac{1}{8}g\Delta\rho(\Lambda)\Lambda$$

It is customary to express $< W(\Delta r) >$ in terms of the average speed v of the convective elements as

$$< W(\Delta r) >= \frac{1}{2}\rho v^2$$

where ρ is the local density of the environment. An extra factor $(1/2)$ is often included in the left-hand side of the previous equation; this factor takes into account the fact that the rising bubbles have to force their way through the surrounding matter, and leaves us with $< W(\Delta r) >= \rho v^2$, whence

$$v^2 = -\frac{1}{8}g\frac{\Delta\rho(\Lambda)}{\rho}\Lambda$$

(notice that $\Delta\rho(\Lambda)$ is negative, because the bubble has a density lower than the environment, therefore v^2 is positive).

In this equation the convective speed is related to $\Delta\rho(\Lambda)$, and we need now to express this quantity as a function of the convective temperature gradient. This is easily done by considering Equation (3.21), that provides

$$\frac{\Delta\rho(\Lambda)}{\rho} = -\frac{\chi_T}{\chi_\rho}\frac{\Delta T(\Lambda)}{T} - \frac{\chi_\mu}{\chi_\rho}\frac{\Delta\mu(\Lambda)}{\mu}$$

Substituting this equation into the expression for the square of the convective velocity provides

$$v^2 = \frac{1}{8}gQ\frac{\Delta T(\Lambda)}{T}\Lambda$$

where

$$Q = \frac{\chi_T}{\chi_\rho} + \frac{\chi_\mu}{\chi_\rho}\left(\frac{d\ln(\mu)}{d\ln(T)}\right)_P$$

In the case of a perfect gas with negligible radiation pressure $Q = 1 - (d\ln(\mu)/d\ln(T))_P$. The temperature difference $\Delta T(\Lambda)$ is given by

$$\Delta T(\Lambda) = \left[\left(\frac{dT}{dr}\right)_{ad} - \left(\frac{dT}{dr}\right)\right]\Lambda$$

By remembering the definition of H_P one can write,

$$\frac{dT}{dr} = -\frac{T}{H_P}\frac{d\ln T}{d\ln P} = -\frac{T}{H_P}\nabla$$

and the equation for the convective velocity can be rewritten as

$$v^2 = \frac{1}{8}g\frac{\Lambda^2}{H_P}Q(\nabla - \nabla_{ad}) \qquad (3.27)$$

where ∇ is the unknown temperature gradient in the convective region. In the same way, the convective flux can be rewritten as

$$F_{conv} = \frac{1}{2}\rho v c_P T\frac{\Lambda}{H_P}(\nabla - \nabla_{ad})$$

Notice how both the convective velocity and the convective flux depend on the free parameter $\alpha_{ml} = \Lambda/H_P$. The condition that the total flux has to be the sum of the radiative plus the convective one can be thus rewritten as

$$\frac{L_r}{4\pi r^2} = \frac{4acgT^4}{3\kappa P}\nabla + \frac{1}{2}\rho v c_P T\frac{\Lambda}{H_P}(\nabla - \nabla_{ad}) \qquad (3.28)$$

Once the total flux is specified by the other equations of stellar structure, Equations (3.27) and (3.28) can be solved to obtain the unknowns ∇ and v. If convection is efficient in the deep stellar interiors, these two relationships provide $\nabla \to \nabla_{ad}$ and velocities of the order of 1–$100\,\mathrm{m\,s}^{-1}$, many orders of magnitude smaller than the local sound speed. On the contrary, in convective layers close to the surface the gradient is strongly superadiabatic and velocities are much larger, of the order of $1-10\,\mathrm{km\,s^{-1}}$, close to the local sound speed.

We conclude this section on the convective energy transport by briefly discussing under what conditions $\nabla_{\mathrm{rad}} > \nabla_{\mathrm{ad}}$ is satisfied inside stars. The value of ∇_{ad} is provided by the equation of state of the stellar matter, and is typically equal to ~ 0.4, with the exception of partially ionized regions where it can drop below 0.1. The radiative gradient is proportional, among others, to the local energy flux and the opacity. When the nuclear energy production has a steep dependence on the temperature, the energy source is concentrated in the very central part of the star, where the local flux is therefore very high and induces a large value of ∇_{rad} that favours the onset of core convection. On the other hand, in the partial ionization regions close to the surface ∇_{ad} is very small and the opacity high (see below) an occurrence that favours the onset of envelope convective regions.

3.1.5 The opacity of stellar matter

As we have discussed before, an evaluation of the opacity κ of the stellar matter is necessary for determining the temperature gradient due to radiative and electron conduction transport. Radiative opacity mechanisms are all those processes that extract photons from the outgoing flux and redistribute them isotropically. There is no net energy loss in the opacity processes; the equation of radiative transport tells us that the effect of the opacity is simply to resist the flow of radiation, analogous to electrical resistance. When the opacity is higher, the temperature gradient has to be steeper in order to force a given flow of photons through the stellar matter. It will be clear from the discussion below that in general κ is a function of the chemical composition of the stellar matter, its temperature and density, and needs an evaluation of the ionization states of the various elements. It goes without saying that an accurate determination of the opacity coefficient rests on the accuracy of the stellar matter EOS.

In the following we summarize briefly the basic processes that contribute to the radiative opacity κ_{rad}. To determine the final value of κ_{rad} one has to sum the contribution of all these processes to the monochromatic opacities κ_ν, and then perform the integration shown in Equation (3.13).

1. **Bound–bound transitions**   Absorption of radiation by an electron bound to an ion that causes the electron to move from one bound state to a more energetic one. After the transition the electron will have to move down to the original bound state, for the ion has to return in equilibrium with the surroundings. The photon emitted during this de-excitation will have the same energy of the absorbed one, but it is re-emitted in a random direction. This process involves only photons of certain frequencies, corresponding to the energy differences between the initial and final bound state of the electron. Bound–bound processes can become important contributors to the total opacity for temperatures below $\sim 10^6\,\mathrm{K}$.

2. **Bound–free transitions** Absorption of radiation by an electron bound to an ion, that moves the electron to a free state. The conservation of energy demands that $h\nu = \chi_0 + (m_e/2)v_e^2$, where ν is the frequency of the absorbed photon, χ_0 the ionization energy and v_e the velocity (relative to the ion) acquired by the free electron. Again, to regain thermodynamical equilibrium an isotropic photon emission will follow with the net effect of depleting the outgoing photon flux. The Rosseland opacity due only to the bound-free process can be approximated by: $\kappa_{bf} = 4.3 \times 10^{25} Z(1 + X)\rho T^{-7/2}$ (the constant is given in cgs units). The dependence $\kappa \propto \rho T^{-7/2}$ is called Kramers' law.

3. **Free–free transitions** Absorption of a photon by an unbound electron that is moving in the field of an ion. These are effectively transitions between unbound electronic states. An approximate formula for the Rosseland opacity due only to this process can be written again as a Kramers' law, i.e. $\kappa_{ff} = 3.7 \times 10^{22} (X + Y)(1 + X)\rho T^{-7/2}$ (constant in cgs units).

4. **Electron scattering** Collisions between photons and electrons scatter both components without loss of energy (Thomson scattering); this process is the dominant source of opacity when the temperature is high and atoms are fully ionized. The scattering cross section is inversely proportional to the square of the electron mass; this explains why nucleon scattering is inefficient, since in this case the cross section would be of the order of 10^6 times smaller than for electrons. For a fully ionized gas the Rosseland opacity due only to electron scattering is given by $\kappa_s \sim 0.2(1 + X)$ in cgs units, i.e. it is independent of temperature and density. This simple formula neglects relativistic corrections (Compton scattering), e.g. the transfer of momentum from the photons to the electrons when photon energies are comparable to the rest mass energy of the electron. The relativistic corrections in principle reduce the value of κ_s given before, but their effect is negligible unless $T > 10^8$ K. Even at $T = 10^8$ K the reduction is only of about 20 per cent.

In the external stellar layers, where temperature and densities are low, the presence of molecules also contributes to the stellar opacity through electronic transitions but also transitions between rotational and vibrational molecular states. The most important molecules that contribute to the Rosseland opacity of the stellar matter (when the temperature is below $\sim$5000 K) at different evolutionary stages are H^-, H_2, TiO, H_2O and CO.

In electron degenerate layers it is also necessary to compute the contribution of the electron conduction opacities κ_{el}. Normally electrons do not transport energy very efficiently (i.e. their mean free path is much shorter than the photon one) but in conditions of electron degeneracy all quantum states with momentum p lower than the Fermi value p_F are filled up, and electrons have difficulty in exchanging momentum when they interact with ions or other electrons. This means that encounters are rare and therefore the mean free path of degenerate electrons is large; degenerate electrons can then transport energy efficiently and the associated opacity is very low (recall that the opacity is inversely proportional to the mean free path) much lower than κ_{rad}. Since the total opacity κ is the harmonic sum of κ_{rad} and κ_{el}, it is the electron conduction opacity which mainly determines the value of κ in condition of electron degeneracy. When the degeneracy is very strong $\kappa_{el} \propto \rho^{-2}T^{2}$. It is interesting to notice that the larger the density the lower κ_{el}, due to the fact that the degree of degeneracy and the electron mean free path both tend to increase; in the case of radiative opacities the opposite is generally true, since higher densities favour the interactions discussed before and therefore increase κ_{rad}. In Figure 3.2, we show in the temperature–density diagram, the regions where the different opacity processes are important.

Updated tabulations of stellar radiative opacity for various chemical compositions and the relevant range of T and ρ are given by [1] and [106]. The former covers only the lower temperatures where molecules (neglected by [106]) are relevant contributors to the Rosseland opacity. Accurate electron conduction opacities are provided by [154].

3.1.6 Energy generation coefficient

In addition to the opacity κ, the energy generation coefficient ϵ (see Equation (3.7)) is needed to solve the equations of stellar structure. As we will see in the following, ϵ is – as in case of the opacity κ – a function of chemical composition, T and ρ.

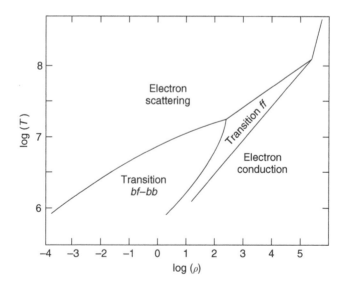

Figure 3.2 The regions in the temperature–density plane where the different opacity processes are important

The energy generation coefficient ϵ contains the following three terms, which we are going to discuss separately

$\epsilon_n \equiv$ energy per unit time and unit mass generated by nuclear reactions;

$\epsilon_g \equiv$ energy per unit time and unit mass generated by the thermodynamical transformations experienced by stellar matter during the star evolution – this contribution is usually (albeit somewhat improperly) named the gravitational energy term;

$\epsilon_\nu \equiv$ energy per unit time and unit mass associated to neutrino production processes, which is effectively subtracted from the stellar energy budget, since neutrinos barely interact with the surrounding stellar matter.

In general $\epsilon = \epsilon_n + \epsilon_g - \epsilon_\nu$.

Nuclear reactions

Most observed stars, including the Sun, are powered by thermonuclear fusion. This term denotes nuclear reactions induced by the thermal motion of the ions, whereby lighter nuclei combine to form a heavier one. If the combined mass of the interacting nuclei (denoted as M_{int}) is larger than the mass of the produced nucleus, M_p, the difference $\Delta M = M_{int} - M_p$ is converted into energy according to the Einstein's relation $E = \Delta M c^2$. As an example, when four hydrogen ions combine to form a helium nucleus, $M_{int} = 4.0324$ atomic mass units and $M_p = 4.0039$ atomic mass units; about 0.7 per cent of the original masses has hence been converted into an energy of about 26.5 MeV. This energy, previously locked into the interacting nuclei is therefore now available to be shared among the various particle species and radiated away.

The mass difference between the interacting protons and the product of their interaction stems from the properties of the nuclear binding energy E_B. The binding energy E_B of a given nucleus is the energy required to separate the nucleons against their mutual attraction due to the nuclear forces, and is essentially a measure of the stability of the nucleus. An approximated semi-empirical formula for E_B is the following

$$E_B(A, Z) = a_1 A - a_2 A^{2/3} - a_3 (Z^2 A^{-1/3}) - 0.25 a_4 (A - 2Z)^2 A^{-1} - a_5$$

where Z is the charge of the nucleus and A its atomic weight. The numerical values of these constants are $a_1 = 16.918$ MeV, $a_2 = 19.120$ MeV, $a_3 = 0.7228$ MeV, $a_4 = 101.777$ MeV and $a_5 = \pm 132 A^{-1}$ MeV where the $+$ and $-$ sign correspond, respectively, to the cases where the number of neutrons and protons are both odd or both even. On the other hand, if the total nuclear mass m_{nuc} is known, then E_B is simply given by

$$E_B(A, Z) = [Z m_p + (A - Z) m_n - m_{nuc}] c^2$$

where m_p is the proton mass (1.672×10^{-24} g) and m_n the neutron mass (1.675×10^{-24} g).

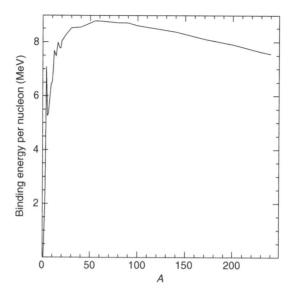

Figure 3.3 Binding energy per nucleon as a function of the atomic weight A

The binding energy per nucleon E_B/A is a most interesting quantity; its value increases with A first steeply – e.g. it is zero for H, 1.11 MeV for deuterium, 7.07 MeV for ^{4}He, 7.98 MeV for ^{16}O – then more gently, until it reaches a maximum of 8.79 MeV around $A = 56$, corresponding to ^{56}Fe, and then drops slowly down to 7.57 MeV for ^{238}U (see Figure 3.3). This means that iron nuclei are the most stable ones. Elements up to ^{56}Fe will therefore produce energy by thermonuclear fusion, whereas for higher values of A it is the splitting of heavier nuclei into lighter ones (nuclear fission) that is able to produce energy.

According to the shell model for the atomic nucleus, protons and neutrons occupy quantized energy levels characterized by three quantum numbers, n (related to the number of nodes of the radial wave function) l (related to the angular momentum vector **l**) and j (the sum of **l** plus the spin vector **s**). For a given value of n, various states of different energies are available, thanks to the spin–orbit interaction, much stronger than in the case of electrons in atoms. The largest gaps between energy levels occur when the number of protons or neutrons arranged in the nucleus is 2, 8, 20, 28, 50, 82, 126. These numbers are called 'magic numbers'. Nuclei with these magic numbers of protons or neutrons are especially stable. Some nuclei, such as ^{4}He and ^{16}O, have magic numbers of both protons and neutrons.

We now focus on the determination of the nuclear energy generation coefficient ϵ_n, and consider a reaction of the type A + b → C + d, in which the target nucleus A interacts with the particle b (typically a proton or a nucleus of another chemical species) and forms nucleus C plus particle d. Three (or more) body interactions have a much smaller probability of happening and can be neglected to a first approximation.

We define a cross section σ as the number of reactions per target nucleus (A) per unit time, divided by the number of incident particles (b) per unit area per unit time (i.e. the flux of incident particles). By denoting with v the relative velocity of species A and b, the number of reactions r per unit volume and time is given by $r = v\sigma(v)N_A N_b$, where N_A and N_b are the number densities of A and b. Since the particles in the stellar gas have a distribution of velocities $n(v)$, one has to integrate the previous expression for r over all the allowed range of relative velocities v

$$r = N_A N_b \int_0^\infty v\sigma(v)n(v)dv = N_A N_b < \sigma v >$$

where $n(v)$ is normalized to one. The product $N_A N_b$ is the number of pairs of interacting particles, $< \sigma v >$ is a measure of the probability that a pair undergoes a reaction of the type described before, and must be determined. If A and b are identical, the number of interacting pairs is $N_A^2/2$, and the previous equation can be generalized as

$$r = N_A N_b \int_0^\infty v\sigma(v)n(v)dv = \frac{N_A N_b}{1+\delta_{Ab}} < \sigma v > \tag{3.29}$$

where δ_{Ab} is equal to one if A and b are the same particle species, otherwise it is zero.

We now define $R \equiv r/\rho$, which corresponds to the number of reactions per unit time and unit mass. By recalling that the mass fraction of a generic element k is given by $X_k = (N_k m_H A_k)/\rho$, where A_k is its atomic weight, we get the following expression for R:

$$R_{Ab} = \rho \frac{X_A X_b}{m_H^2 A_A A_b} \frac{< \sigma v >_{Ab}}{1+\delta_{Ab}}$$

and the nuclear energy generation coefficient can be therefore written as

$$\epsilon_n = R_{Ab} Q_{Ab}$$

with Q_{Ab} being the amount of energy released by a single reaction. The value of Q_{Ab} can be determined from the difference between the sum of the masses of the interacting particles and the sum of the masses of the products. Often nuclear masses are expressed in atomic mass units, and it is worth recalling that 931.494 MeV is the amount of energy associated with one atomic mass unit (1.6605×10^{-24} g).

If a positron is produced by the nuclear reaction, it is customary to add to Q_{Ab} its annihilation energy corresponding to $2m_e c^2 = 1.022$ MeV. In the case where neutrinos are produced, their energy has to be subtracted from the total energy production budget since neutrinos cross the stellar structure without interacting with the rest of the stellar matter, hence effectively taking away energy from the star. Typical cross sections for the interaction of neutrinos with stellar matter are $\approx 10^{-18}$ times smaller than the cross section for the photon interaction with matter.

As an example, we consider the reaction $H + H \rightarrow {}^2D + e^+ + \nu_e$, that is the first stage of hydrogen thermonuclear fusion in stars with masses of the order of the Sun or lower. The difference between the masses of two hydrogen nuclei and the mass of the deuterium nucleus plus the annihilation energy of the positron is equal to 1.442 MeV. The energy associated with the neutrino produced by the reaction (an electron neutrino) is 0.263 MeV and must be subtracted from the previous contribution in order to provide the effective Q value of this reaction.

In order to determine ϵ_n we have still to discuss how $< \sigma v >$ can be determined. Matter involved in nuclear reactions is well approximated by a fully ionized perfect gas, and the particle velocities and kinetic energies therefore follow the Maxwell distribution. If the individual particles have a Maxwell velocity distribution, their relative velocities v also follow the same distribution, and the corresponding kinetic energy E is given by $E = (1/2)mv^2$, where m is the reduced mass $m = (m_A m_b)/(m_A + m_b)$. The expression for $< \sigma v >$ thus becomes:

$$< \sigma v >= \frac{2^{3/2}}{(m\pi)^{1/2}(K_B T)^{3/2}} \int_0^\infty \sigma(E)E \; e^{-E/K_B T} dE \qquad (3.30)$$

In order to calculate $< \sigma v >$ we need further information about the dependence of $\sigma(E)$ on the energy E; this can be obtained by studying the process of thermonuclear fusion in more detail.

Whenever a pair of particles A and b interacts (i.e. they come close enough that the attractive nuclear force tends to fuse them together) a compound nucleus C' is formed. The atomic and mass number of this compound nucleus is the sum of, respectively, the atomic and mass numbers of A and b. The nucleus C' has no memory of how it was formed, since its component nucleons are mixed together independently of their origin, and the energy of the interacting particles is shared among all of them. This compound nucleus is in a particular excited state for a given time τ, and will then break up in many possible ways, producing, among various possibilities, the particles C and d, the same compound nucleus in a stable state with the emission of a photon, or even the interacting particles A and b themselves. Conservation of energy, momentum, angular momentum and nuclear symmetries has to be ensured when determining the possible outcomes of the the decay of the intermediate compound nucleus C'.

The outgoing particles then obtain an amount of kinetic energy that will be shared with the surroundings in thermal equilibrium, with the exception of neutrinos. As an example, we consider an interaction that produces a compound ^{14}N nucleus, with an excitation energy of, for example, 15 MeV. This compound nucleus can then decay according to these channels

$$^{14}N \rightarrow {}^{13}N + n$$
$$^{14}N \rightarrow {}^{13}C + p$$
$$^{14}N \rightarrow {}^{12}C + {}^2H$$
$$^{14}N \rightarrow {}^{10}B + {}^4He$$

Coming back to our generic reaction $A + b \rightarrow C + d$, the probability that it happens depends on the product of the probability that the interacting particles come close enough to experience the effect of the attractive nuclear force, and the probability that this interaction produces the particles C and d (what we will denote as the nuclear part of the reaction).

We now examine separately these two probabilities. In order to obtain the fusion of charged particles like protons, they have to be able to get close enough that the short-range attractive nuclear forces dominate over the long-range repulsive Coulomb forces. Nuclear attraction dominates for distances between A and b smaller than $r_2 = 1.4 \times 10^{-13}(A_A^{1/3} + A_b^{1/3})$ cm (see Figure 3.4). For distances larger than r_2 the repulsive electrostatic force prevails, and the potential changes according to $V_{\text{Coul}} = (Z_A Z_b e^2)/r$, where Z_A, Z_b are the electric charges of A and b. In the stationary reference frame of A, a particle b can classically overcome the electrostatic potential only if its kinetic energy is larger than the Coulomb potential at a distance r_2.

However, due to quantum effects first investigated by G. Gamow, there is a small but finite probability of 'tunnelling' through the Coulomb barrier even for particles with energy E lower than $V_{\text{Coul}}(r_2)$. In terms of the energy E of the relative motion of the colliding particles, this probability is $P = p_0 E^{-1/2} e^{-2\pi\eta}$ where $\eta = (m/2)^{1/2}(Z_A Z_b 2\pi e^2 h^{-1} E^{-1/2})$ (not to be confused with the degeneracy parameter discussed in the section about the EOS) and p_0 a constant that depends on the value of the relative angular momentum and the reduced mass of the two interacting particles. This means that nuclear fusion can happen at temperatures much lower than predicted by classical physics. For a given pair of nuclei the probability of tunnelling increases with increasing E (for two protons it is about 10^{-20} at $T = 10^7$ K typical

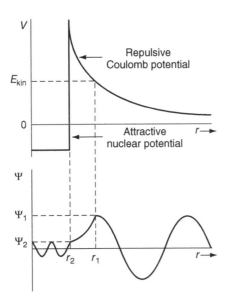

Figure 3.4 Illustration of the potential seen by particle b when approaching particle A with a kinetic energy E_{kin}, and the corresponding wavefunction Ψ; classically, particle b would reach only a distance r_1 from particle A before being repelled by the Coulomb force

of the centre of the Sun, and, due to the large number of reacting particles, it is sufficient to generate the energy needed to support the solar structure). At a given E the probability decreases for increasing Z_A, Z_b, and one needs progressively higher temperatures for nuclear 'burnings' as the various nuclear fusion phases are usually called – involving heavier nuclei.

In astrophysics it is customary to define the quantity

$$S(E) = \sigma(E)E \ e^{+2\pi\eta}$$

called astrophysical factor or S-factor, that is mainly sensitive to the nuclear contribution to the cross section. We may now rewrite Equation (3.30) by including $S(E)$

$$< \sigma v > = \frac{2^{3/2}}{(m\pi)^{1/2}(K_B T)^{3/2}} \int_0^\infty S(E) \ e^{-E/K_B T - 2\pi\eta} dE \qquad (3.31)$$

which is the form usually found in the astrophysical literature. In many relevant astrophysical cases, i.e. the so-called non-resonant reactions, $S(E)$ is a slowly varying function of E in the energy range of interest. An examination of Equation (3.31) (see Figure 3.5) shows that the first term of the product $e^{-E/K_B T} e^{-2\pi\eta}$ represents the high energy wing of the Maxwell distribution – which decreases rapidly for higher energies – and the second one represents the energy dependence of the tunnelling probability – which decreases rapidly for small energies. The product produces a strongly peaked curve, with a maximum at the so-called Gamow peak, which is the most efficient energy for the nuclear reaction to occur. The energy E_G of the Gamow peak is

$$E_G \sim 0.122(m/m_H)^{1/3}(Z_A Z_b)^{2/3}(T/10^9)^{2/3} \text{MeV}$$

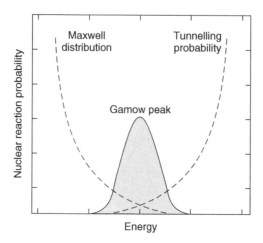

Figure 3.5 Illustration of the location and shape of the Gamow peak (not to scale) as compared with the functions $e^{-E/K_B T}$ and $e^{-2\pi\eta}$

and the full width at half maximum ΔE_G is given by

$$\Delta E_G \sim 0.237(Z_A^2 Z_b^2 m/m_H)^{1/6}(T/10^9)^{5/6} \text{MeV}$$

Both E_G and ΔE_G increase for increasing temperature, and the rate of increase with T is very similar (only slightly higher for ΔE_G); the ratio $\Delta E_G/E_G$ is always below unity and the shape of the peak region remains nearly the same. It is in the Gamow peak energy range that $S(E)$ needs to be known in order to derive the reaction rate. This energy range is at least a factor of about 10 lower than the energies at which the stellar reactions can be produced in laboratories; therefore one has to extrapolate the measured $S(E)$ values to much lower energies if an empirical determination of $S(E)$ is sought.

As mentioned above, $S(E)$ is often a slowly varying function of E and $<\sigma v>$ is usually obtained from Equation (3.31) by considering a second-order Taylor expansion of $S(E)$ and $e^{-E/K_B T - 2\pi \eta}$ around E_G, and neglecting all terms of order higher than $K_B T/E_G$. This provides

$$<\sigma v> \sim S_{\text{eff}} \left(\frac{32 E_G}{3m}\right)^{1/2} \left(\frac{1}{K_B T}\right)^{3/2} e^{-3E_G/K_B T} \tag{3.32}$$

with

$$S_{\text{eff}} = S(E_G)\left(1 + \frac{5K_B T}{36 E_G}\right) + S'(E_G)\left(E_G + \frac{35 K_B T}{36}\right) + \frac{1}{2}S''(E_G)E_G\left(E_G + \frac{89 K_B T}{36}\right) \tag{3.33}$$

S' and S'' being the first and second derivatives of S with respect to the energy. The values for the astrophysical factor and its derivatives are obtained either from laboratory measurements (extrapolated down to the stellar energies) or from theoretical quantum mechanical computations.

Contrary to the previous assumption of a slowly varying $S(E)$ with respect to E, in some cases $S(E)$ happens to show very rapid variations, with narrow peaks at specific values of E (within the Gamow peak range) corresponding to excited energy levels within the nucleus. In this case a resonance is said to occur, and the approximation discussed before breaks down. The cross section is generally greatly enhanced in the presence of a resonance, and it is dominated by the resonance, which occurs at the resonance energy E_R, independent of T (remember that the Gamow peak energy E_G does depend on T). The width Γ of the resonance is proportional to the total lifetime τ of the excited state according to $\Gamma = h/(2\pi\tau)$, and $<\sigma v>$ can be written, in this case ([166])

$$<\sigma v> = \left(\frac{2\pi}{mK_B T}\right)^{3/2} \left(\frac{h}{2\pi}\right)^2 \omega\gamma e^{-E_R/K_B T} \tag{3.34}$$

where $\gamma = \Gamma_{Ab}\Gamma_{Cd}/\Gamma$ and $\omega = (2J+1)/[(2J_A+1)(2J_b+1)]$, and J is the total angular momentum of the excited energy level that gives origin to the resonance, J_A, J_b the angular momenta of the interacting particles; Γ_{Ab}, Γ_{Cd} are defined as Γ, but considering the lifetime for the decay of the compound nucleus into the original interacting particles $A + b$, and into the products $C + d$, respectively. In the case of heavy compound nuclei many resonances are present in the range of stellar energies. In this case one has to sum the contribution of overlapping resonances.

Other nuclear reactions effective in stars are neutron captures, whereby a nucleus of mass A and electric charge Z captures a neutron and becomes a nucleus with mass $A + 1$. Since the electric charge of neutrons is zero, the cross section for neutron captures is much larger than for charged particles, because neutrons do not have to overcome the Coulomb potential. Usually the compound nuclei formed by this process have many resonances in the range of stellar energies, and the resonances are broad, so that they tend to overlap and make the cross section almost independent of T. In fact, experimentally one finds that $<\sigma v>$ is almost constant with temperature. If, however, the target nucleus is light or is magic numbered, there are few resonances and $<\sigma v>$ is no longer a smooth function of T.

Stars can synthesize chemical elements heavier than iron through two kinds of processes involving neutron captures, in those evolutionary phases when appreciable neutron fluxes are produced (see Chapter 7): the s- and r-processes.

Suppose that a sufficient neutron flux is available. This triggers a chain of reactions with nuclei capturing neutrons and producing heavier isotopes of the same element. A generic reaction chain is the following

$$E(A, Z) + n \rightarrow E_1(A + 1, Z)$$

$$E_1(A + 1, Z) + n \rightarrow E_2(A + 2, Z)$$

$$E_2(A + 2, Z) + n \rightarrow E_3(A + 3, Z) \ldots$$

As long as element E_i is stable, the chain can proceed. Otherwise E_i will decay as

$$E_i(A + i, Z) + n \rightarrow F(A + i, Z + 1) + e^- + \bar{\nu}$$

If the element F is stable, a new neutron capture chain can start. Otherwise, multiple β decays follow until a stable element is produced. In this way, increasingly heavier stable elements are created.

The comparison between the neutron capture and the β decay timescales marks the distinction between s- and r-processes. When the timescale for neutron capture is shorter – which means that a very large neutron flux is available – the unstable isotopes will capture more neutrons before decaying and the process is called r-process ('r' stands for rapid). If the opposite is true, the unstable isotope decays before capturing a further neutron. This is the so-called s-process ('s' stands for slow). The practical difference between r- and s-processes is that via the former mechanism one can produce chemical elements heavier than in case of s-processes.

Extensive tabulations of reaction rates for all the relevant reactions happening in stars are listed in [4,51].

It is important to notice that in the previous discussion and the cross sections provided in the literature refer to nuclear reactions happening in vacuum. Instead, ions in stellar interiors are surrounded by a sea of free electrons that tend to cluster near the nuclei and reduce their Coulomb potential. Therefore an approaching particle will feel a Coulomb potential different from the case of an isolated positive charge. This shielding effect of the electrons increases the reaction rates $<\sigma v>$ by a so-called screening factor f that will necessarily be a function of the matter density. Just to give a brief qualitative introduction to the problem of electron screening, we may consider a nucleus of charge Ze that causes a polarization in its surroundings. Negative electrons are attracted by the nucleus and will have a density n_e in its neighbourhood, larger than the average value $<n_e>$ without the electric field of the nucleus. The other positive ions will be repelled and their local density n_i will be lower than the average value $<n_i>$ without the field of the ion. In general terms, given the electrostatic potential u generated by an isolated ion, the surrounding density of particles of charge q will be changed according to

$$n = <n> e^{-qu/K_B T}$$

By considering the electrons and the various species of ions present in the stellar matter, one can determine the effective electrostatic potential u' at a distance r from an ion of charge Ze, assuming spherical symmetry for the surrounding charge distribution; the result is

$$u' = \frac{Ze}{r} e^{-r/R_D}$$

where R_D is the Debye radius defined in Section 2.1.3. For $r \to 0$ the potential becomes the unscreened Coulomb potential. One can say that R_D is a measure of the radius of the electron distribution that shields part of the ion potential to an external observer. A lower electrostatic potential means a lower Coulomb barrier and a larger cross section. By considering again the generic reaction $A + b \to C + d$ and defining $E_D = (Z_A Z_b e^2)/R_D$, the screening effect is considered weak when $E_D \ll K_B T$, and strong when $E_D \gg K_B T$. In general, the effective cross section for a given reaction will be given by the unscreened one discussed previously, multiplied by a factor f that accounts for the screening effects and depends on the charge of the interacting particles and gas chemical composition. It is possible to estimate if the screening effect is substantial by comparing R_D with the closest distance to which the ions can classically approach each other if they have the Gamow energy, i.e. $r_c = (Z_A Z_b e^2)/E_G$. In the case of the solar core and most stellar evolutionary phases $R_D \gg r_c$ and screening effects are relatively small (less than 10 per cent for the Sun) because the electrostatic potential at distances below the Debye radius is only slightly different from the Coulomb unscreened one.

For non-resonant reactions, the weak screening result provides

$$f \approx e^{0.188 Z_A Z_b (\delta \rho)^{1/2} (T/10^6)^{-3/2}}$$

(the factor δ has been defined in the section devoted to the treatment of ionization in the EOS) whereas in the case of strong screening

$$f \approx e^{0.205[(Z_A+Z_b)^{5/3}-Z_A^{5/3}-Z_b^{5/3}](\rho/\mu_e)^{1/3}(T/10^6)^{-1}}$$

The intermediate case as well as screening factors for resonant reactions are more complicated; a full treatment of screening effects can be found in [186,86].

We conclude this section by briefly noticing that the screening factors shown above both predict an increase of f when the density increases and temperature decreases. At extremely high densities and low temperatures typical of astrophysical compact objects, the screening corrections dominate the evaluation of the cross section. In this case one talks of pycnonuclear reactions, whose rate is strongly dependent on ρ but weakly sensitive to the temperature. Pycnonuclear reactions might play a role in very advanced stages of stellar evolution, when they may constitute an energy source for very compact objects (see, for example, [31, 186]).

Gravitational energy

Matter inside a star can in principle experience a series of thermodynamical transformations due to changes in the local radius, pressure and temperature, as well as local chemical composition. Even in the absence of nuclear reactions these thermodynamical transformations can generate energy due to the first principle of thermodynamics, that relates the heat dQ added to the star (per unit mass) to the internal energy per unit mass U and the specific volume $v = 1/\rho$ (i.e. the volume corresponding to the unit mass) through the following relationship

$$dQ = dU + Pdv$$

Dividing both sides by dt one obtains

$$\frac{dQ}{dt} = \frac{dU}{dt} + P\frac{dv}{dt} = -\epsilon_g$$

The minus sign arises from the fact that a positive dQ/dt means energy added to the mass layer, whereas the energy generation coefficient is positive if energy is released by the mass layer. One can write the total differential of the internal energy U as

$$dU = \left(\frac{dU}{dv}\right)_{T,\mu} dv + \left(\frac{dU}{dT}\right)_{v,\mu} dT + \left(\frac{dU}{d\mu}\right)_{T,v} d\mu$$

and the gravitational energy generation coefficient can be rewritten as

$$-\epsilon_g = \left(\frac{dU}{dv}\right)_{T,\mu} \frac{dv}{dt} + \left(\frac{dU}{dT}\right)_{v,\mu} \frac{dT}{dt} + \left(\frac{dU}{d\mu}\right)_{T,v} \frac{d\mu}{dt} + P\frac{dv}{dt}$$

The term $(dU/d\mu)_{T,v}(d\mu/dt)$ gives the variation of U at constant temperature and volume due to the change of chemical abundances. Its contribution to the stellar energy budget is usually negligible when nuclear reactions are efficient, but it is important in the case of white dwarfs where nuclear burnings are inactive. It will be neglected in the rest of this derivation, but we will discuss its effect in Section 7.4 dealing with white dwarf stars.

Our aim is now to rewrite ϵ_g in terms of only P, T, ρ, μ, c_P and ∇_{ad} and derivatives of these quantities. In this way one minimizes the number of variables needed as an input from the EOS and, moreover, ϵ_g will be described by a relationship easy to implement in stellar evolution codes. In the following we will omit the suffix μ for the derivatives of thermodynamical quantities taken at constant chemical composition. From elementary thermodynamics it is known that $(dU/dv)_T = T(dP/dT)_v - P$ and $c_v = (dU/dT)_v$; by substituting these results into the previous equation for ϵ_g we obtain

$$-\epsilon_g = c_v \frac{dT}{dt} + T\left(\frac{dP}{dT}\right)_v \frac{dv}{dt}$$

By recalling the definition of specific volume this equation becomes

$$-\epsilon_g = c_v \frac{dT}{dt} - \frac{T}{\rho^2}\left(\frac{dP}{dT}\right)_v \frac{d\rho}{dt}$$

We can again make use of basic thermodynamics and write $(dP/dT)_v = (p\delta')/(T\alpha)$, where $\delta' = -(d\ln(\rho))/(d\ln(T))_P$ and $\alpha = (d\ln(\rho)/d\ln(P))_T$; moreover, neglecting the differentiation with respect to μ, one can also write $(d\rho/\rho) = \alpha(dP/P) - \delta'(dT/T)$. By substituting these equations into the expression for $-\epsilon_g$ one obtains, after some algebra

$$-\epsilon_g = \left(c_v + \frac{P\delta'^2}{\rho\alpha T}\right)\frac{dT}{dt} - \frac{\delta'}{\rho}\frac{dP}{dt}$$

We now use from thermodynamics the additional relationships $\nabla_{ad} = (P\delta')/(T\rho c_P)$ and $(c_P - c_v) = (P\delta'^2)/(\rho T\alpha)$ to reach the sought result:

$$\epsilon_g = -c_P\left(\frac{dT}{dt} - \nabla_{ad}\frac{T}{P}\frac{dP}{dt}\right) \tag{3.35}$$

It is now natural to ask ourselves what is the energy source responsible for the ϵ_g contribution to the total energy budget. To answer this question we write ϵ_g again in its basic form

$$\epsilon_g = -\frac{dU}{dt} - P\frac{dv}{dt} = -\frac{dU}{dt} + \frac{P}{\rho^2}\frac{d\rho}{dt}$$

where we have used the relationship $v = 1/\rho$. It is evident that the first term in ϵ_g arises from the variation of the star internal energy. As for the second term, let us start by considering the total gravitational potential (Ω) of the star; as we will show in Section 3.1.8,

$$\Omega = -3 \int_0^M \frac{P}{\rho} dm$$

Differentiation of this equation with respect to time does provide

$$\frac{d\Omega}{dt} = -3 \int_0^M \frac{dP}{dt} \frac{1}{\rho} dm + 3 \int_0^M \frac{P}{\rho^2} \frac{d\rho}{dt} dm \qquad (3.36)$$

We now differentiate with respect to time the hydrostatic equilibrium equation (with mass as an independent variable) hence

$$\frac{d(dP/dt)}{dm} = 4 \frac{Gm}{4\pi r^4} \frac{dr}{dt} \frac{1}{r}$$

then multiply both sides by $4\pi r^3$ and integrate over m, obtaining:

$$\int_0^M 4\pi r^3 \frac{d(dP/dt)}{dm} dm = 4 \int_0^M \frac{Gm}{r} \frac{dr}{dt} \frac{1}{r} dm \qquad (3.37)$$

The right-hand side of this equation is $4(d\Omega/dt)$ whereas the integration by part of the left-hand side provides:

$$\left[4\pi r^3 \frac{dP}{dt} \right]_0^M - 3 \int_0^M 4\pi r^2 \frac{dr}{dm} \frac{dP}{dt} dm$$

The term in square brackets vanishes at both the centre ($m = 0$) and the surface ($m = M$) since $r = 0$ at $m = 0$ and $P \sim 0$ independent of time at $m = M$; using the equation of continuity of mass the integral can be rewritten as

$$-3 \int_0^M \frac{dP}{dt} \frac{1}{\rho} dm$$

After these transformations Equation (3.37) can be rewritten as

$$-3 \int_0^M \frac{dP}{dt} \frac{1}{\rho} dm = 4 \frac{d\Omega}{dt}$$

which, once inserted in place of the first term in the right-hand side of Equation (3.36) gives

$$\frac{d\Omega}{dt} = - \int_0^M \frac{P}{\rho^2} \frac{d\rho}{dt} dm \qquad (3.38)$$

The integrand of the right-hand side is the second term contributing to ϵ_g (in addition to the time variation of the internal energy) that now appears explicitly as due to the time derivative of the gravitational potential of the star.

Neutrino production

As mentioned before, neutrinos do not practically interact with the stellar matter (apart for extreme conditions encountered during some type of supernova explosion, as we will see later on) therefore their energy makes a negative contribution to the total stellar energy budget. We should also point out that only electron neutrinos are produced in a stellar environment.

In addition to being created during nuclear burnings by the β decays of unstable nuclei (the negative energy contribution of nuclear reaction neutrinos is already accounted for in the ϵ_g term) neutrinos can be produced in stellar interiors by additional processes, as shown below. The first four of them are purely leptonic processes, i.e. a consequence of the electron–neutrino coupling predicted by the unified electroweak interaction theory. In fact, due to this coupling, a neutrino–antineutrino pair can be produced whenever an electron changes its momentum or an electron–positron pair annihilates. The last two processes are nuclear processes, however, usually independent of the nuclear burnings happening in the stellar centres.

- Pair annihilation process: $e^- + e^+ \rightarrow \nu + \bar{\nu}$. In very hot environments, e.g. for $T > 10^9$ K, the energy of a large fraction of photons is high enough to produce electron–positron pairs $e^- - e^+$, that are annihilated to produce photons and, approximately once in 10^{19} times, a neutrino–antineutrino pair.

- Photoneutrino process: $\gamma + e^- \rightarrow e^- + \nu + \bar{\nu}$. When a Compton scattering between a photon and an electron happens, there is a small probability that after scattering the photon is replaced by a neutrino and antineutrino pair.

- Plasma neutrino process: $\gamma_p \rightarrow \nu + \bar{\nu}$. When a photon propagates in a dense ionized gas, it is coupled with the collective motions of the electrons, and behaves like having an effective mass given by $(h\omega_p)/(2\pi c^2)$, where $\omega_p = (4\pi n_e e^2/m_e)^{1/2}$ in case of non-electron degenerate matter, and $\omega_p = (4\pi n_e e^2/m_e)^{1/2}[1 + (h/(2\pi m_e c))^2 (3\pi^2 n_e)^{2/3}]^{-1/4}$ when matter is electron degenerate (n_e being the electron number density). Under these conditions photons are referred to as plasmons, and they can decay into a neutrino–antineutrino pair. In normal conditions this is not possible because, having the photon a zero rest mass, momentum and energy cannot be both conserved. This plasma neutrino process is particularly effective in dense and electron degenerate stellar cores.

- Bremsstrahlung neutrino process: inelastic scattering (e.g. deceleration) of electrons in the Coulomb field of a nucleus usually leads to the emission of so-called Bremsstrahlung radiation, i.e. photons. However, at very high densities and low temperatures the photons can be replaced by neutrino–antineutrino pairs. This process shows a strong dependence on the charge of the nucleus involved and can only be important for nuclei with a large atomic number.

- Recombination neutrino process: $e^- + (Z, A) \rightarrow (Z - 1, A) + \nu$. In this process neutrinos are produced when a free electron is captured into the K-shell around a fully ionized nucleus of charge Z and atomic weight A. The efficiency of this channel for the neutrino production is restricted mainly to part of the white dwarf star evolution.

- URCA process: $(Z, A) + e^- \rightarrow (Z - 1, A) + \nu, (Z - 1, A) \rightarrow (Z, A) + e^- + \bar{\nu}$. In this process an electron capture on a stable nucleus (Z, A) is followed by a β decay which results in neutrino production but no change of the chemical composition. In order for the URCA process to happen, the nucleus $(Z - 1, A)$ has to be unstable to β decay. If the appropriate nuclei are present in the stellar matter, the efficiency of this process increases with both ρ and T. In general one needs degenerate matter at high densities in order for the electrons to have energies high enough to be captured by nuclei. If only the first part of the URCA process happens (the electron capture) neutrinos are still produced.

Other possible mechanisms for neutrino production like inelastic electron–electron scattering or photon–photon scattering (both producing a neutrino–antineutrino pair) have been found to be completely negligible in stellar conditions. Detailed tabulations and analytic expressions for the various individual energy loss rates due to these mechanisms can be found in [8] and [108]. A simple set of equations that provides the energy loss rate due to the three dominant mechanisms of pair-, photo- and plasma processes is given by [8]

$$\iota = \left(\frac{\rho/\mu_e}{10^9 \mathrm{g/cm^3}} \right)^{1/3} \varphi^{-1}$$

$$\varphi = \frac{T}{5.9302 \times 10^9 \mathrm{K}}$$

$$g(\varphi) = 1 - 13.04\varphi^2 + 133.5\varphi^4 + 1534.0\varphi^6 + 918.6\varphi^8$$

$$f_{\mathrm{pl}} = \frac{(6.002 \times 10^{19} + 2.084 \times 10^{20}\iota + 1.872 \times 10^{21}\iota^2)e^{-5.5924\iota}}{\iota^3 + 9.383 \times 10^{-1}\varphi^{-1} - 4.141 \times 10^{-1}\varphi^{-2} + 5.829 \times 10^{-2}\varphi^{-3}}$$

$$f_{\mathrm{ph}} = \frac{(4.886 \times 10^{10} + 7.580 \times 10^{10}\iota + 6.023 \times 10^{10}\iota^2)e^{-1.5654\iota}}{\iota^3 + 6.290 \times 10^{-3}\varphi^{-1} + 7.483 \times 10^{-3}\varphi^{-2} + 3.061 \times 10^{-4}\varphi^{-3}}$$

$$f_{\mathrm{pa}} = \frac{(2.320 \times 10^{-7} + 8.449 \times 10^{-8}\iota + 1.787 \times 10^{-8}\iota^2)e^{-0.56457\iota}}{\iota^3 + 2.581 \times 10^{-2}\varphi^{-1} + 1.734 \times 10^{-2}\varphi^{-2} + 6.990 \times 10^{-4}\varphi^{-3}}$$

$$\epsilon_\nu = \frac{\rho^2}{\mu_e^3} f_{\mathrm{pl}} + \frac{1}{\mu_e} \varphi^5 f_{\mathrm{ph}} + g(\varphi)e^{(-2/\varphi)} f_{\mathrm{pa}}$$

3.1.7 Evolution of chemical element abundances

The chemical composition of the stellar matter at a given point within a star is subject
to changes with time due to the effect of nuclear reactions and convection. Equations
describing the time evolution of the various chemical species have to be added to
the equations of the stellar structure, since the physical inputs (e.g. EOS, opacities,
nuclear reaction rates) needed for the computation of stellar models are themselves
influenced by the local chemical composition. We first discuss the effect of nuclear
reactions, and find the connection between the number of reactions and the variation
of the abundances of chemical elements.

We consider a generic reaction between two nuclei i and j: $i + j \rightarrow k$. As already
discussed before, the number of reactions per unit volume and unit time is given by

$$r_{ij} = \frac{N_i N_j}{1 + \delta_{ij}} < \sigma v >_{ij}$$

where $< \sigma v >_{ij}$ is the cross section for the reaction between i and j and N_i, N_j are the
numbers of nuclei i and j per unit volume. By recalling that $R_{ij} \equiv r_{ij}/\rho$ is the number
of reactions per unit time and unit mass, the temporal variation of the abundance of
element i (that is destroyed in this reaction) is given by

$$\frac{dX_i}{dt} = -A_i m_H R_{ij} = -\rho \frac{X_i X_j}{m_H A_j} \frac{< \sigma v >_{ij}}{1 + \delta_{ij}}$$

We can now envisage a most general situation where element i is produced by w
reactions of the following kind

$$n_h h + n_k k \rightarrow n_p i$$

and it is destroyed by l reactions of the following kind

$$n_d i + n_j j \rightarrow n_z z$$

By a simple extension of the previous relationships one finds that the variation of the
abundance (by mass fraction) of i will be given by

$$\frac{dX_i}{dt} \frac{1}{A_i} = \sum_w \rho^{n_h + n_k - 1} n_p \frac{X_h^{n_h} X_k^{n_k}}{A_h^{n_h} A_k^{n_k}} \frac{< \sigma v >_{hk}}{m_H^{n_h + n_k - 1} n_h! n_k!} - \sum_l \rho^{n_d + n_j - 1} n_d \frac{X_i^{n_d} X_j^{n_j}}{A_i^{n_d} A_j^{n_j}} \frac{< \sigma v >_{ij}}{m_H^{n_d + n_j - 1} n_d! n_j!}$$

$$(3.39)$$

The first term in Equation (3.39) accounts for all reactions which produce element
i, while the second term involves only the reactions responsible for the destruction
of the same element. A system of algebraic equations of this kind can be written for
all elements involved in the nuclear burnings, and has to be added to the equations

of stellar structure. In actual stellar evolution calculations Equation (3.39) is usually much simpler, since in most cases only one particle of each species is involved in the reactions.

In the case of spontaneous decays involving a generic element i one can write

$$\frac{1}{X_i}\frac{dX_i}{dt}=\frac{1}{\tau_d} \tag{3.40}$$

where τ_d is the decaying constant of the process considered. This equation comes straightforwardly from the well-known equation of radioactive decay.

A very interesting phenomenon happens when we are dealing with a cyclic chain of reactions, where two elements a and c are both produced and destroyed by this kind of reaction chain (an example of this is the *CN cycle* – see Section 5.2.2 – in the hydrogen burning phase)

$$a + b \rightarrow c$$

$$c + b \rightarrow a$$

It is common use to define as 'primary elements' the ones which in a cyclic reaction are just produced or destroyed, as element b in our example; the so-called 'secondary elements' are both produced and destroyed, as elements a and c.

The change of the abundance of element a with time can be written as

$$\frac{dX_a}{dt}=\rho\left(\frac{X_c X_b A_a}{m_H A_c A_b}<\sigma v>_{cb}-\frac{X_a X_b}{m_H A_b}<\sigma v>_{ab}\right)$$

and the change of c is given by

$$\frac{dX_c}{dt}=\rho\left(\frac{X_a X_b A_c}{m_H A_a A_b}<\sigma v>_{ab}-\frac{X_c X_b}{m_H A_b}<\sigma v>_{cb}\right)$$

We assume that ρ and X_b are approximately constant during the process, and introduce the constants $A\equiv(\rho X_b A_a)/(m_H A_c A_b)$, $B\equiv(\rho X_b)/(m_H A_b)$. With these new definitions the equation for X_a becomes

$$\frac{dX_a}{dt}=X_c A<\sigma v>_{cb}-X_a B<\sigma v>_{ab}$$

Being secondary elements, the sum of the abundances by number of a and c must be conserved, i.e.

$$X_a/A_a+X_c/A_c=X_a^0/A_a+X_c^0/A_c=K^0$$

where X_a^0 and X_c^0 are the initial values at time $t = 0$. We can therefore write $X_c = A_c K^0 - (A_c X_a)/A_a$ and transform the equation for the evolution with time of X_a into

$$\frac{dX_a}{dt} = K^0 A_c A <\sigma v>_{cb} - X_a \left(\frac{A_c A}{A_a} <\sigma v>_{cb} + B <\sigma v>_{ab} \right)$$

If the reaction rates are approximately constant during the process we can define the constant $P \equiv \frac{A_c A}{A_a} <\sigma v>_{cb} + B <\sigma v>_{ab}$, and the previous differential equation (with constant coefficients) can be rewritten as

$$\frac{dX_a}{dt} + PX_a = K^0 A_c A <\sigma v>_{cb}$$

with solution given by

$$X_a = X_a^0 e^{-Pt} + X_a^\infty (1 - e^{-Pt})$$

with

$$X_a^\infty = \frac{K^0 A_c A <\sigma v>_{cb}}{P}$$

In the case of element c one obtains an analogous solution with

$$X_c^\infty = \frac{K^0 A_c B <\sigma v>_{ab}}{P}$$

This means that for cyclic reactions, within the conditions described above and given enough time, the elements that are both produced and destroyed tend to converge to abundances constant with time, called equilibrium abundances. These abundances can be different from the initial ones and are related to the rates of the nuclear reactions involved. If the elements a and c start from initial abundances different from the equilibrium values, they will converge to X_a^∞ and X_c^∞ after a time $t \gg \tau_{eq}$, where $\tau_{eq} = 1/P$. The conditions of constant ρ and X_b may be satisfied when, for example, $X_b \gg X_a, X_c$ and the timescale τ_{eq} is much shorter than the timescale for the variation of the physical structure of the star.

Another process able to change the local chemical composition of the stellar matter is convection. Whenever the convective energy transport mechanism is efficient, large-scale motions of the stellar gas are excited; these motions transport energy and matter effectively from one point to another of the stellar interior. In a region where previous (or ongoing) nuclear reactions have established a chemical stratification, the chemical abundance profile is completely erased if convection sets in, as matter becomes fully mixed within the entire convection zone and the timescale for the convective mixing is very short, much shorter than the timescale for abundance changes due to nuclear reactions. This latter timescale can be quantified – in analogy with the radioactive decay constant – by the product $(1/X_i)(dX_i/dt)$, where X_i is the mass fraction of a given element i, determined by Equation (3.39). Typically, the timescale of convective motions (that can be estimated from the convective velocity and typical stellar sizes) is much shorter compared with the inverse of the quantity $(1/X_i)(dX_i/dt)$ due to nuclear reactions. This means that even if nuclear burning is effective in part of the convective region, the chemical composition within the region is always uniform, because at a given time t the local abundance variation due to nuclear reactions is instantaneously redistributed all over the convective region.

With this knowledge in hand we can finally provide a general solution to the problem of the chemical abundance evolution within the stellar structure. In radiative regions, where there is no exchange of matter between neighbouring mass layers, the chemical abundances are changed by the effect of nuclear reactions and their evolution is described by Equations (3.39) and (3.40).

In the case of convective regions, one has to take into account that the newly created nuclei (if nuclear reactions are effective) and all other elements are dispersed homogeneously within the entire convective region. Let us suppose that the convective region extends from mass layer m_1 (inner boundary) to mass layer m_2 (outer boundary) within the star. Inside this region the abundance X_i of a generic element i is constant (and will be denoted as $< X_i >$). At the boundaries (one or both of them) one usually has a discontinuity between the homogeneous convective chemical profile and the profile in radiative regions, that is usually affected by the nuclear reactions (previous and/or present). Due to these effects, the time evolution of $< X_i >$ is to a first order given by

$$\frac{d < X_i >}{dt} = \frac{1}{\Delta m}\left[\int_{m_1}^{m_2}\frac{dX_i}{dt}dm + \frac{dm_2}{dt}(X_{i2}- < X_i >) - \frac{dm_1}{dt}(X_{i1}- < X_i >)\right]$$

(3.41)

where $\Delta m = m_2 - m_1$, X_{i1}, X_{i2} are the abundances on the radiative side of the discontinuities at, respectively, the inner and outer boundary of the convective region. The integral in the right-hand side describes the variation due to the nuclear burnings, whereas the other two terms in brackets describe the change in composition when the boundaries of the convective zone move into surrounding regions of – in principle – inhomogeneous composition.

3.1.8 Virial theorem

The virial theorem plays a fundamental role in stellar evolution theory and here we provide a derivation suited to our purposes; more general derivations are given elsewhere (see, for example, [62], [123]).

Consider a bound spherical gas system of mass M in hydrostatic equilibrium, which can be described by Equation (3.5), and assume that the temperatures are not high enough to start nuclear reactions. Multiplying both sides of Equation (3.5) by $4\pi r^3$ and integrating over dm from the centre to the surface provides

$$\int_0^M \frac{dP}{dm}4\pi r^3 dm = -\int_0^M \frac{Gm}{r}dm$$

An integration by part of the left-hand side of this equation gives

$$[4\pi r^3 P]_0^M - \int_0^M 12\pi r^2 \frac{dr}{dm} P dm = -\int_0^M \frac{Gm}{r} dm$$

Using the fact that $P \approx 0$ at the surface and $r = 0$ at the centre, and applying Equation (3.2) to the second term on the left-hand side of this equation, we obtain

$$3 \int_0^M \frac{P}{\rho} dm = \int_0^M \frac{Gm}{r} dm \qquad (3.42)$$

The right-hand side of Equation (3.42) is $-\Omega$, where Ω denotes the total gravitational energy of the system, i.e. the potential energy of a mass element dm integrated over all the system. The left-hand side is clearly related to the thermodynamics of the system. In the assumption of a perfect monatomic gas the internal energy per unit mass E is equal to (see Equations (2.6) and (2.7)) $(3/2)(P/\rho)$ and Equation (3.42) can be rewritten as

$$E = -\frac{\Omega}{2} \qquad (3.43)$$

which represents the virial theorem.

The total energy E_t of the system is given by $E_t = E + \Omega$; given that $\Omega = -2E$ (Equation (3.43)) we obtain $E_t = -E$ (or $E_t = \Omega/2$, which is negative because the gravitational energy is negative). The total energy is negative, in agreement with the hypothesis that the system is bound. In the specific case of stars (with no active nuclear reactions) an initial hydrostatic equilibrium state is perturbed when energy is radiated away from surface (because stellar surfaces are hotter than the surrounding interstellar space) hence the total energy E_t decreases according to

$$L = -\frac{dE_t}{dt}$$

Following the result of the virial theorem, this is equivalent to

$$L = -\frac{1}{2}\frac{d\Omega}{dt}$$

and Ω has to decrease in order to produce a positive luminosity, i.e. the star has to contract. At the same time the internal energy will increase by an amount $\Delta E = -\Delta E_t = -\Delta\Omega/2$, and thus, somewhat surprisingly, the star heats up (E is proportional to T) when energy is lost from its surface. This fact can be described by saying that stars have a negative specific heat. Summarizing, the loss of energy from the surface causes the contraction of the system; half of the gain in gravitational energy goes into internal energy, the other half is radiated away.

We can use the virial theorem to put constraints on the average internal temperature and density of the star of mass M. The gravitational energy of a sphere of mass M and radius R is given by

$$\Omega = -\alpha \frac{GM^2}{R}$$

where α is a numerical constant of the order of unity, whose exact value depends on the density profile within the system. In the case of constant density $\alpha = 3/5$. If the gas is a fully ionized monatomic perfect gas, the internal energy E is given by

$$E \sim \frac{3}{2} K \bar{T} \frac{M}{\mu m_\mathrm{H}}$$

where $\bar{T}$ is the mean temperature within the star. Introducing now the average density $\bar{\rho} \propto M/R^3$, and expressing R as a function of $\bar{\rho}$, the virial theorem provides

$$\bar{T} \propto M^{2/3} \bar{\rho}^{1/3}$$

This means that two stars of different mass must have different mean densities in order to achieve the same mean temperature, i.e. less massive objects have to be denser.

The virial theorem also provides a criterion for the stability of a star. Considering the equation of state for a generic perfect gas, we can write the internal energy as

$$E = \frac{1}{(\gamma - 1)} \frac{P}{\rho}$$

where the quantity γ has been already defined as $\gamma \equiv c_P/c_v$; it is equal to 5/3 for a perfect monatomic gas. Substituting this equation into Equation (3.42) one obtains a more general expression for the virial theorem

$$E = -\frac{\Omega}{3(\gamma - 1)} \qquad (3.44)$$

By using this equation, and remembering that $E_\mathrm{t} = E + \Omega$, one obtains

$$E_\mathrm{t} = \frac{(3\gamma - 4)}{3(\gamma - 1)} \Omega \qquad (3.45)$$

Equation (3.45) provides a fundamental criterion for the stability of stellar structures: a star will be in hydrostatic equilibrium only as long as γ is larger than 4/3.

In the case where the pressure at the surface of the gas system (P_0) is non-negligible, it is easy to show that the virial theorem becomes

$$3(\gamma - 1)E + \Omega = 4\pi R^3 P_0 \qquad (3.46)$$

3.1.9 Virial theorem and electron degeneracy

We consider now the case of a gas made of degenerate electrons and a perfect monatomic ion gas (an analogous result holds if Coulomb corrections are included). A detailed derivation of the virial theorem for electron degenerate configurations, including the effect of Coulomb corrections can be found in [117].

For highly degenerate electrons (large values of the degeneracy parameter η) $3(\gamma - 1) = 2$ in the non-relativistic case and $3(\gamma - 1) = 1$ in the relativistic case; in the case of the ions $3(\gamma - 1) = 2$. For a mixture of these three cases within the same object, the average $3(\gamma - 1)$ is between 1 and 2 and therefore the total energy is negative and the star can evolve in hydrostatic equilibrium. Let us consider the specific case of non-relativistic degenerate electrons, which corresponds to $3(\gamma - 1) = 2$, as for the perfect ion gas; this implies (assume $P_0 = 0$) that $\Omega = -2E = 2E_t$ and $L = -(1/2)d\Omega/dt$.

The gravitational energy Ω is proportional to $1/R$, hence to $\rho^{1/3}$, so that $(d\Omega/dt)(1/\Omega) = (1/3)(d\rho/dt)(1/\rho)$, where ρ is an average value of the star density. A contraction increases ρ hence the internal energy E_e of the electrons that is given by $E_e \approx \rho^{2/3}$ (see Chapter 2). This means that $(dE_e/dt)(1/E_e) = (2/3)(d\rho/dt)(1/\rho)$, which gives

$$\frac{dE_e}{dt} \approx 2\frac{E_e}{\Omega}\frac{d\Omega}{dt} \tag{3.47}$$

Because of the virial theorem $\Omega = -2E$, where the internal energy E is given by $E = E_e + E_i$, E_i being the energy of the ions. When the electron degeneracy is high $E_e \gg E_i$, hence $\Omega \sim -2E_e$ and from Equation (3.47) we obtain

$$\frac{dE_e}{dt} \approx -\frac{d\Omega}{dt}$$

Using this relationship for the total energy balance provides

$$L = -\frac{dE_t}{dt} = -\frac{dE_i}{dt} - \frac{dE_e}{dt} - \frac{d\Omega}{dt} \sim -\frac{dE_i}{dt} \tag{3.48}$$

This discussion shows that the half of the gravitational energy that is not radiated away increases the internal energy of the electrons, i.e. their Fermi energy, whereas the thermal energy of the ions, proportional to T, decreases; in particular, the energy lost by radiation is nearly equal to the rate of decrease of the thermal energy of the ions. In brief, the virial theorem tells us that the star contracts, the temperature decreases and E_F increases. The contraction of electron degenerate structures is, however, hard to detect – and they can be safely assumed to evolve at constant radius – since the equilibrium configuration of these objects is at small radii (see the section about white dwarf stars) hence $|\Delta\Omega| = |\Delta R|/R^2$ is comparable to L (as prescribed by the virial theorem) for extremely small values of ΔR.

This qualitative picture does not appreciably change when considering a more realistic EOS for the degenerate electrons and the ions.

3.2 Method of solution of the stellar structure equations

We start by summarizing the stellar structure equations derived above (we drop the index r in the notation)

$$\frac{dr}{dm} = \frac{1}{4\pi r^2 \rho} \tag{3.49}$$

$$\frac{dP}{dm} = -\frac{Gm}{4\pi r^4} \tag{3.50}$$

$$\frac{dL}{dm} = \epsilon_n - \epsilon_\nu - c_P \left(\frac{dT}{dt} - \nabla_{ad} \frac{T}{P} \frac{dP}{dt} \right) \tag{3.51}$$

$$\frac{dT}{dm} = -\frac{T}{P} \nabla \frac{Gm}{4\pi r^4} \tag{3.52}$$

$$\frac{dX_s}{dt} \frac{1}{A_s} = \sum_w \rho^{n_h + n_k - 1} n_p \frac{X_h^{n_h} X_k^{n_k}}{A_h^{n_h} A_k^{n_k}} \frac{<\sigma v>_{hk}}{m_H^{n_h + n_k - 1} n_h! n_k!}$$

$$- \sum_l \rho^{n_d + n_j - 1} n_d \frac{X_s^{n_d} X_j^{n_j}}{A_s^{n_d} A_j^{n_j}} \frac{<\sigma v>_{sj}}{m_H^{n_d + n_j - 1} n_d! n_j!} \tag{3.53}$$

Equation (3.53) is actually a set of I equations ($s = 1, \ldots, I$) for the change of the mass fraction of the chemical elements considered; the meaning of the right-hand side of Equation (3.53) has been discussed in the previous section (in case of convective regions – if present – one has to employ Equation (3.41)). These equations and the corresponding physical and chemical variables refer to a generic stellar layer located at a mass coordinate m (m runs from zero to the value of the total stellar mass M) and have to be applied to all mass layers from the centre to the stellar surface.

To solve these equations one needs to know the local values of the auxiliary functions (dependent on the local chemical composition) ϵ_n (and the related nuclear cross sections) ϵ_ν, the equation of state of the stellar matter $P = P(\rho, T)$ and related thermodynamical quantities, the opacities κ and the appropriate temperature gradient ∇ (radiative, adiabatic or superadiabatic). If these auxiliary functions are known we have at each stellar layer a set of $4 + I$ differential equations for the $4 + I$ variables r, P, T, L, X_s (with $s = 1, \ldots, I$); the independent variables are m and t. In general one needs to specify the total mass M of the star and its initial chemical composition; then the equations are solved to determine the structure of the star at the initial instant t_0 and its evolution with time.

The equations must be integrated with numerical techniques, since there is no analytical solution. This means that one can only determine an approximation of the real solution in a discrete number of points (corresponding to discrete values of the mass, that is used as independent variable) within the star and at discrete intervals of time. Equations (3.49) to (3.52) describe the mechanical structure of the star for a given chemical composition, while Equation (3.53) describes the chemical structure and its time evolution.

Time derivatives appear explicitly only in Equations (3.51) and (3.53), i.e. the time evolution is driven by the energy generation and losses, and the consequent chemical transformations of the stellar matter. It is important to realize that the timescales associated with the time derivatives in Equation (3.51) are different from the case of Equation (3.53). The former derivatives are associated with the rate of change of the gravitational and internal energy (the two contributors to ϵ_g) with time; the corresponding timescale is called the Kelvin–Helmholtz timescale and can be estimated by assuming that no nuclear reactions are efficient, and the star physical conditions evolve according to the virial theorem. In this case gravitational and internal energy change with the same speed, and the corresponding timescale is defined as $\tau_{KH} \equiv |\Omega|/L$, where L is the star luminosity and Ω its total gravitational potential energy. In the case of the Sun, with the approximation $\Omega \sim GM_\odot^2/R_\odot - M_\odot$ being the solar mass and $R_\odot$ its Radius – one obtains $\tau_{KH} \approx 10^7$ years. This would approximately be the solar lifetime in the case where nuclear reactions are not active and the solar evolution is driven by the virial theorem.

The time derivatives in Equations (3.53) are instead associated with the characteristic nuclear reaction timescale $\tau_n \equiv E_n/L$, where E_n is the nuclear energy reservoir; in the case of the Sun the nuclear reactions transform H into He. If the Sun were made of pure H and only the central 10 per cent of its mass is hot enough to undergo nuclear reactions, $\tau_n \approx 10^{10}$ years (which is close to a more accurate estimate of the solar lifetime). A comparison between these two timescales shows that $\tau_n \gg \tau_{KH}$; this holds not only for the present Sun, but for the main part of the lifetime of stars with different masses.

A third timescale that is useful to consider (and that we have briefly discussed in Section 3.1) is the so-called free-fall timescale τ_{ff}; τ_{ff} is the timescale for the collapse of the star in the case where the gravitational force is not balanced by the pressure, i.e. the star is not in hydrostatic equilibrium (see Equation (3.50)). In the case of the Sun τ_{ff} is of the order of 10 minutes, and in general $\tau_n \gg \tau_{KH} \gg \tau_{ff}$.

The inequality between τ_n and τ_{KH} means that Equation (3.53) can be decoupled from the other four equations. This is no longer true in the most advanced evolutionary phases of massive stars, where the speed of the chemical evolution of the stellar matter becomes comparable to the evolution of the physical variables. Decoupling Equation (3.53) from the other equations means that one can solve the mechanical part of the star at a given instant t with a given set of chemical abundances, then apply a time step Δt, solve Equation (3.53) to determine the new chemical abundances using the value of the physical variables determined at the previous instant t, then integrate the equation of the mechanical structure again at time $t + \Delta t$ using these updated chemical abundances, and so on.

In the following we will first deal with the problem of solving Equations (3.49) to (3.52); the first step is to determine the appropriate boundary conditions. We need four boundary conditions for r, P, T and L, respectively. As for the chemical composition, one has to specify the initial abundances of the various elements, assumed at the beginning to be homogeneous throughout the star and corresponding to the chemical composition of the protostellar cloud (nuclear reactions have not yet modified the stellar chemical abundances).

Boundary conditions need to be given at both the stellar centre ($m=0$) and surface ($m = M$). We recall that the surface for the stellar structure equations is located at a layer where the diffusion approximation for the energy transport starts to be applicable, e.g. a layer where $\tau \sim 1$ (photosphere). The value of the stellar mass at this layer is practically equal to the total mass M, since the overlying stellar atmosphere does not contain an appreciable amount of mass, due to its extremely low density. At $m = 0$ it is easy to recognize that $r = 0$ and $L = 0$. However, we cannot know in advance the central density and temperature ρ_c and T_c; therefore the other two boundary conditions must come from the surface.

Care has to be devoted to the treatment of the star centre. The central conditions described before produce a singularity, and therefore the integration has to start from a value m very small but larger than zero. It is not difficult (but slightly laborious) to demonstrate that a first-order Taylor expansion of the equations provides the following set of boundary conditions at $m = m'$

$$r = \left(\frac{3}{4\pi\rho_c}\right)^{1/3} m'^{1/3}$$

$$P = P_c - \frac{3G}{8\pi}\left(\frac{4\pi\rho_c}{3}\right)^{4/3} m'^{2/3}$$

$$L = \epsilon_c m'$$

$$T^4 = T_c^4 - \frac{1}{2ac}\left(\frac{3}{4\pi}\right)^{2/3} \kappa_c \epsilon_c \rho_c^{4/3} m'^{2/3} \quad \text{(radiative)}$$

$$\ln(T) = \ln(T_c) - \left(\frac{\pi}{6}\right)^{1/3} G\frac{\nabla_{ad,c}\rho_c^{4/3}}{P_c} m'^{2/3} \quad \text{(convective)} \qquad (3.54)$$

where the suffix c denotes the central values of the physical variables. These relationships are a two parameter set of boundary conditions; by assuming arbitrary values of P_c and T_c one obtains (using the equation of state and the auxiliary functions ϵ and Rosseland opacity) all other boundary values.

At the surface $m = M$ and $L = L_s$, where L_s is the unknown total stellar luminosity, but we still need a relationship for the temperature, pressure and radius. In fact, it is true that at the surface $P \sim 0$ and $T \sim 0$ (the so-called zero conditions) at least when compared with the deeper stellar layers, yet they are not exactly zero because of the gradual and extended transition to the values of the interstellar medium happening in the atmospheric layers.

A more rigorous procedure is to use results from model atmosphere computations to obtain the outer boundary conditions. In general model atmospheres are computed considering a plane–parallel geometry, and solving the hydrostatic equilibrium equation together with the frequency dependent (no diffusion approximation is allowed in these rarefied layers) equation of radiative transport (and convective transport when necessary) plus the appropriate equation of state. Due to the plane–parallel geometry (some models include the effect of sphericity) a given atmosphere is completely defined by the chemical composition, acceleration of gravity $g = GM/R^2$ (assumed constant throughout the atmosphere) and the so-called effective temperature T_{eff}, e.g. the temperature corresponding to the black body that yields the same surface flux of energy as the star. A given pair of T_{eff} and L_s values is linked to the total radius R by $L_s = 4\pi R^2 \sigma T_{eff}^4$ (the so called Stefan–Boltzmann law) and to g (since the stellar mass is given). With the known chemical composition of the stellar atmosphere (assumed to be the same as the composition of the stellar outer layers) T_{eff} and g, model atmosphere computations provide the pressure P_s at the appropriate value of τ where the diffusion approximation starts to be valid (e.g. the layer that defines the stellar surface from the point of view of the stellar evolution equations).

A simplified yet generally used approach is – for a given chemical composition, g and T_{eff} – to integrate the atmospheric layers using the following equations

$$\frac{dP}{d\tau} = \frac{g}{\kappa} \tag{3.55}$$

$$T^4 = \frac{3}{4} T_{eff}^4 \left(\tau + \frac{2}{3} \right) \tag{3.56}$$

plus the equation of state. The first equation is simply the equation of hydrostatic equilibrium written in the case of constant mass using the optical depth τ (computed from the Rosseland opacity) as independent variable, while the second equation is an approximation of the atmospheric temperature stratification as a function of τ. The integration is carried out from $\tau = 0$ (where $T \sim 0$ and $P \sim 0$) down to $\tau = 2/3$ where $T = T_{eff}$, using the shooting method described at the end of this section. This approximation for the atmosphere computation is called grey atmosphere; the $T(\tau)$ relationship given by Equation (3.56) is obtained from the Eddington approximation of the grey radiative transfer. In principle this procedure it is not strictly correct, since one is using Rosseland opacities when a frequency dependent treatment should be employed; however, it is sufficient to provide reasonable boundary conditions in most (albeit not all) cases. Some authors use this same procedure, but instead of Equation (3.56) they employ a $T(\tau)$ relationship determined empirically on the Sun, i.e. a so-called solar $T(\tau)$ relationship.

With the surface boundary conditions described above we have now determined a relationship among the four surface conditions T_{eff}, P_s, L_s and R; as in the case of the central values they are a two-parameter set of boundary conditions. By assuming arbitrary values of, e.g., R and L_s all other boundary values are determined.

Having fixed the boundary conditions for four of the equations describing the mechanical structure, we now briefly present the standard method employed to solve the system of Equations (3.49) to (3.52), the so-called Henyey method, first described in [96]. The star is divided into N mass shells with boundaries at mass m_j, where $m_1 = 0$ and $m_N = M$; the structure equations are then replaced by finite difference equations, that establish relations between the solution at different mesh points m_j. As an example, Equation (3.49) is replaced by the following

$$\frac{r_{j+1} - r_j}{m_{j+1} - m_j} = \frac{1}{4\pi r_{j+1/2}^2 \rho_{j+1/2}} \tag{3.57}$$

where $r_{j+1/2} = (r_j + r_{j+1})/2$ and $\rho_{j+1/2} = (\rho_j + \rho_{j+1})/2$. The relation between the physical quantities at m_1 and m_2 is given by the series expansion discussed before.

As for the time derivatives on the right-hand side of Equation (3.51) one has to use the following replacements

$$\left(\frac{dP}{dt}\right)_{j+1/2} = \frac{1}{\Delta t}(P_{j+1/2}^{t+\Delta t} - P_{j+1/2}^t)$$

$$\left(\frac{dT}{dt}\right)_{j+1/2} = \frac{1}{\Delta t}(T_{j+1/2}^{t+\Delta t} - T_{j+1/2}^t)$$

We can now write for each mesh j the four finite difference equations for the mechanical structure in the following compact way

$$E_j^i \equiv \frac{y_{j+1}^i - y_j^i}{m_{j+1} - m_j} - f_i(y_{j+1/2}^1, \ldots, y_{j+1/2}^4) = 0 \tag{3.58}$$

where f_i is a generic function of the four unknowns $y^1 = r$, $y^2 = P$, $y^3 = T$, $y^4 = L$ computed at $j + 1/2$, i runs from 1 to 4 and j runs from 2 (the equations show a singularity at the centre $i = 1$) to $N - 1$.

The boundary conditions at the surface $(j = N)$ can be rewritten, in the same vein as Equation (3.58), as

$$S_1 \equiv y_N^2 - f_s(y_N^1, y_N^4) = 0$$

$$S_2 \equiv y_N^3 - g_s(y_N^1, y_N^4) = 0$$

where f_s and g_s are two functions of the star total radius and total luminosity. As for the central boundary conditions Equation (3.54) one can write

$$C_i(y_2^1, y_2^2, y_2^3, y_2^4, y_1^2, y_1^3) = 0$$

with i running from 1 to 4, $j = 1$ being the centre and $j = 2$ the first mass layer after the centre, i.e. $m = m'$, as discussed before. The lack of dependence of the functions C_i on y_1^1 and y_1^4 reflects the fact that $y_1^1 = y_1^4 = 0$.

After the transformation of the stellar structure equations to finite difference equations it is wise to check the number of equations available and unknowns. The total number of unknowns is $4N - 2$, since we know that at the centre $y_1^1 = y_1^4 = 0$. The E_j^i equations provide $4N - 8$ (j runs from 2 to $N - 1$) relations, to which we have to add the two equations provided by S_1 and S_2, and the four additional relations provided by the C_i equations. The final budget is $4N - 2$ equations for $4N - 2$ unknowns.

Consider now a generic instant t; the run of the chemical abundance is given by the previous solution of the chemical evolution part, and we now have to solve for each mesh point j the system of equations $E_j^i = 0$ using the boundary conditions $C_i = 0$ and $S_1 = S_2 = 0$. We will see that the technique discussed below is very similar to the standard Newton–Raphson method for the solution of systems of algebraic equations. Suppose we have a first approximation of the solution $(y_j^i)_1$ (the suffix 1 here denotes the first approximation to the solution), e.g. given by the solution of the mechanical structure at the previous timestep $t - \Delta t$, where Δt is a known quantity, used for the computation of ϵ_g. We use these approximated values as a first guess for the arguments in the functions E_j^i, S_i and C_i (we notice that in case of the equation of energy transport and the corresponding central boundary condition one has first to test the radiative stability by using the trial solution, and then select the appropriate transport mechanism); since they will not represent the correct solution, we find that the equations are not satisfied, i.e. $(S_i)_1 \neq 0$, $(C_i)_1 \neq 0$ and $(E_j^i)_1 \neq 0$. The next step is to determine the correction to this approximate solution, i.e. $(y_j^i)_2 = (y_j^i)_1 + \delta y_j^i$, so that the equations are satisfied. Assuming the corrections are small, we can use a first-order Taylor expansion to express the variations δS_i, δC_i and δE_j^i as functions of the unknown corrections δy_j^i applied to $(y_j^i)_1$. A solution for the δy_j^i values can be then found by requiring that $(S_i)_1 + \delta S_i = 0$, $(C_i)_1 + \delta C_i = 0$ and $(E_j^i)_1 + \delta E_j^i = 0$. With this linearization the latter three conditions can be written as

$$\frac{dS_i}{dy_N^1}\delta y_N^1 + \frac{dS_i}{dy_N^2}\delta y_N^2 + \frac{dS_i}{dy_N^3}\delta y_N^3 + \frac{dS_i}{dy_N^4}\delta y_N^4 = -(S_i)_1 \quad i = 1, 2$$

$$\frac{dC_i}{dy_2^1}\delta y_2^1 + \frac{dC_i}{dy_2^2}\delta y_2^2 + \frac{dC_i}{dy_2^3}\delta y_2^3 + \frac{dC_i}{dy_2^4}\delta y_2^4 + \frac{dC_i}{dy_1^2}\delta y_1^2 + \frac{dC_i}{dy_1^3}\delta y_1^3 = -(C_i)_1 \quad i = 1, \ldots, 4$$

$$\frac{dE_j^i}{dy_j^1}\delta y_j^1 + \frac{dE_j^i}{dy_j^2}\delta y_j^2 + \frac{dE_j^i}{dy_j^3}\delta y_j^3 + \frac{dE_j^i}{dy_j^4}\delta y_j^4 + \frac{dE_j^i}{dy_{j+1}^1}\delta y_{j+1}^1 + \frac{dE_j^i}{dy_{j+1}^2}\delta y_{j+1}^2$$

$$+ \frac{dE_j^i}{dy_{j+1}^3}\delta y_{j+1}^3 + \frac{dE_j^i}{dy_{j+1}^4}\delta y_{j+1}^4 = -(E_j^i)_1 \quad i = 1, \ldots 4 \quad j = 2, \ldots, N - 1$$

$$(3.59)$$

Here, the S_i, C_i and E_j^i and their derivatives are evaluated using the first approximation as arguments. In practical terms the derivatives of these three functions are determined by varying one at a time the $(y_j^i)_1$ values (keeping the others fixed) by a small amount δ (usually a small fraction of $(y_j^i)_1$) and determining the variation of S_i, C_i and A_j^i due to the variation of the given (y_j^i).

The system of $4N - 2$ linear algebraic equations for the $4N - 2$ unknowns δy_j^i ($i = 1, \ldots 4$ and $j = 1, \ldots N$, with $\delta y_1^1 = 0$ and $\delta y_1^4 = 0$) given by Equations (3.59) has usually a matrix of the coefficients (the so-called Henyey matrix, with non-zero elements only near the diagonal) with non-zero determinant, and can be solved by standard mathematical methods. Because Equations (3.59) have been obtained by a linear approximation, their solution will provide second trial values $(y_j^i)_2$ that may not yet satisfy the equations of stellar structure. The procedure described above has therefore to be repeated in order to determine $(y_j^i)_3$ trial values and so on, until $S_i < \epsilon$, $C_i < \epsilon$ and $E_j^i < \epsilon$ where ϵ is the accuracy (to be specified a priori) with which one wants to satisfy the stellar structure equations.

The Henyey method sketched briefly above has the great advantage of being extremely stable, since small local errors do not propagate to other mesh points and do not affect appreciably the general solution. In most cases the solution is found with only a handful of iterations and if a few dozens of iterations are allowed for, extremely complicated structures, typical of advanced evolutionary phases, can be also very accurately modelled.

After the mechanical structure at time t has been derived, a new timestep Δt is selected, and the equations of the chemical evolution (Equations (3.53)) are solved again using the mechanical structure determined at the previous timestep. In general, Equations (3.53) are solved after a transformation to finite difference equations, in the same vein as the equations of the mechanical structure; we will then have at each mesh point a system of I algebraic equations, where I denotes the number of chemical elements taken into account. We remember again that within convective regions one has to use Equation (3.41). The equations of the chemical evolution can be solved in a manner similar to the case of the mechanical structure. One uses a trial solution that is then improved following the same procedure described for the Henyey method.

It is clear, from the previous description, that there is a problem with the computation of the first model of any evolutionary sequence. The method needs a trial solution that usually comes from the solution at the previous timestep, and the computation of ϵ_g requests at each mesh point the values of P and T at the previous timestep. The latter problem can be overcome by computing a first model with ϵ_g artificially set to zero, and then gradually switching on ϵ_g during the next few models. Provided the timestep is suitably small, this inconsistency does not affect the rest of the evolutionary sequence. The initial chemical stratification is assumed to be homogeneous (and equal to the specified initial chemical abundances) since stars are expected to be fully convective at birth. As for the the lack of a suitable trial solution, it is customary to evaluate the initial model of an evolutionary sequence using a different technique, the so-called shooting method. This method is based on the following idea. Let's start from the outer mesh point; here we have four boundary conditions T_{eff}, P_s, L_s, R (T_{eff} and P_s can be expressed in terms of L_s and R, as discussed before). The value of a generic physical variable y^i at mesh j, y_j^i, can be obtained from the value at the adjacent mesh $j-1$ or $j+1$ by a first-order expansion $y_j^i = y_{j-1}^i + (dy_{j-1}^i/dm)dm$, or $y_j^i = y_{j+1}^i + (dy_{j+1}^i/dm)dm$, where dm is the mass step between two adjacent mesh points, and the derivative is evaluated at mesh point $j-1$ in the first case, and at mesh point $j+1$ in the second case. Starting from the surface mesh point $j = N$ we can therefore determine the values of y_{j-1}^i at mesh $j = N-1$ by using $y_{N-1}^i = y_N^i + (dy_N^i/dm)dm$. The derivative in the right-hand side of this equation is computed at $j = N$ and is given by the right-hand side of the corresponding equation of the mechanical structure, evaluated at $j = N$. After y_{N-1}^i is known, one can determine y_{N-2}^i in the same way, and move further inward, down to a given mesh point within the star, for example half way between the surface and the centre. It is not wise to try to integrate the equations down to the centre, since the approximate solution that we are seeking can diverge more and more from the real one, the larger the integration path. Instead, we stop the integration at a given mesh point $j = f$, and denote with $(y^i_f)_{surf}$ the values of the physical variables at this point, usually called the fitting point. We then integrate the equations from the centre, using the central boundary conditions as starting values, up to $j = f$, and denote with $(y^i_f)_{centre}$ the values of the y^i variables at $j = f$. Since the boundary conditions are just trial values, $(y^i_f)_{centre} \neq (y^i_f)_{surf}$, and suitable corrections have to be applied to these trial boundary values. To obtain these corrections we first write the dependence of the $(y^i_f)_{centre}$ and $(y^i_f)_{surf}$ values on the boundary trial values as first-order Taylor expansions, i.e.

$$\Delta(y^i_f)_{surf} = \frac{d(y^i_f)_{surf}}{dy^1_N} \delta y^1_N + \frac{d(y^i_f)_{surf}}{dy^4_N} \delta y^4_N$$

$$\Delta(y^i_f)_{centre} = \frac{d(y^i_f)_{centre}}{dy^2_1} \delta y^2_1 + \frac{d(y^i_f)_{centre}}{dy^3_1} \delta y^3_1$$

The numerical values of the derivatives on the right-hand sides of these equations can be obtained by trial inward and outward integrations applying, one at a time, small corrections δ to the relevant boundary values.

After defining $\Delta_f \equiv (y^i_f)_{centre} - (y^i_f)_{surf}$, we can find the corrections to the four boundary trial values solving the following system of four algebraic equations $\Delta(y^i_f)_{centre} - \Delta(y^i_f)_{surf} = -\Delta_f$. We then follow the previously described procedure using these corrected boundary values; at the fitting point we will probably still have $(y^i_f)_{centre} \neq (y^i_f)_{surf}$, since the linear approximation described before. All the procedure has thus to be repeated until $(y^i_f)_{centre} - (y^i_f)_{surf} < \epsilon$, where ϵ is the prescribed accuracy of the solution.

This shooting method is generally used to determine the first model of an evolutionary sequence, or at most the first simpler evolutionary phases, where the run of the physical variables is reasonably smooth. It completely fails in advanced evolutionary phases where the physical variables show strong gradients, and the Henyey method must be the technique of choice. The reason is that the values of the parameters at the fitting point are too heavily dependent on the assumed boundary conditions, causing instabilities in the computation of the corrections; also unavoidable small errors due to the approximations made in the numerical solution tend to add up when moving from the surface (or the centre) to the fitting point.

3.2.1 Sensitivity of the solution to the boundary conditions

In this section we highlight the sensitivity of the solution to the surface boundary conditions and consider first the case of radiative envelopes. Dividing the equation of radiative transport by the hydrostatic equilibrium gives

$$\frac{dT}{dP} = \frac{3\kappa_{\text{rad}}L}{16\pi acT^3 Gm}$$

where in the external layers we can assume m and L are constant and equal to the total stellar mass and luminosity, respectively. By approximating κ_{rad} with $\kappa_{\text{rad}} = \kappa_0 \rho T^{-3.5}$ (κ_0 being a constant) and using the perfect gas EOS to express ρ in terms of P, we obtain an equation of the form

$$T^{7.5} dT = APdP$$

with A being a positive constant. The integration of this equation provides

$$T^{8.5} = A'(P^2 + C) \qquad (3.60)$$

where A' is $(8.5/2)A$ and C is a constant of integration that depends on the boundary conditions. It is easy to see that, due to the dependence of T on P, the solution becomes rapidly independent of the boundary value C. In fact, moving toward the interior P^2 gets rapidly much larger than any possible realistic value of C, and the solution converges fast to the case of $C = 0$ that corresponds to $P = 0$ and $T = 0$ at the surface. Moreover, by differentiating Equation (3.60) one obtains

$$\nabla_{\text{rad}} = \frac{2}{8.5} \frac{A'P^2}{T^{8.5}}$$

which has to be lower than $\nabla_{\text{ad}} \sim 0.4$ (in the approximation of fully ionized monatomic gas) for the envelope to be radiative starting from the surface. Dividing both sides of Equation (3.60) by $A'P^2$ we realize that the ratio $A'P^2/T^{8.5}$ is always smaller than unity when $C \geq 0$, hence the hypothesis of radiative envelope is satisfied, because $\nabla_{\text{rad}} \leq 0.235$.

   If the boundary conditions provide $C < 0$, one can obtain $\nabla_{\text{rad}} > 0.4$ right at the surface, so that the external layers are convective. In this case the precise value of C is important; in fact, for increasing P the contribution of C in Equation (3.60) becomes progressively negligible – as in the radiative case – and at some point, depending on the initial value of C, the solution will converge to $\nabla_{\text{rad}} < 0.4$. This means that the depth of the envelope convection is sensitive to the surface boundary conditions.

3.2.2 More complicated cases

Massive stars evolve very fast, especially in advanced evolutionary phases, and the acceleration term in the equation of the hydrostatic equilibrium cannot be neglected any longer. Moreover, due to the short nuclear timescales, it is not possible to compute separately mechanical and chemical structure, since the involved timescales are now comparable.

In this case one has to solve simultaneously (using the Henyey method) the equations of the mechanical and chemical structure of the star, including the velocity as an additional variable, so that the acceleration term in the equation of hydrostatic equilibrium can be rewritten in terms of the differential of the velocity with respect to time.

The system of algebraic equations to be solved at a given time t now contains the mechanical structure equations plus the equations related to the evolution of the chemical elements considered. When the solution is found, a timestep Δt is applied, the previous solution (for all the physical and chemical variables) is used as trial solution, and the Henyey method is applied again to determine the new solution at $t + \Delta t$.

One could ask at this point how to deal with convective mixing if the associated timescale is also comparable to the nuclear timescale for some particular evolutionary phase and/or mass range. We will discuss this point below when dealing briefly with non-canonical physical processes. We just notice here that if nuclear burning timescales are comparable to convective ones, the chemical composition of a convective region containing an active nuclear burning region will not be homogeneous.

3.3 Non-standard physical processes

Stellar evolution, like any other field of science, uses the Occam's razor as guideline, i.e. the explanation relying on the smallest number of hypotheses is the one to be preferred. Standard stellar evolution models are very successful at explaining the main properties of stars neglecting magnetic fields, mass loss, rotation (which are all observed properties of stars) and other chemical element transport mechanisms that are predicted by physics; their contribution is somewhat a second-order effect, although some specific observed properties of stars can be fully explained only with the inclusion of one or more of these processes in the model computation.

The presence of magnetic fields at the surface of stars is directly revealed by the Zeeman splitting of spectral lines and the polarization of the light emitted by many types of stars. The potential importance of magnetic fields for the structure and evolution of stars is due to the fact that they exert a pressure which could, in principle, reach a substantial fraction of the gas pressure, and in combination with rotation can affect the transport of angular momentum in stars (see below). Unfortunately we are not able to determine empirically the internal magnetic field configuration of stars, and from the theoretical point of view the modelling of the evolution of magnetic fields in stellar interiors is a very complicated three-dimensional problem way beyond present computational capabilities.

One important process not included in the definition of the standard stellar model is mass loss, although it is routinely included in stellar model computations. The effect of mass loss processes, whereby mass is lost from the external layers of stars, is important to explain some observations, although the observed rates are usually low and do not alter significantly (apart from specific cases) the structural, chemical and evolutionary properties of stars. We will show when relevant, the effect of mass loss on the star evolution. As for the practical implementation of mass loss processes in stellar model computations, a simple and good approximation is the following. When solving the mechanical structure equations at time $t + \Delta t$ the total mass of the star is reduced according to the chosen mass loss rate and the value of Δt; the value of the actual mass corresponding to the more external mass layers (i.e. the layers containing two or three times the mass lost) is then rescaled (for example linearly) in such a way that the total mass of the star is reduced by the right amount. The chemical and physical profiles at time t are unchanged, but of course their distribution in the most external layers is now altered because of the rescaling of the independent variable m. These profiles (with the rescaled m) are then used as trial solutions for the new model at time $t + \Delta t$ and then adjusted by the Henyey method until the equations are satisfied. As long as the mass change between two consecutive models is kept small (by adjusting Δt appropriately) this procedure converges rapidly.

In the following we will discuss briefly two other main groups of non-standard physical processes that affect to various degrees the evolution of stars, i.e. atomic diffusion plus radiative levitation, and rotation plus associated mixings. Our presentation of stellar evolution and related techniques developed in the following chapters is based purely on standard stellar models with inclusion of mass loss effects. The efficiency and the interplay between atomic diffusion, radiative levitation and rotational mixings is still under debate and we will not enter into details on this subject. Introductions to this subject can be found in [46], [52] and [161].

3.3.1 Atomic diffusion and radiative levitation

Atomic diffusion (some authors use the term atomic diffusion to include the radiative levitation process as well) is a basic physical transport mechanism driven by collisions of gas particles; if several chemical species are present in the gas, the net effect is to change the chemical stratification and therefore it is potentially important for the study of stellar evolution.

Pressure, temperature and chemical abundance gradients are the driving forces behind atomic diffusion. A pressure gradient and a temperature gradient tend to push the heavier elements in the direction of increasing pressure and increasing temperature, whereas the resulting concentration gradient tends to oppose the above processes. The speed of the diffusive flow depends on the collisions with the surrounding particles, as they share the acquired momentum in a random way. It is the extent of these collision effects that dictates the timescale of atomic diffusion once the physical and chemical profiles are specified. Comprehensive and detailed mathematical presentations of the theory of atomic diffusion are given in [27] and [55].

The chemical evolution of the mass fraction X_i of an element i at a given mass layer m due to atomic diffusion can be written as:

$$\frac{dX_i}{dt} = \frac{d}{dm}(4\pi r^2 \rho X_i w_i) \tag{3.61}$$

where w_i is the diffusion velocity of element i in the centre-of-mass reference frame. In general Equation (3.61) is treated as an additional term to the equation of the evolution of chemical species (Equation (3.53)). The velocity w_i is determined by the diffusion coefficients $A_P(i)$, $A_T(i)$ and $A_X(i)$ associated to the pressure, temperature and chemical abundance gradients, respectively; they are obtained, for a given chemical and physical profile, by imposing the conditions of mass, momentum and energy conservation, together with charge neutrality. General formulae for the computation of the diffusion coefficients are given by [197] and [217].

It can be useful to provide the rough dependence of the mean diffusion coefficients associated with the pressure and temperature gradients on the thermal properties of the stellar matter

$$< A > \propto \frac{T^{5/2}}{\rho}$$

In the centre of the Sun, $< A >$ is of the order of $5 \, \text{cm}^2 \text{s}^{-1}$. The timescale of atomic diffusion is given by the relation

$$\tau_{\text{diff}} \sim \frac{R^2}{< A >}$$

where R is the characteristic length of the system, i.e. the stellar radius in the case of stars. Therefore, in the case of the Sun, by substituting R with $R_\odot$ and using the approximate value of $< A >$, one obtains $\tau_{\text{diff}} \approx 6 \times 10^{13}$ years. This means that the effects of atomic diffusion are significant only when the evolutionary lifetimes are of the order of at least $\approx 10^9 - 10^{10}$ years.

Radiative levitation is an additional transport mechanism caused by the interaction of photons with the gas particles, which acts selectively on different atoms and ions. In simple terms, since within the star a net energy flux (albeit locally small, to preserve local thermodynamical equilibrium) is directed towards the surface, photons provide an upward 'push' to the gas particle with which they interact, effectively reducing the gravitational acceleration. Since we are dealing with interactions of photons with gas particles, it is clear that the efficiency of radiative levitation is related to the opacity of the stellar matter, in particular to the monochromatic opacity, and increases for increasing temperature (more energetic photons).

The effect of levitation on the chemical abundances is simply an additional contribution to w_i in Equation (3.61) coming from the computation of the radiative acceleration, i.e. the acceleration felt by an individual atomic species through the absorption of a photon and before the momentum change is shared with the other species of the gas through collisions. The determination of the radiative accelerations involves the knowledge of the cross sections for absorption and scatter of photons, and how the momentum of photons is distributed among species and ionization states, since this modifies the distribution in the rest of the gas. In addition one must also know how the momentum is shared between the electron and a given ion undergoing photoionization, since this determines whether the ion is pushed forward or backward by the photon.

Equation (3.61) is also suited to treat convection in a time-dependent way, whenever this is necessary, as discussed before. In this case one includes in w_i the contribution due to the convective velocity estimated from the mixing length theory. In general, within a convective region, the convective velocity is always much larger than the contributions of atomic diffusion and radiative levitation, so that the rehomogenization due to convection always prevails over atomic diffusion and radiative levitation.

3.3.2 Rotation and rotational mixings

Spectroscopic observations of stars tell us that stars rotate, with rotation velocities being generally higher the higher the stellar mass. In rotating stars centrifugal forces act on the gas reducing the effective gravity, as a consequence the condition of hydrostatic equilibrium is achieved at a lower pressure, and hence a lower temperature, with respect to non-rotating stars. Therefore, for a given stellar mass, the main effect of rotation is to slightly cool down the stellar interiors in each of the evolutionary phases.

An additional, sometimes very important effect (depending, of course, on the rotational velocity) is that rotation leads to deviations from spherical symmetry, although for small and moderate rates the deformations are not large. Inclusion of stellar rotation into one-dimensional spherical models using the same stellar structure equations presented before is discussed at length by [70], [114] and [136]. In the case of a conservative effective potential Φ – where Φ is the sum of the gravitational potential plus the centrifugal force contribution – spherical surfaces are replaced by equipotential surfaces where P, T and ρ are constant. This hypothesis of conservative effective potential is justified for solid-body rotation or for constant rotation on cylinders centred on the axis of rotation.

In the equations of stellar structure one can then reinterpret the mass dm between two spherical shells as dm_Φ, i.e. the mass enclosed between two equipotential surfaces with Φ and $\Phi + d\Phi$ that are in principle non-spherical. The radius r becomes r_Φ, defined as $V_\Phi = (4/3)\pi r_\Phi^3$, where V_Φ is the volume enclosed within the equipotential surface Φ. With this reinterpretation of r and m the equations look exactly the same as the ones without rotation, with the exception of a multiplicative factor f_p in the right-hand side of the equation of hydrostatic equilibrium, and f_t/f_p on the right-hand side of the equation of radiative transport. These factors f_p and f_t described in [70] and [114] are functions of Φ, and hence of the rotational velocity.

As discussed in [136] a conservative Φ is usually not a realistic approximation since during time the internal rotation generally evolves towards non-conservative rotation laws. A more realistic picture for the rotation is the so-called shellular rotation ([239]) whereby the rotation rate is constant on isobars (surfaces of constant pressure); exactly the same equations discussed before can also be used in this more realistic case, provided that the equipotential surfaces are reinterpreted as the isobars, and the quantities describing the mechanical structure are reinterpreted as mean values over the isobars (the chemical abundances are supposed to be homogeneous over isobars).

An additional equation has then to be added for the transport of angular momentum within the star due to mixing of chemical elements, contraction and/or expansion of the stellar layers and eventual mass loss.

In general the hydrostatic effects of rotation have only a very small influence of the order of a few per cent on the internal evolution of rotating stars; however, the extra mixing processes associated with rotation can have a more relevant impact. We cannot discuss here all possible mixing processes associated with rotation, and refer the reader to [95] for a good introduction and references. At least five processes (the so-called dynamical shear instability, Solberg–Hoiland instability, secular shear instability, Eddington–Sweet circulation and Goldreich–Schubert–Fricke instability) should be taken into account, their efficiency (poorly known) evaluated and the corresponding mixing velocity w included as additional terms in Equation (3.61), in the hypothesis that they can be treated as diffusive processes.

4 Star Formation and Early Evolution

4.1 Overall picture of stellar evolution

Condensation of interstellar matter forms protostars in hydrostatic equilibrium and at low temperatures. Due to the virial theorem (since nuclear reactions are not yet efficient) these objects contract and increase their central temperature until the fusion of hydrogen into helium becomes effective (the so-called H-burning phase). These burning processes effectively halt the contraction due to the extra energy input that is enough to keep the star in hydrostatic equilibrium. When hydrogen is exhausted in the central hot regions (whose chemical composition is now essentially pure helium) the stellar core starts to contract again and increases its central temperature until nuclear reactions transforming He (produced by the previous H-burning phase) into C and O become effective. The evolution is driven by the He-burning until He in the centre is exhausted, and the sequence repeats again leading to the burnings of progressively heavier elements which were essentially the products of the previous burning stage. The burning stages end when Fe is left in the core. Since fusion of Fe subtracts energy instead of producing it, at this stage the central stellar layers collapse and the envelope is expelled, giving origin to the supernova phenomenon. What is left of the iron core is a neutron star or a black hole, depending on the mass of the remnant.

If the initial stellar mass is below a given threshold, the onset of electron degeneracy at the end of either the helium or carbon burning phase prevents (according to the virial theorem) the increase of central temperature that leads to the next burning stage. Instead, the energy of the non-degenerate ions is radiated away with a consequent progressive decrease of the star temperature (and luminosity) producing a so-called white dwarf star.

Evolution of Stars and Stellar Populations Maurizio Salaris and Santi Cassisi
© 2005 John Wiley & Sons, Ltd

Some further general and useful rules of stellar evolution are given below.

- The higher the mass the shorter the lifetime of the star will be.

- The lower the mass, the higher will be the central density and the lower the central temperature in a given evolutionary phase.

- The higher the metallicity – keeping the initial He abundance fixed – the lower the luminosity and T_{eff} will be, and the longer the evolutionary timescale of a star of a given mass during the major evolutionary phases.

- The higher the initial He content – keeping the metallicity fixed – the higher the luminosity and T_{eff} will be, and the shorter the evolutionary timescale of a star of a given mass in the major evolutionary phases

This broad sketch of stellar evolution highlights the pivotal role played by the stellar mass (and to a lesser extent the initial chemical composition) in determining the evolutionary properties of stars. In the following sections we will start to give more details of this general picture, a necessary step if we want to use stars as tools to build a coherent picture of the evolution of the universe as a whole.

4.2 Star formation

The previous chapter has shown that stars can be modelled as gravitating gas spheres in hydrostatic equilibrium, fuelled by nuclear reactions and gravitational energy release. Before reaching the hydrostatic equilibrium stage, stars must have formed out of interstellar matter (ISM) previously in a diffuse form. The enormous volume of space between stars in our galaxy – and in other galaxies undergoing star formation activity – is filled with ISM, which is heated and ionized by the photons emitted by all kinds of stars, and the mechanical energy released during the explosion of massive stars. The ISM is made of gas plus dust; the relative number of chemical elements in the gas phase is about 90 per cent of hydrogen, 10 per cent of helium and traces of heavy elements. Dust is in the form of grains that make approximately 1 per cent of the total ISM mass, mostly silicate and graphite grains. The ISM of the Milky Way contains several gas phases that exist in rough pressure equilibrium, at temperatures ranging from ≈ 100 to $\approx 10^6$ K, and densities between 10 and 10^{-3} particles/cm^3. Within the ISM one can find large molecular clouds (made mainly of molecular hydrogen plus dust) of masses in the range 10^5–10^6 solar masses, temperature between 10 and 100 K and densities between 10 and 10^2 particles/cm^3, these clouds are the star forming regions. Without entering into the details of a very complex problem (with many questions still unanswered) we just sketch the main ideas behind our understanding of the process of star formation.

In order to produce energy by means of nuclear reactions, stars have to become much hotter and denser (at least in the centre) than the typical density of the molecular clouds. This suggests that some process of gravitational contraction had to be effective during the star formation phase. We now assume that rotation and magnetic fields are negligible and make the approximation of spherical symmetry; in this case one can use the virial theorem to derive easily an approximate criterion for the collapse of an interstellar cloud. According to the virial theorem, the hydrostatic equilibrium of a gas sphere requires that the relationship between the internal energy E (equal to the kinetic energy of the gas particles in the approximation of perfect gas) and its gravitational potential Ω is $E = -(1/2)\Omega$. If Ω is larger than the value prescribed by the virial theorem the cloud will collapse under its own gravity. For a spherical gas cloud of total mass M, made of a perfect monatomic gas of approximately constant density ρ, temperature T and molecular weight μ, Equation (3.42) becomes

$$\frac{3K_{\mathrm{B}}TM}{\mu m_{\mathrm{H}}} = \int_{0}^{M} \frac{Gm}{r} \, dm$$

The integral on the right-hand side (i.e. the gravitational potential of the cloud) can easily be computed if the density is constant. In fact, in this case $m/r^3 = M/R^3$, where R and M are the total radius and mass of the cloud, and r and m their couterparts at a generic point within the cloud. Using this relationship we can express r as $r = (m/M)^{1/3}R$ and substitute this expression into the integral whose solution is $3/5(GM^2/R)$. The criterion for gravitational collapse can then be written as

$$\frac{3K_{\mathrm{B}}TM}{\mu m_{\mathrm{H}}} < \frac{3}{5}\frac{GM^2}{R}$$

We can now replace R with $R = [(3/4)(M/\pi\rho)]^{1/3}$ and after some rearrangement one obtains that the cloud will collapse when

$$M > M_{\mathrm{J}} \equiv \left(\frac{3}{4\pi\rho}\right)^{1/2} \left(\frac{5K_{\mathrm{B}}T}{G\mu m_{\mathrm{H}}}\right)^{3/2} \tag{4.1}$$

The minimum value of the cloud mass to undergo a gravitational collapse – for the given chemical composition, ρ and T – is called Jeans mass (denoted by M_{J}). It is very important to notice, for what follows, that once the chemical composition is fixed, M_{J} scales as $M_{\mathrm{J}} \propto T^{3/2}\rho^{-1/2}$. For typical temperatures and densities of large molecular clouds the value of M_{J} is $\approx 10^5 \, M_{\odot}$. The collapse timescale t_{ff} when $M > M_{\mathrm{J}}$ is given by the time a mass element at the cloud surface needs to reach the centre. A mass element at the surface is subject to an acceleration $g = GM/R^2$ and, in the approximation of constant g during the collapse, the time needed to cover the distance R is given by $(1/2)(GM/R^2)t_{\mathrm{ff}}^2 = R$. By approximating R using $R^3 \approx (M/\rho)$, we obtain (neglecting the numerical factor $(1/2)$)

$$t_{\mathrm{ff}} \approx (G\rho)^{-1/2}$$

We now follow the collapse of a giant molecular cloud with $M > M_J$. During the collapse the density has to increase (because the volume gets smaller and mass is conserved) whereas the evolution of T is affected by the energy exchange between the cloud and the surroundings. If, as a first approximation, the evolution of the cloud is adiabatic, the temperature changes as $T \propto \rho^{2/3}$ for a perfect monatomic gas; with this temperature evolution the value of M_J increases during the collapse phase as $\sqrt{\rho}$ and the evolution will always reach a point when Equation (3.42) is satisfied. From this moment on the cloud would start to evolve in hydrostatic equilibrium following the virial theorem, at which point we assume that a star is born.

This kind of adiabatic evolution is, however, just an idealization; more realistically, the matter within the cloud is heated and cooled by various mechanisms. Heating can be due to cosmic rays penetrating the cloud and ionizing the particles, and the high-energy electrons produced by these interactions may transmit their kinetic energy to other particles – and eventually ionize them – by collisions. Also X-rays and eventually the radiation field of the surrounding stars may penetrate the gas and ionize the cloud particles. On the other hand the cloud is cooled by molecular or dust radiation, and the more efficient mechanism (i.e. with the shorter timescale) will prevail. The efficiency (hence the timescale) of cooling and heating is determined, for a given environment, by the physical conditions of the cloud.

In most cases during the cloud collapse cooling processes are very efficient and their timescale t_{cool} is much shorter than the collapse timescale t_{ff}. Under these conditions the collapse of the cloud will be to a good approximation isothermal (the gravitational energy acquired during the collapse is immediately lost due to the cooling processes) and, since ρ increases and T stays constant, M_J decreases. If there are within the clouds inhomogeneities with mass larger than the actual value of M_J, they will collapse by themselves with their local t_{ff}, different from the initial t_{ff} of the whole cloud. This process, called fragmentation, will continue as long as t_{cool} is shorter than the local t_{ff}, and produces increasingly smaller collapsing subregions. Eventually the density of the various subunits will become so large that the matter becomes optically thick (i.e. photons are not able to leave the cloud easily) and the evolution becomes adiabatic. The adiabatic evolution will then lead all these collapsing subregions to hydrostatic equilibrium. We see that in this way a giant molecular cloud can originate a group of stars with various masses, the mass distribution being determined by the fragmentation process, i.e. by the cloud physical conditions and the efficiency of the undergoing cooling processes.

The typical minimum stellar mass produced by the fragmentation process, that is the minimum mass of the collapsing subregions when the matter becomes optically thick, is of the order of $0.01\,M_\odot$ (i.e. much larger than the mass of planets that therefore cannot be formed by this process).

We now cover the evolution of a single collapsing subregion in a bit more detail. During the collapse the density does not stay uniform in any realistic case, but increases inwards. Since the core is denser, the optically thick phase is reached first in the central regions that eventually reach the hydrostatic equilibrium. This leads to the formation of a core with free falling gas surrounding it; during this phase the energy released by the core (now obeying the virial theorem) is absorbed by the envelope and radiated away as infrared radiation. After the accretion is essentially completed the whole protostar, as it is called in this phase, increases its temperature by virtue of the virial theorem. The steady increase of the central temperature causes first the dissociation of the H_2 molecule, then the ionization of hydrogen and the first and second ionization of helium. During these processes the energy gained by the contraction mainly goes into the dissociation and ionization of these species (which almost make the totality of the protostar chemical composition). The protostar collapses (free-fall timescales) again during these phases until all dissociations and ionizations are completed and hydrostatic equilibrium is restored. The total energy involved in a single dissociation or ionization process is given by the product of how many particles have to be dissociated/ionized times the dissociation/ionization energy of a single particle. The sum of the energies involved in all four processes listed above has to be at most equal to the energy available to the star through the virial theorem, i.e. half the gravitational energy increase due to the contraction between the beginning of the protostar phase and the moment when the major dissociation/ionizations listed above are completed. This simple estimate tells us that the maximum initial radius R_{max}^i of a protostar of mass M has to be

$$\frac{R_{max}^i}{R_\odot} \approx 50 \frac{M}{M_\odot}$$

We conclude this section by posing this question. In the light of our knowledge of the star formation mechanisms, is it possible to predict theoretically the number of stars with a given mass to be born in a given molecular cloud (the so-called Initial Mass Function – IMF) ? The answer is – up to now – no, and we have to apply, for example, stellar evolution based techniques (discussed in later chapters) to study empirically the IMF in different environments. For example, the effect of pre-existing magnetic fields can be relevant if the cloud is partially ionized and therefore the gas is a good conductor (if not, the magnetic lines of force can slide through the gas without affecting the collapse). In this case the magnetic lines of force are trapped by the gas and during the cloud collapse the magnetic lines also contract and increase the strength of the magnetic field proportional to R^{-2}. If at the beginning the gravitational force (which also increases during the collapse as R^{-2}) was larger than the magnetic force, it will remain larger during the whole contraction phase and the picture sketched above is unchanged. If the magnetic forces at the beginning were larger than the gravitational ones, they will prevent the contraction of the cloud perpendicularly to the direction of the magnetic lines; therefore the contraction can only go on along the magnetic lines and the cloud forms a disk that can then fragment into smaller disks for which cooling processes are more efficient than for a spherical cloud.

Particularly difficult is the case of primordial matter with the chemical composition typical of the cosmological nucleosynthesis. In this case the metal content is essentially zero and without metals cooling processes are very ineffective; it is therefore

difficult to have efficient fragmentation that produces stars with masses comparable
to the solar mass. The general belief is that star formation in primordial matter produces only very massive, short-lived stars, of the order of $100M_\odot$ Due to current
uncertainties in the physics of star formation it is, however, not completely clear if
this was really the case. We will discuss this point again briefly in the next chapter.

The classical expression for the IMF – that we will often use in the rest of this
book – determined empirically for the solar neighbourhood, is the so-called Salpeter's
law ([185]), i.e.

$$dn/dM = CM^{-x}$$

where dn is the number of stars born with mass between M and $M + dM$, $x = 2.35$,
and C is a normalization constant. Hence, star formation appears to be biased towards
low mass stars, at least in the solar vicinity. More recent evaluations ([120]) confirm
the Salpeter IMF for $M \geq 0.5M_\odot$, whereas for $0.1 \leq M/M_\odot < 0.5$, x appears to be
smaller, i.e. $x = 1.3$.

4.3 Evolution along the Hayashi track

The evolutionary timescales from the first equilibrium configuration (i.e. from when
we assume that a star is born) until the end of the ionization processes are negligible
with respect to the rest of the star lifetime. At this stage the protostar (which we will
call star from now on) is in a physical state that can be described by the equations of
stellar evolution we have derived above. Stars in this phase are fully convective, T_{eff}
is low, the radius large and the luminosity high relative to subsequent evolutionary
phases.

Traditionally the evolution of the surface (bolometric) luminosity L and T_{eff} of a
star is described by the so-called stellar evolutionary track, i.e. the path described
in the $\log(L/L_\odot)$ vs $\log(T_{eff})$ diagram, the so-called Hertzsprung–Russel diagram
(HRD). By convention the quantity displayed along the horizontal axis (T_{eff}) increases
towards the left.

During this fully convective phase stars evolve in the HRD along an almost vertical
line, i.e. at approximately constant T_{eff} and decreasing luminosity (since the radius
is decreasing) as shown in Figure 4.1. This almost vertical line is called the Hayashi
track.

4.3.1 Basic properties of homogeneous, fully convective stars

The location of the Hayashi track is sensitive to the chemical composition of the star,
i.e. the track of a given mass moves towards lower T_{eff} when the metallicity increases
or the helium abundance decreases (the dependence on helium is, however, very
small). At a fixed chemical composition the Hayashi track is shifted to lower T_{eff} when

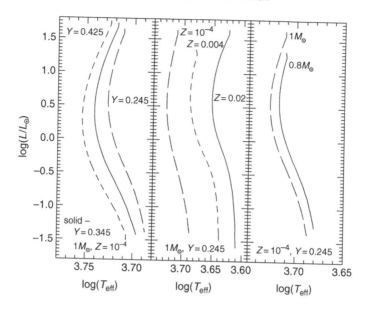

Figure 4.1 HRD location of Hayashi tracks for various assumptions about the chemical composition and the stellar mass

the mass decreases, again with a mild dependence. As for the value of α_{ml}, the Hayashi track moves towards higher T_{eff} when α_{ml} increases; in fact, an increase of α_{ml} makes convection more efficient, the superadiabatic gradient approaches the adiabatic one and overall the gradient in the superadiabatic region decreases, producing a higher T_{eff}.

The effect of the total mass and the shape of the Hayashi track on the HRD can be understood with the following analysis. Consider the stellar structure equations using r as the independent variable. For a star with a given total mass M, total luminosity L, total radius R and homogeneous chemical composition with molecular weight μ one can define the following dimensionless variables

$$x \equiv \frac{r}{R}, \quad q \equiv \frac{M_r}{M}, \quad t \equiv \frac{T}{T_0}, \quad p \equiv \frac{P}{P_0}$$

where T_0 and P_0 are constants given by

$$T_0 = \frac{\mu G M m_H}{R K_B}, \quad P_0 = \frac{G M^2}{4 \pi R^4}$$

Using these dimensionless variables the first two equations of stellar structure can be rewritten as

$$\frac{dq}{dx} = \frac{p}{t} x^2 \quad \text{(continuity of mass)}$$

$$\frac{dp}{dx} = -\frac{p}{t} \frac{q}{x^2} \quad \text{(hydrostatic equilibrium)}$$

In the approximation of adiabatic temperature gradient for the fully convective star we can add the equation of an adiabatic transformation for a monatomic perfect gas, $P = CT^{5/2}$, with C being a constant that determines the adiabat followed by the temperature stratification. In terms of the dimensionless variables this equation can be rewritten as

$$p = Et^{2.5}$$

where the constant E is given by

$$E = 4\pi G^{3/2} C M^{1/2} R^{3/2} \left(\frac{\mu m_H}{K_B} \right)^{5/2}$$

These three dimensionless structure equations are sufficient to describe the mechanical structure of the star since there are three unknowns (p, q, t) and three equations (x is the independent variable). With these new variables the surface is located at $x = 1$ and the corresponding boundary conditions can be taken as $q_s = 1$, $t_s = T_s/T_0$, $p_s = Et_s^{2.5}$.

In general t_s and E are two free parameters that have to be appropriately adjusted for the integration to satisfy the boundary condition at the centre $x = 0$, $q = 0$. When this condition is satisfied by the right choice of t_s and E, the other two central values for t and p are also obtained. To solve these equations one starts the numerical integration from the surface with trial values of t_s and E, moving towards the centre using the shooting method. If the boundary condition $q = 0$ at the centre is not satisfied the integration is repeated with updated t_s and E until agreement is found. In this way one determines t_s and E (hence p_s) and the run of q, p, t from the surface to the centre. It has been shown by [93] that a star can be fully convective (in the adiabatic approximation) only when the dimensionless constant E is equal to

$$E = 4\pi G^{3/2} C M^{1/2} R^{3/2} \left(\frac{\mu m_H}{K_B} \right)^{5/2} = 45.48 \qquad (4.2)$$

We now want to determine the properties of the Hayashi line on the HRD using the previous result. Let us define the constant C as

$$C = C' M^{-1/2} R^{-3/2}$$

with C' being another constant equal to

$$C' = \frac{45.48}{4\pi G^{3/2}} \left(\frac{K_B}{\mu m_H} \right)^{5/2}$$

whose value depends on the chemical composition of the star, and integrate the atmosphere using Equation (3.55). This treatment of the atmosphere assumes radiative transport, a reasonable approximation for the more external atmospheric layers (because a star radiates energy into space). Even if the convection within the star reaches layers above $\tau = 2/3$, one can always perform the atmospheric integration down to the point where convection sets in and take the boundary conditions there. For the discussion that follows, this would simply change a constant entering the solution.

We assume an average constant value of the opacity throughout the atmosphere and a constant acceleration of gravity $g = GM/R^2$ (the mass contained in the atmosphere is negligible). The straightforward integration from $\tau = 0$ (where $P = 0$) down to $\tau = 2/3$ provides

$$P_s = \frac{2}{3} \frac{GM}{R^2} \frac{1}{\kappa}$$

If we now assume that the constant value of κ throughout a given atmosphere is equal to $\kappa = \kappa_0 P^a T^b$ where κ_0, a and b are constants, and P and T are the values at $\tau = 2/3$ ($P = P_s$ and $T = T_{eff}$) the boundary condition on the pressure becomes

$$P_s = \left(\frac{2G}{3}\right)^{1/(a+1)} \left(\frac{MT_{eff}^{-b}}{\kappa_0 R^2}\right)^{1/(a+1)} \tag{4.3}$$

In the interior we assumed that $P = CT^{5/2}$ that gives, at the interface with the stellar atmosphere

$$\log(T_{eff}) = 0.4 \log(P_s) + 0.4 \log(C^{-1})$$

$$= 0.4 \log(P_s) + 0.4 \left(\frac{3}{2} \log(R) + \frac{1}{2} \log(M) - \log(C')\right)$$

while from Equation (4.3) (after integration of the atmospheric layers) we can also write

$$(a+1) \log(P_s) = \log(M) - 2 \log(R) - \log(\kappa_0) - b \log(T_{eff}) + \log\left(\frac{2G}{3}\right)$$

These two equations must be satisfied simultaneously along the Hayashi track. For a fixed total mass M, a generic value of R fixes T_{eff} and P_s through these two equations. From T_{eff} and R one then gets the luminosity $L = 4\pi R^2 \sigma T_{eff}^4$. By continuously varying the free parameter R one can determine the locus on the HRD occupied by an homogeneous fully convective star of a given mass and chemical composition that is contracting along the Hayashi track. This procedure yields an equation of this type

$$\log(T_{eff}) = \alpha \log(L) + \beta \log(M) + \delta \tag{4.4}$$

α, β and δ being constants determined by the values of a, b, κ_0 and C' (which contains the constant $E = 45.48$ and the molecular weight μ). In particular, α and β depend on a and b only. Using realistic values $a \sim 1$ and $b \sim 3$ (the main opacity source is the $^-$H ion, where the necessary electrons are provided by the partially ionized heavy elements in the external layers) one obtains $\alpha < 0.1$ and $\beta \sim 0.2$.

The slope of the Hayashi line in the HRD is therefore $d\log(L)/d\log(T_{\text{eff}}) > 10$, i.e. the Hayashi line is basically vertical; also, since $d\log(T_{\text{eff}})/d\log(M) \sim 0.2$, its location has a mild dependence on the stellar mass, and it moves towards higher T_{eff} for increasing M. The effect of the chemical composition is included in the constant δ through the value of κ_0 and μ, and one can demonstrate that an increase of the metallicity (that causes an increase of the opacity) shifts the track to lower effective temperatures, whereas an increase of helium at constant metallicity (that increases μ and decreases the opacity, because the opacity of He is lower than the hydrogen opacity) has the opposite (and less relevant) effect.

Two important points may be noticed here. The first one is that this result is independent of the stellar energy source (we did not make use of the equation of energy generation). The second point is that the properties of the Hayashi track are determined by the surface boundary conditions. In the more realistic case of a superadiabatic gradient in the external layers, the Hayashi track properties are determined by both the boundary conditions and the superadiabatic zone. The bottom line is that the external layers where the temperature gradient is either radiative or superadiabatic, fix the adiabat that describes the inner temperature stratification.

4.3.2 Evolution until hydrogen burning ignition

During the evolution along the Hayashi track the star increases its temperature due to the virial theorem, a radiative core will form at the centre and grow in mass as the star evolves. This evolutionary phase is the so called Pre-Main Sequence (PMS). When the radiative core appears the star is no longer fully convective, and it has to depart from its Hayashi track; this departure moves the track towards higher T_{eff} (to the left of the Hayashi track in the HRD) so that the Hayashi track acts as a rightmost boundary to the evolution of stars in the HRD. In fact, a fully convective (and consequently homogeneous) star has to lie on its Hayashi track, but whenever a radiative core is present, the star will have to stay on the left side of its Hayashi track.

When stars are not fully convective $E < 45.48$, decreasing when the extension of the convective region gets smaller. The constant E enters Equation (4.4) through the term δ, and it can be shown that δ – hence T_{eff} – increases for decreasing values of E. The path on the HRD is almost horizontal when a sizable radiative core is established, with a constant increase of the central temperature due to the virial theorem.

During this early evolution some nuclear burnings take place inside the star, although they are not relevant from the point of view of energy production. When temperatures reach the order of 10^6 K deuterium is transformed into ^{3}He by proton captures and the nuclear energy released through these reactions temporarily slows down the evolution. When $T \sim 2.5 \times 10^6$ K lithium is destroyed by proton captures and produces mainly ^{4}He. Also much less abundant elements like beryllium and boron are going to be destroyed during these early stellar evolution phases.

The initial abundances of these light elements are typically very small and are practically reduced to zero where the reactions are efficient, although they do not appreciably affect the abundances of the elements they produce (mainly helium). If stars were fully convective during these first nuclear burnings we would have, for example, a complete depletion of lithium and deuterium at the surface, since matter in a convective star is always well mixed and all lithium and deuterium would be destroyed. However, the convective region will start to shrink at some point. If the convective region is confined to the external and cooler layers before all the, i.e. lithium, reservoir has been destroyed in the fully convective phase, some lithium nuclei will still be present in the convective envelope. This means that the resulting surface abundances of these light elements depend on the interplay between the increase of the stellar temperature, the resulting efficiency of the nuclear reactions involved and the timescale for the shrinking of the convective envelope. In general, once the chemical composition is fixed, larger masses have lower surface depletion of the light elements. For a given mass the depletion increases with increasing metallicity. This result depends on the fact that increasing mass or decreasing metallicity is equivalent to lower (or non-existent) convective envelopes at the beginning of the hydrogen burning, but also to an earlier retreat of convection from the central regions.

The HRD evolution of selected masses until the start of central hydrogen burning is displayed in Figure 4.2. The lifetime and selected surface chemical abundances

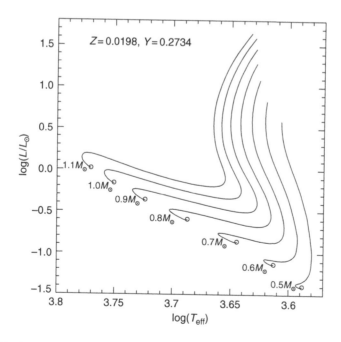

Figure 4.2 Evolutionary tracks of stars with different masses and solar initial chemical composition, until the onset of core H-burning (marked with an open circle)

Table 4.1 Age and surface chemical abundances of light elements (in mass fractions) at the start of central hydrogen burning, for selected masses and the two initial chemical compositions. In the case of $Z = 0.02$, the initial chemical abundances of the light elements are: $X^0_{2D} = 4.78E - 05$, $X^0_{3He} = 2.94E - 05$ and $X^0_{7Li} = 1.07E - 08$; while for $Z = 0.01$, the following values have been adopted: $X^0_{2D} = 4.99E - 05$, $X^0_{3He} = 2.72E - 05$ and $X^0_{7Li} = 5.35E - 09$ (from [201])

Mass ($M_\odot$)	Age (Myr)	X_{2D}	X_{3He}	X_{7Li}
	$Z = 0.02$, $Y = 0.277$			
0.2	730	$7.82E - 18$	$1.90E - 04$	$6.50E - 17$
0.5	210	$9.88E - 18$	$1.05E - 04$	$5.41E - 17$
1.0	62	$1.64E - 17$	$1.00E - 04$	$3.97E - 09$
2.0	2.9	$1.52E - 17$	$1.00E - 04$	$1.07E - 08$
4.0	0.5	$3.63E - 18$	$1.00E - 04$	$1.07E - 08$
	$Z = 0.01$, $Y = 0.256$			
1.0	50	$2.08E - 17$	$1.02E - 4$	$4.72E - 9$

of selected stellar models are shown in Table 4.1. The effect of changing chemical composition on a $1M_\odot$ star is also shown.

When the central temperature reaches $\sim 10^7$ K, hydrogen burning is ignited in the stellar core and the so called Main Sequence (MS) phase starts.

When the mass of the contracting star along the Hayashi track is below $\sim 0.08 M_\odot$, the contraction and temperature increase due to the virial theorem are halted before reaching the hydrogen burning temperatures due to the onset of electron degeneracy. These objects, that will never produce energy from nuclear reactions, are called brown dwarfs, and their evolution is similar to the much more advanced white dwarf evolutionary phase typical of stars like the Sun (see Chapter 7).

5 The Hydrogen Burning Phase

5.1 Overview

One of the most important phases during the evolution of a star is that corresponding
to the H-burning phase (which is actually being experienced by the Sun). The reasons
for its importance are as follows.

- It is the longest evolutionary phase; as a consequence the number of H-burning
  stars which can be observed is much larger than the number of stars in any other
  phase.

- The structural and evolutionary properties of a star during the central and shell
  H-burning phases determine its evolutionary properties through all successive phases.

- The most important astrophysical 'clock' is that related to the termination of the
  central H-burning phase.

- The final portion of the shell H-burning phase in low-mass, metal-poor stars
  provides an accurate distance indicator for old stellar populations.

- The analysis of the central H-burning phase, and in particular the counts of stars
  evolving through this phase, offers a unique opportunity for deriving the Initial
  Mass Function (IMF) of stellar systems.

We define a star as being on the Main Sequence if its evolutionary rate is controlled
by the timescale of the H-burning process occurring in the core.

Before discussing the main evolutionary and structural properties of low- and
intermediate-mass stars during the central and shell H-burning phases, it is essential
to understand the thermonuclear reactions involved in the H-burning process.

Evolution of Stars and Stellar Populations   Maurizio Salaris and Santi Cassisi
© 2005 John Wiley & Sons, Ltd

5.2 The nuclear reactions

The H-burning mechanism is essentially the nuclear fusion of four protons into one ^{4}He nucleus. By adopting the well-known Einstein's relation for the equivalence between energy and mass, $E = mc^2$, and accounting for a mass deficit of ≈ 0.7 per cent, it can easily be verified that the energy produced by the whole set of reactions is equal to 26.731 MeV. This amount of energy is almost a factor of 10 larger than that produced in any other nuclear reaction process occurring in stars – the nuclear conversion of H into He is a very efficient, from the point of view of the stellar energy budget, mechanism. This occurrence has the consequence that the amount of fuel, i.e. hydrogen, that is available for the burning process, is consumed at a lower rate than in any other evolutionary phase. Therefore, for each fixed star mass, the central H-burning lifetime is longer than that for any other evolutionary phase. For example, the central H-burning phase is longer by a factor of 100 than the central He-burning stage.

The fusion of H nuclei can be achieved through two reaction chains, namely the *p–p chain* and the *CNO cycle*. These usually occur simultaneously but with relative efficiencies depending on the total stellar mass.

5.2.1 The p–p chain

The reaction networks involved in the p–p chain are the following:

pp I

^{1}H $+^1$H $\rightarrow ^2$D $+$e$^+ + \nu_e$

^{2}D $+^1$H $\rightarrow ^3$He $+ \gamma$

^{3}He $+^3$He $\rightarrow ^4$He $+^1$H $+^1$H

pp II

^{3}He $+^4$He $\rightarrow ^7$Be $+ \gamma$

^{7}Be $+$e$^- \rightarrow ^7$Li $+ \nu_e$

^{7}Li $+^1$H $\rightarrow ^4$He $+^4$He

pp III

^{3}He $+^4$He $\rightarrow ^7$Be $+ \gamma$

^{7}Be $+^1$H $\rightarrow ^8$B $+ \gamma$

^{8}B $\rightarrow ^8$Be $+$e$^+ + \nu_e$

^{8}Be $\rightarrow ^4$He $+^4$He

The first reaction in the pp I chain requires that protons experience a β^+ decay; since it is impossible to form a bound system with two protons, this reaction can occur only if the protons are brought together by a nuclear collision. During the short timescale of this encounter, one of the protons has a chance to β^+ decay, thereby becoming a neutron, a positive electron, and a neutrino. The neutron can then be captured by the other proton to form a deuteron. This process is governed by weak interactions and therefore has a low probability of occurring; its nuclear cross section is in fact quite low ($\approx 10^{-23}$ barn; 1 barn $= 10^{-24}$ cm^2). The pp I chain therefore begins to be important only when the core temperature is of the order of 5×10^6 K. Until a temperature of the order of 8×10^6 K is reached, the reactions producing ^{3}He are more frequent than those consuming ^{3}He and as a consequence the abundance of ^{3}He increases. When this temperature is achieved, the nuclear reactions ^{3}He $+^3$He and ^{3}He $+^4$He become effective, so decreasing the ^{3}He abundance. In a short time this element reaches an 'equilibrium' abundance (this point will be discussed in more detail later).

The relative frequency of pp II and pp III depends strongly on the temperature. In particular, the ^{3}He $+^4$He reaction becomes competitive with respect to ^{3}He $+^3$He for $T \approx 15 \times 10^6$ K. As a general rule, with increasing temperature the importance of pp II and pp III increases with respect to pp I if there is a sufficient concentration of ^{4}He (either produced by pp I or primordial). In addition, pp III gradually becomes more important than pp II.

Concerning the dependence of the nuclear energy generation of the p–p chain as a function of the temperature, it holds the relation: $\epsilon_{pp} \propto T^\nu$, with $\nu \approx 6$ for $T \approx 5 \times 10^6$ K and $\nu \approx 3.5$ for $T \approx 20 \times 10^6$ K. The commonly adopted average relation is $\epsilon_{pp} \propto T^4$, which is the smallest temperature sensitivity of all nuclear fusion reactions of astrophysical interest. At the centre of the Sun, $T \approx 15 \times 10^6$ K, thus $\epsilon_{pp} \propto T^{3.83}$ and the mean liberated energy per reacting proton is ≈ 6.54 MeV.

5.2.2 The CNO cycle

The CNO cycle is the combination of two independent cycles: the CN cycle and the NO cycle. The presence of some isotopes of C, N or O are necessary for either cycle to begin. Being both produced and destroyed during these cycles, these elements act as catalysts. The reaction networks involved are the following:

CN cycle

$$^{12}\text{C} +^1\text{H} \rightarrow {}^{13}\text{N} + \gamma$$

$$^{13}\text{N} \rightarrow {}^{13}\text{C} + e^+ + \nu_e$$

$$^{13}\text{C} +^1\text{H} \rightarrow {}^{14}\text{N} + \gamma$$

$$^{14}\text{N} +^1\text{H} \rightarrow {}^{15}\text{O} + \gamma$$

$$^{15}\text{O} \rightarrow {}^{15}\text{N} + e^+ + \nu_e$$

$$^{15}\text{N} +^1\text{H} \rightarrow {}^{12}\text{C} +^4\text{He}$$

NO cycle

$$^{15}N + {}^1H \rightarrow {}^{16}O + \gamma$$

$$^{16}O + {}^1H \rightarrow {}^{17}F + \gamma$$

$$^{17}F \rightarrow {}^{17}O + e^+ + \nu_e$$

$$^{17}O + {}^1H \rightarrow {}^{14}N + {}^4He$$

In the CN cycle the isotopes of C and N act as catalysts, and so behave as 'secondary elements'. As a consequence the cycle can start almost from any reaction if the involved isotope is present, and during a complete loop around the cycle the isotope is consumed and then produced again. However, this does not mean that the concentrations of the different isotopes will be unchanged as the final abundances depend strongly on the relative rates of the nuclear reactions in the cycle. Only at a high enough temperature ($T \approx 15 \times 10^6$ K) will the isotopes achieve their equilibrium abundance, i.e. the rate of production is exactly equal to the rate of destruction. At this point, the abundance of each isotope is inversely proportional to the nuclear cross section of the reaction by which it is destroyed. Since the slowest reaction of the CNO cycle is $^{14}N(p, \gamma)^{15}O$, the most abundant element in the CNO cycle processed material is ^{14}N.[1]

The branching ratio between the proton captures on ^{15}N, producing respectively ^{16}O and ^{12}C, is of the order of 10^{-4}, thus the amount of ^{16}O produced by proton captures on ^{15}N is regligible. Nevertheless, the small production of ^{16}O through this channel is important because it allows the ^{16}O nuclei originally present to take part in the cycle as well, as they are transformed into nitrogen via the NO cycle.

In this context it should be noted that for a typical Population I chemical composition (i.e. about solar) the global amount of CNO elements is of the order of $X_{CNO} \approx 0.75Z$ and $X_O/X_{CNO} \approx 0.7$, where X_O, X_{CNO} and Z are the abundance by mass of oxygen, of all CNO elements and of all metals in the stellar matter, respectively. This means that the final (equilibrium) ^{14}N abundance depends not only on the initial CN abundances but on all CNO element abundances.

In general the NO cycle only becomes efficient for temperatures larger than $\approx 20 \times 10^6$ K. As a rule of thumb, the change of the CNO element caused by the different burning channels is:

- CN cycle processed matter – C↓ N↑

- CNO cycle processed matter – C↓ N↑ O↓

where the symbols ↓ and ↑ mean that the final abundance of the element is lower or larger than the initial one, respectively.

---

[1] Here we have introduced the compact notation $^{14}N(p,\gamma)^{15}O$ instead of $^{14}N + {}^1H \rightarrow {}^{15}O + \gamma$. This example explains clearly the relationship between the two notations; in the rest of the book we will often use this compact notation for nuclear reactions.

The temperature sensitivity of the complete CNO cycle is much larger than that of the p–p chain, where $\epsilon_{CNO} \propto T^{18}$ at $T \approx 10 \times 10^6$ K. This means that the p–p chain dominates at low temperatures – $T \leq 15 \times 10^6$ K – i.e. in stars with mass lower than $\approx 1.3 M_\odot$, while the CNO cycle dominates at higher temperatures, thus for larger stellar masses. In Figure 5.1, the trends of both ϵ_{pp} and ϵ_{CNO} with temperature are shown. At the centre of the Sun $\epsilon_{pp}/\epsilon_{CNO} \approx 10$, so that the contribution of the CNO cycle to the whole energy budget is of the order of 10 per cent.

The quite different temperature sensitivities of the p–p chain and CNO cycle have an important consequence: if the H-burning process is dominated by the CNO cycle, it is confined to the central regions of the core. This results in a large central energy flux, an occurrence which favours the presence of a central convective region (we discuss this point in more detail later).

5.2.3 The secondary elements: the case of ^{2}H and ^{3}He

It has already been noted that in both the p–p chain and the CNO cycle, there are some elements which are simultaneously involved in destruction and production processes. These elements are usually called 'secondary elements'.

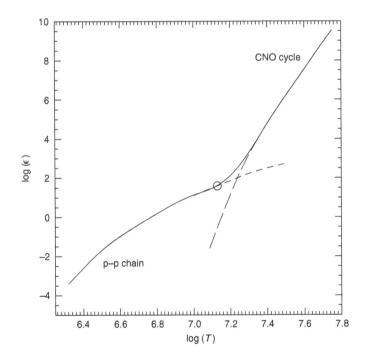

Figure 5.1 The nuclear energy generation in units of erg g^{-1} s^{-1} as a function of the temperature for the p–p chain and the CNO cycle. A chemical composition and density at the centre of the Sun have been assumed. The open circle marks the location of the Sun in this diagram

It is instructive to compute the equilibrium abundances of these elements, in order to understand their changes with time, i.e. with the evolution of the thermodynamical properties (temperature and density) and chemical composition. In the following, we analyse the case of ^{2}H and ^{3}He but the same consideration can be adopted for the treatment of the isotopes of the CNO elements involved in the CNO cycle.

In Section 3.1.7 it was shown how the variation of the abundance (by mass) of the element i can be written down by accounting for all the reactions of production and destruction (Equation (3.39)). By applying this equation to the case of ^{2}H, and using the relation between the abundance by mass of element i and the number of the nuclei of element i per volume unit, i.e. $N_i = (\rho X_i)/(m_H A_i)$, one obtains (in the following the numbers 1 and 2 refer to ^{1}H and ^{2}H, respectively):

$$\frac{dN_2}{dt} = \frac{N_1^2}{2} <\sigma v>_{11} - N_1 N_2 <\sigma v>_{12}$$

The first term accounts for the reaction which produces ^{2}H (^{1}H + ^{1}H) while the second term corresponds to the destruction process (^{2}H + ^{1}H). At equilibrium, by definition $\frac{dN_2}{dt} = 0$, therefore

$$\left(\frac{N_2}{N_1}\right)_{eq} = \frac{1}{2}\frac{<\sigma v>_{11}}{<\sigma v>_{12}}$$

It is evident that if the abundance of ^{2}H is larger than its abundance at equilibrium, then the destruction process prevails over the production one, and the chemical abundance of ^{2}H tends to evolve quickly towards its equilibrium configuration. It is instructive to determine the timescale for equilibrium to be reached, assuming that $N_2 \gg (N_2)_{eq}$. In this case, the destruction process dominates over the production one and

$$\frac{dN_2}{dt} = -N_1 N_2 <\sigma v>_{12}$$

whose solution is of the type $N_2 = N_{2,0} e^{-\frac{t}{\tau}}$ with $\tau = 1/(N_1 <\sigma v>_{12})$. For typical temperatures at which the p–p chain is fully efficient, the value of τ is of the order of 1 second. This means that the equilibrium configuration is achieved in a very short time.

Now consider the case of the ^{3}He, once again using Equation (3.39) altered to account for the abundance by number per unit volume

$$\frac{dN_3}{dt} = N_1 N_2 <\sigma v>_{12} - 2\frac{(N_3)^2}{2} <\sigma v>_{33} - N_3 N_4 <\sigma v>_{34}$$

by imposing $\frac{dN_3}{dt} = 0$ and noting that if ^{3}He is at the equilibrium then deuterium will also be (the timescale for equilibrium of ^{2}H is shorter than the ^{3}He one) one obtains

$$(N_3)_{eq} = \frac{-<\sigma v>_{34} N_4 + \sqrt{(<\sigma v>_{34} N_4)^2 + 2 <\sigma v>_{11} <\sigma v>_{33} (N_1)^2}}{2 <\sigma v>_{33}}$$

If we assume that $N_3 \gg (N_3)_{eq}$ (see the discussion in Section 5.3) then the destruction process is dominant and one can easily estimate the ^{3}He equilibrium timescale. We find

$$\frac{dN_3}{dt} \approx -2 \frac{(N_3)^2}{2} < \sigma v >_{33} - N_3 N_4 < \sigma v >_{34}$$

So the timescale for the ^{3}He equilibrium to be reached is:

$$\tau_3 = \frac{1}{N_3 < \sigma v >_{33} + N_4 < \sigma v >_{34}}$$

It is worth noting that these relations are strictly valid only in a radiative region. They can still be applied in a convective zone, but the various quantities have to be derived as averages over the whole convective region.

5.3 The central H-burning phase in low main sequence (LMS) stars

In Section 4.3.2 the evolutionary and structural properties of stars during the Pre-Main Sequence phase were discussed. During this phase the evolutionary rate of the stars – neglecting the small contribution to the energy budget provided by the burning of light elements such as Li and Be – is dictated by the gravitational energy release. Also, according to the previous discussion on the secondary elements involved in the H-burning processes, the first model that is fully supported by nuclear burning is not yet a *true* Main Sequence (MS) model, because the secondary elements have not yet reached the equilibrium configuration. The Zero Age Main Sequence (ZAMS) is the *first* MS model fully supported by H-burning in which the secondary elements are at their equilibrium configuration. Therefore, there is an 'intermediate' phase between the Pre-Main Sequence and the ZAMS during which the star attains the chemical equilibrium of the secondary elements.

Due to the dependence on temperature of the different H-burning processes, in low-mass stars with $M \leq 1.3 M_{\odot}$ (the exact value depending on the initial chemical composition) the main H-burning mechanism is the p–p chain. It is common to define these stars as belonging to the *Low Main Sequence*.

It is worth briefly discussing the approach of these stars to the ZAMS. Near the end of the Pre-Main Sequence phase, due to the lack of ^{3}He nuclei, the dominant reaction is that of ^{3}He production. As the nuclear processing proceeds, the amount of this element increases, thus increasing the number of ^{3}He $+ ^3$He $\rightarrow ^4$He $+ ^1$H $+ ^1$H reactions. The large number of these reactions allows the star to complete the ppI branch and to produce more energy per burnt H nucleus. During the previous phases, with only a partially efficient p–p chain, the star was forced to reach a higher temperature and density in order to satisfy its energy needs. Now, with a more efficient p–p chain, the star slightly decreases its core density and temperature, fixing the number of nuclear reactions to the exact value required by its energy needs. This situation lasts until ^{3}He equilibrium is achieved.

During the phases preceding the ^{3}He equilibrium, a small convective core is present, a consequence of the fact that for a short time ^{3}He production is larger in the central portion of the star, so energy production is more concentrated in the centre. Due to the large energy flux, F (we recall that $\nabla_{rad} \propto F$) the central region becomes convective. This convective core vanishes as soon as the abundance of ^{3}He is increased – by the nuclear burning – in more external regions, as this causes a redistribution of the energy generation over a larger area.

In general, for any hydrostatic evolutionary phase not affected by electron degeneracy, the star regulates its thermonuclear burning rate so that the nuclear energy production is just enough to enforce the hydrostatic equilibrium condition. If, for any reason, the number of nuclear reactions is larger than needed, the star reacts by expanding, thus decreasing the temperature and density, so that the burning rate decreases and equilibrium is re-established.

Beginning on the ZAMS, a star burns H in a *radiative region* whereas, due to the large opacity associated with the presence of partially ionized H and He, the *outer regions are convective*. Due to the small temperature dependence of the p–p chain, H-burning involves a relatively large mass fraction of the star. In Figure 5.2 the H chemical profile is shown in a $1M_\odot$ star at various stages, from the ZAMS to the exhaustion of H at the centre.

During the conversion of H into He the total number of free particles decreases and so does the pressure. In order to remain in equilibrium, the star-slightly contracts and heats up. The combined effect of the change in the opacity[2] in the stellar interior, the increase in the mean molecular weight and the temperature increase, causes a slow but monotonic increase of the star's luminosity. The evolutionary path in the HRD of low-mass stars during the central H-burning phase is shown in Figure 5.3. From this figure, one can notice that during this evolutionary phase, the surface luminosity and effective temperature increase. The central H-burning phase continues until the H at the centre is completely exhausted; this stage corresponds roughly to the hottest point on the evolutionary tracks shown in tracks shown in Figure 5.3, called the *turn off* (TO). It marks the end of the central burning phase in low-mass stars. The TO is the most important 'clock' provided by stellar evolution, and its use in comparison with observations for dating stellar populations will be outlined in Section 9.2.

Evolutionary lifetimes during the central H-burning phase are quite long for low MS stars, being of the order of ~ 10 Gyr. This lifetime is a strong function of the stellar mass, decreasing as the mass increases.

At the exhaustion of H at the centre as the burning process already extends to a significant portion of the surrounding regions (see Figure 5.2) the core slightly contracts while H-burning continues in a shell around an He core.

---

[2] The opacity of He is less than that of H in the same thermal conditions.

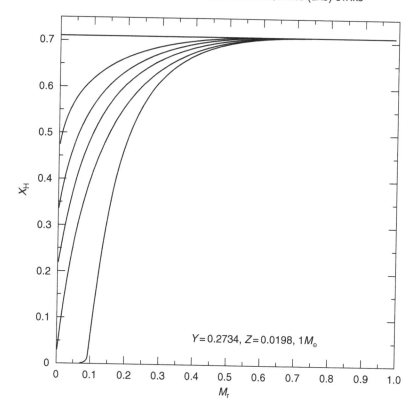

Figure 5.2 The chemical profile of hydrogen in a $1M_\odot$ star at different stages during the core H-burning phase

5.3.1 The Sun

The importance of studies of the Sun's internal structure lies not only in its intrinsic interest, but also in the possibility of testing some of the fundamentals of theory, both physical and astrophysical. The Sun is the star for which, through helioseismology and spectroscopic measurements, we can collect the largest amount of empirical data which can be directly compared to theoretical predictions.

Without going into detail, from the analysis of the spectra associated with solar non-radial oscillations, it is possible to determine accurately the depth of the envelope convective zone and the speed of sound, $c_b = (0.221-0.225)\,\mathrm{Mm\,s^{-1}}$, at the transition radius, $R_b = (0.710-0.716)R_\odot$, between convective and radiative regions.

Several determinations of the photospheric helium abundance have been derived from helioseismology, yielding the estimate $Y_{phot} = 0.233-0.268$. The interpretation of data on solar oscillations provides the strongest evidence for microscopic diffusion in the Sun and, presumably, in any low-mass stars. From spectroscopic measurements,

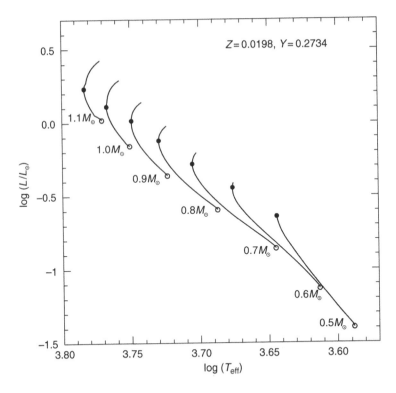

Figure 5.3   The HRD for low-mass stars of different mass during the core H-burning phase. The solid dot marks the location of the turn off along each track

we know that the present ratio of the heavy element abundance over hydrogen in the solar photosphere is $Z/X = 0.0245$. Recently, on the basis of new analysis ([6]) performed by adopting realistic time-dependent, three-dimensional models of the solar atmosphere, the estimated solar heavy element abundance has been significantly reduced ($Z_\odot \sim 0.013$ so by almost a factor of two with respect to the commonly adopted value). However, these measurements have to be confirmed by additional investigations.

Coupling this empirical evidence with the other known properties of the Sun, i.e. its luminosity, $L_\odot = 3.842 \times 10^{33}$ erg·s^{-1}, radius, $R_\odot = 6.9599 \times 10^{10}$ cm and mass, $M_\odot = 1.9891 \times 10^{33}$ g, one can work at obtaining the best stellar evolution model, based on the most up-to-date physics, able to reproduce the whole set of observational properties of the Sun at its present age ($t_\odot \approx 4.5$ Gyr): this would be a *Solar Standard Model* (SSM).

When computing an SSM, rotation and magnetic fields are assumed to have a negligible effect on the evolution. Furthermore, mass loss is almost universally neglected during the central H-burning phase of solar-like stars because the estimated mass loss rate is of the order of $(dM/dt) \sim 10^{-14} M_\odot$yr^{-1}. Thus, the free parameters which, aside from some practical limits, are varied in solar ZAMS models until 4.5 Gyr worth of evolution yields the present-day Sun, are as follows.

- The abundance of heavy elements Z, which affects the structural evolution through the effect on the stellar opacity and on the efficiency of the CNO cycle (that also contributes to the energy generation in the Sun).

- The He content, affecting the structure through the change in the mean molecular weight and, in turn, in the nuclear energy release, and in the opacity. Note that in constructing the SSM, the adopted initial values of Z and Y (Z_{ini} and Y_{ini}) are not independent. This is because (i) the relation $X = 1 - Y - Z$ holds, and (ii) after 4.5 Gyr and the work performed by atomic diffusion, the Z/X ratio in the SSM must reproduce the value observed at the surface of the Sun.

- The value adopted for the free parameter α_{ml} in the mixing length theory. This affects the efficiency of convection in the envelope and, in turn, fixes the radius of the solar model.

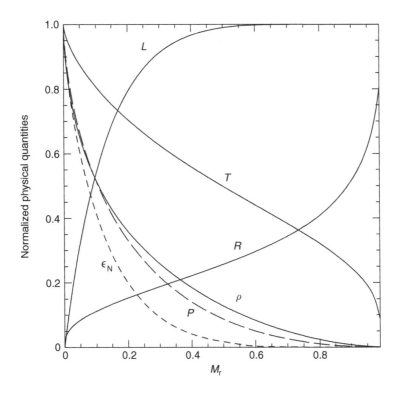

Figure 5.4 The trend with mass coordinate of different physical quantities within a Solar Standard Model. Each physical quantity has been normalized to its maximum value within the star

To obtain the appropriate SSM one has to change (see the discussion in Section 5.5) the values of these free parameters until the model properties agree with the solar values. Different SSMs computed with similar, up-to-date physical inputs, predict similar values for these free parameters. Typical values are: $Z_{ini} = 0.0198$, $Y_{ini} = 0.269$, $\alpha_{ml} \sim 1.7$.

Due to the effect of atomic diffusion, predicted values for the metallicity and He content in the photosphere of the SSM are: $Z_{phot} = 0.0182$, $Y_{phot} = 0.238$. In Figure 5.4, the behaviour of the most important physical quantities such as temperature, pressure, density, luminosity, radius and nuclear energy production coefficient (ϵ_N) in the interior of an SSM are shown.

The calibration of the SSM is a very important topic in stellar evolution. In fact, it allows not only the prediction of the structural properties of the Sun, which can be directly compared with helioseismology and spectroscopic measurements, but also the calibration of the mixing length parameter. It is quite common to use the value of the mixing length efficiency, α_{ml}, obtained by the calibration of the SSM, for computing the evolution of stars of any mass, chemical composition and evolutionary stage. Even if there is no a priori reason why this assumption should be correct for other stars, the comparison between stellar evolution models and empirical evidence clearly shows that this approach works quite well.

5.4 The central H-burning phase in upper main sequence (UMS) stars

In the previous section, we discussed the main structural and evolutionary properties of Lower Main Sequence stars. Here, we address the same topic but for more massive stars, i.e. structures with a mass larger than $\sim 1.2-1.3 M_\odot$ (the *Upper Main Sequence*). The evolutionary paths in the HRD during the central H-burning phase for these stars are shown in Figure 5.5: it is characterized by a monotonic increase of the luminosity and an almost steady decrease of the effective temperature.

The main effect of an increase in the stellar mass is a significant increase in the interior temperature (in Table 5.1 are listed the main thermal properties in the centre and evolutionary properties for different stellar masses). The most important effect of this temperature increase is that the CNO cycle becomes the dominant energy production mechanism: in a $1.5 M_\odot$ star the CNO cycle contributes 70 per cent of the nuclear energy budget at the centre and 50 per cent to the total luminosity, whereas a $1.8 M_\odot$ star is almost completely under the control of the CNO cycle.

The strong dependence of ϵ_{CNO} on temperature has two important effects.

- The nuclear burning process is more concentrated to the centre: in a $10 M_\odot$ star, more than the 90 per cent of the total luminosity is produced in the innermost 10 per cent of the mass, compared to 70 per cent of the mass in a $1 M_\odot$ star.

- The concentration of the energy production causes a steep increase of the radiative gradient towards the centre, as there is a very high energy flux in the innermost portion of the structure. Due to this, according to the Schwarzschild criterion, the interior of the star becomes convective and it will remain unstable against convection throughout the central H-burning phase.

The existence of a convective core means that H is converted into He in a region which is fully mixed. As a consequence, the chemical gradient inside these stars at various times during the MS phase is completely different (see Figure 5.6) from that in less massive stars.

With increasing stellar mass, an additional effect related to the temperature increase is the increasing contribution to the total pressure provided by the radiation pressure ($P_{rad} \propto T^4$): in a $1M_\odot$ star $P_{rad}/P_{tot} \approx 0.0001$, whereas this ratio is of the order of 0.30 in a $50M_\odot$ star.

As the stellar mass increases, the same is true for the mass size of the convective core, a consequence of the larger temperature in the interior which leads to a larger energy flux. In addition, for massive stars, the decrease of the adiabatic gradient due

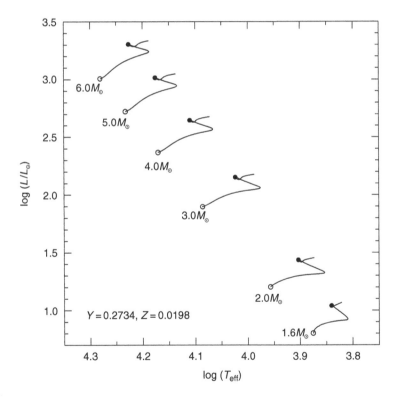

Figure 5.5 Evolutionary tracks of different intermediate-mass stars during the core H-burning phase. The solid dot marks the evolutionary stage equivalent to the turn off point in less massive stars

Table 5.1 Selected properties of core H-burning stars with different masses and solar chemical composition: surface luminosity, effective temperature, central temperature, central density and mass-size of the convective core at the ZAMS, and the core H-burning lifetime. The units are K for temperatures and g cm^{-3} for the density

$M(M_\odot)$	$\log(L/L_\odot)$	$\log(T_{\text{eff}})$	$\log(T_c)$	$\log(\rho_c)$	$M_{cc}(M_\odot)$	t_{Hc} (Myr)
0.5	−1.397	3.588	6.948	1.857	–	128543.7
0.6	−1.133	3.613	6.984	1.872	–	77015.7
0.7	−0.860	3.645	7.021	1.887	–	44031.9
0.8	−0.595	3.687	7.062	1.898	–	26952.3
0.9	−0.363	3.723	7.101	1.900	–	17360.2
1.0	−0.161	3.751	7.133	1.890	–	11513.3
1.1	0.018	3.771	7.158	1.874	0.000	7913.1
1.2	0.213	3.790	7.184	1.887	0.000	5640.7
1.3	0.400	3.813	7.217	1.928	0.085	4138.5
1.4	0.556	3.831	7.248	1.932	0.111	3043.3
1.5	0.691	3.851	7.268	1.918	0.136	2312.5
1.6	0.812	3.874	7.284	1.900	0.168	1844.3
1.7	0.923	3.897	7.297	1.879	0.199	1520.8
1.8	1.019	3.919	7.306	1.851	0.232	1333.3
1.9	1.113	3.939	7.315	1.829	0.266	1142.9
2.0	1.207	3.956	7.325	1.807	0.299	946.5
2.1	1.290	3.973	7.332	1.786	0.335	836.0
2.2	1.371	3.988	7.339	1.762	0.358	727.1
2.3	1.448	4.003	7.346	1.742	0.394	637.8
2.5	1.591	4.030	7.357	1.700	0.447	502.3
2.6	1.657	4.042	7.362	1.679	0.492	447.9
2.8	1.783	4.065	7.372	1.640	0.542	360.9
3.0	1.898	4.086	7.380	1.603	0.603	296.6
4.0	2.368	4.171	7.414	1.446	0.922	138.1
5.0	2.724	4.232	7.437	1.325	1.242	78.5
6.0	3.007	4.280	7.455	1.227	1.549	51.0
7.0	3.238	4.320	7.470	1.147	1.887	36.3
8.0	3.433	4.352	7.482	1.079	2.309	27.7
9.0	3.601	4.380	7.493	1.022	2.672	22.1
10.0	3.749	4.404	7.502	0.972	3.159	18.3

to the increasing contribution of radiation pressure to the total pressure also has a significant effect on the dimension of the convective core. These stars, at odds with low-mass stars, have a convective core but a radiative envelope. This is because the regions of partial ionization of H and He (with high opacity and hence large radiative gradient) are located too high up in the atmosphere – thus at too low a density – to affect the thermal properties of the envelope.

A remaining open problem for the stars with convective cores during the central H-burning phase is related to the location of the 'true' boundary of the convective

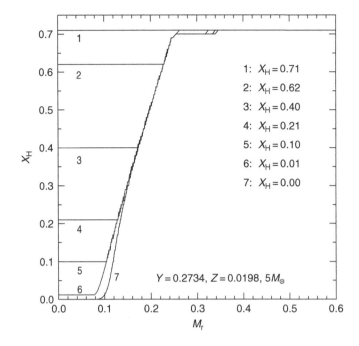

Figure 5.6 The chemical profile of hydrogen in a $5M_\odot$ star at different stages during the core H-burning phase

region: the problem of overshoot (see discussion in Section 3.1.4). The case for a significant amount of overshoot, i.e. for an increase of the size of the convective core with respect the value predicted by standard (canonical) models, has been made many times, theoretically and observationally, but the results have so far been contradictory. From a purely theoretical point of view, there are at least three physical 'mechanisms' which could induce an increase in the size of the convective core: (1) any change to the physical inputs (opacity, equation of state) which affect the radiative gradient, (2) stellar rotation, and (3) true mechanical overshoot. Whatever the physical reason, the effects on the evolutionary and structural properties can be easily predicted from basic theory as follows.

- The star becomes brighter, because the increase of the mean molecular weight (the H-burning luminosity depends on μ as $L_H \propto \mu^7$, everything else being constant) related to the conversion of H into He, now involves a larger fraction of the structure.

- The central H-burning lifetime increases, as there is more H available for nuclear burning.

- At the end of the MS phase the mass size of the He core is much larger and, as a consequence, during the central He-burning phase the star will be brighter and the He-burning lifetime shorter.

Figure 5.7 shows the HRD of two stellar models computed with and without overshoot. The more relevant properties of the these models are given in Table 5.2.

The evidence that stars belonging to the Upper MS burn H inside a convective core has an important effect at the end of central H-burning phase: when the abundance of H inside the convective core becomes lower than $X \sim 0.05$ (which corresponds to point B in Figure 5.7) the amount of H is not sufficient for providing the energy

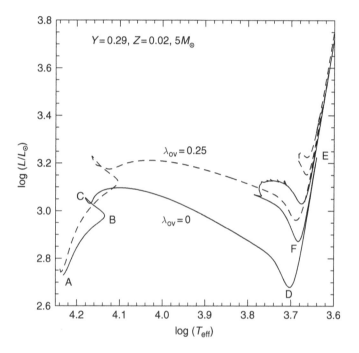

Figure 5.7 The HRD of a $5M_\odot$ star for two different assumptions about the efficiency of core overshoot during the central H-burning phase

Table 5.2 Selected properties of two $5M_\odot$ models with ($\lambda_{OV} = 0.25$) and without core convective overshoot (see also Figure 5.7); the mass of the He core at the core H-exhaustion, the core H-burning lifetime, the mass of the He core at the onset of core He-burning and the core He-burning lifetime are listed

Model	$M_{\mathrm{He}}(X_\mathrm{H}=0)(M_\odot)$	$t_{\mathrm{Hc}}(\mathrm{Myr})$	$M_{\mathrm{He}}(M_\odot)$	$t_{\mathrm{Hec}}(\mathrm{Myr})$
Canonical	0.573	85.1	0.660	19.9
Overshoot	0.777	100.1	0.854	10.0

necessary to maintain the structure in equilibrium, and so the star begins to contract. This evolutionary phase, called *overall contraction*, corresponds to the path between points B and C in Figure 5.7. It is important to note that during the overall contraction phase the dominant energy source is gravitational and so the evolutionary timescale is shorter than the nuclear one.

Point C marks the end of the central H-burning phase. After this point, while the central regions contract, the outer layers expand and this occurrence produces an increase in the stellar radius and a cooling of the external layers. This drastically increases the opacity of the envelope.

In massive stars, i.e. $M \geq 10 M_{\odot}$, there is another phenomenon related to convection which makes their evolutionary properties near the end of the central H-burning phase uncertain. When the abundance of H in the core is strongly decreased, the convective core retreats leaving behind a chemical abundance gradient, with the He abundance decreasing moving outwards. In the region outside the contracting core, the radiative flux also increases as a consequence of the large energy flux produced by gravitation, and therefore exceeds the adiabatic gradient. According to the Schwarzschild criterion this region should become convective. However, if this region is mixed with the original convective core, its resulting He abundance should be larger than the initial one. As a consequence the opacity, hence the radiative gradient of these layers would decrease, inhibiting convection. We are thus facing a dilemma: if these layers are not mixed they are unstable against convection; if they are mixed they appear to be stable and should not have been mixed.

The solution to this problem is related to the choice of the most appropriate thermal gradient in a chemically inhomogeneous region (see discussion in Section 3.1.4). In stellar evolution computations, this is commonly reduced to the choice between the Schwarzschild and the Ledoux criterion. For our purposes, it is sufficient to note that in models for massive stars, it is a common procedure to adopt the Ledoux criterion. This usually inhibits the convection in the layers around the convective core. These models seem better at reproducing the empirical evidence.

5.5 The dependence of MS tracks on chemical composition and convection efficiency

In previous sections, the main evolutionary and structural properties of Main Sequence (MS) stars were discussed and it was often emphasized that these properties depend mainly on the stellar mass. However, in order to better understand how theoretical predictions allow us to extract from empirical evidence, information on single stars, stellar clusters and galaxies, one needs to know how stellar evolution models depend on the adopted inputs.

- *The He content*: the initial abundance of helium affects the MS evolution through the changes induced in the radiative opacity and the mean molecular weight. Any

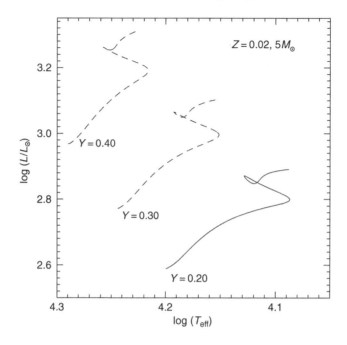

Figure 5.8 The evolutionary tracks of a $5M_\odot$ star, computed for different values of the initial He abundance

increase of He causes a decrease of the opacity and an increase of the nuclear H-burning reaction rates ($L_H \propto \mu^7$). As a consequence, the structures become brighter and hotter, as shown in Figure 5.8. The evolutionary lifetime is consequently affected, decreasing with increasing He content.

- *The metallicity*: a change in metallicity affects the radiative opacity much more than the nuclear energy generation. This occurrence is due to the fact that the p–p chain efficiency is not dependent on the abundance of heavy elements. This is not the case for the CNO cycle given that CNO isotopes act as catalysts. However, due to the very strong dependence of the CNO cycle efficiency on the temperature, even a low CNO abundance is enough to ensure its full efficiency (this is not true for extremely metal-poor stars as Population III objects – see Section 5.11). Any increase in the metallicity produces an increase of the stellar radiative opacity and the structure becomes fainter and cooler. When discussing the evolutionary effect of metallicity, it is usually assumed that the distribution of the heavy elements in the mixture is close to solar (the so-called *scaled solar mixture*). Nevertheless, empirical evidence shows that the distribution of α-elements, i.e. the elements synthesized by nuclear α-capture reactions (O, Ne, Mg, Si, S, Ca, Ti, etc.) are enhanced with respect to iron in Population II stars, compared to the solar mixture.

For a fixed iron abundance, the enhancement of the α-elements has two effects: (1) the CNO cycle efficiency is increased because the sum (C+N+O) is larger due to the increase of the O abundance, and (2) the opacity is increased, and two opacity bumps are created. The first is at $\log(T) \sim 6$, and is due to absorption processes involving the K shell electrons of O, and the second is at $\log(T) \sim 5.5$, and is due to the combined L edges of Mg, Si and Ne. As a consequence, for a fixed iron content, accounting properly for α-elements enhancement produces MS tracks which are fainter and cooler with respect to the *scaled solar* one, with longer central H-burning lifetimes ($\sim$5 per cent for $X_{Fe} \sim 4 \times 10^{-5}$). These effects continue to increase as the metallicity does. The impact of the inclusion of α-elements enhancement on theoretical isochrones will be discussed further in Section 9.2.1).

- *The efficiency of convection*: the effect of core overshoot has already been discussed. Stellar models are also affected by the treatment of superadiabatic convection. The most common approach for treating the convection in the outer layers is the mixing length formalism, which has a free parameter, α_{ml}. It has already been shown that the calibration of the SSM allows one to fix the value of α_{ml}. This notwithstanding, consider the effect of a change of α_{ml} on the stellar models: a change in the value of α_{ml} has no effect on the surface luminosity, but it does affect

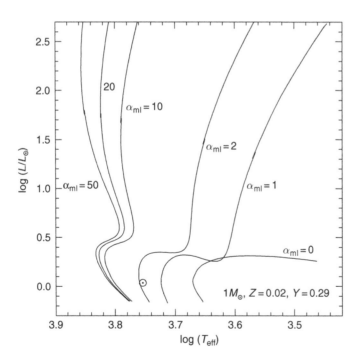

Figure 5.9 The HRD for a $1M_{\odot}$ star computed adopting different values for the mixing length. The $\odot$ symbol marks the location of the Solar Standard Model

the stellar radius and in turn the effective temperature. An increase of α_{ml} corresponds to a larger efficiency of convection – a bubble moves further before losing its excess energy – so a lower thermal gradient is required, and as a consequence the effective temperature increases and the radius decreases (see Figure 5.9).

5.6 Very low-mass stars

The evolutionary and structural properties of low-mass stars change significantly when entering into the realm of very low-mass (VLM) stars, i.e. objects with mass less than $0.3–0.4M_\odot$. Stars below this mass limit are fully convective all along their MS lifetime. With the exception of the PMS evolution for all stars, this is peculiar to VLM stars and is due to the huge opacity values characteristic of the matter in these objects. To point out how peculiar these stars are in the field of stellar evolution theory, Table 5.3 lists the values of density and temperature at the centre and the base of the photosphere for three stellar masses: $(0.1, 0.6, 1.0)M_\odot$, with solar composition and an age of about 10 Gyr.

Within the temperature/density range listed in this table, molecular H and atomic He are stable in the outermost part of the stellar envelope, while most of the star ($\sim$90 per cent in mass) is a fully ionized H/He plasma. Due to these peculiar conditions, the thermodynamical properties are very sensitive to the treatment of pressure ionization and dissociation, non-ideal Coulomb interactions. In addition, electron degeneracy becomes very important for masses less than $\sim0.12M_\odot$.

Theoretical and observational analyses have demonstrated that the emergent radiative flux of VLM stars shows an incredible number of features related to the presence of several opacity sources in the stellar atmosphere. Due to the large density and pressure at the photosphere, collision effects become significant and induce molecular dipoles on H_2. This produces the so-called *Collision Induced Absorption* (CIA) which provides an important contribution to the opacity. The CIA of H_2 suppresses the flux longward of $2\mu m$ and causes the redistribution of the emergent radiative flux towards shorter wavelengths. Around and below $T_{eff} \approx 4000$ K, the presence of molecules such as H_2O, CO, VO and TiO is very important: TiO and VO govern the energy flux in the optical wavelength range, while H_2O and CO control the spectrum in the infrared. For effective temperatures lower than $\sim$2800 K, the presence of grains also becomes important.

Table 5.3 Selected properties at the centre and at the base of the photosphere of low- and very low-mass stars with different masses

$M(M_\odot)$	$T_c(10^6$ K$)$	$\rho_c(\mathrm{g\,cm}^{-3})$	$T_{phot}(K)$	$\rho_{phot}(\mathrm{g\,cm}^{-3}$
1.0	16	100	6000	10^{-7}
0.6	10	150	4000	10^{-6}
0.1	5	500	2800	10^{-5}

For masses less than $\sim 0.4 M_\odot$, only pp I provides a contribution to the energy production, due to the low temperatures in their interiors. This means that the destruction process of ^{3}He is negligible, and this element behaves as a pseudo-primary element, for which the timescale required for equilibrium is longer than the Hubble time. It is evident that in such a situation, the definition of ZAMS as given previously is completely meaningless. In addition, the evolutionary lifetimes of these stars are so long that their location on the HRD, after they start to burn hydrogen, does not significantly change within a Hubble time.

Figure 5.10 shows the HRD of VLM stars for various metallicities and for an age of 10 Gyr. One can immediately recognize the usual dependence of the MS location on the metallicity. It is important to note the sinuous shape of the MS, with two well-defined bending points:

- the first, located at $T_{\rm eff} \sim 4500$ K and $M \sim 0.5 M_\odot$, is due to changes in the adiabatic gradient related to the recombination process of H_2;

- the second, located at $T_{\rm eff} \sim 2800$ K and $M \sim 0.15 M_\odot$, is related to the increasing level of electron degeneracy when the stellar mass decreases.

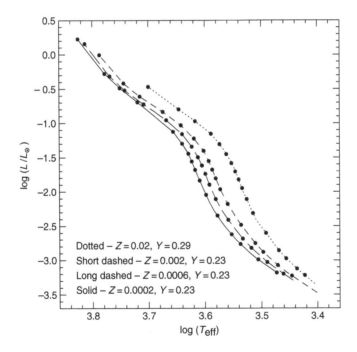

Figure 5.10 The MS location of very low-mass stars with different initial chemical compositions at an age of 10 Gyears. The solid dots along each sequence mark the location of the following masses (from top to bottom): $M/M_\odot = 0.80, 0.70, 0.65, 0.60, 0.50, 0.45, 0.40, 0.35, 0.30, 0.25, 0.20, 0.15, 0.12, 0.11, 0.10, 0.095$

The role played by electron degeneracy in VLM stars is very important. For any fixed chemical composition, it determines the value of the Minimum Mass for H-burning (MMHB). With decreasing stellar mass, in order to achieve the central temperatures necessary for the p–p chain to be effective, stars contract more and more, and so larger and larger central densities are attained. In VLM stars, this results in such a large density that electron degeneracy becomes important, and for masses around $0.1M_\odot$, the cooling associated with conductive opacity is so efficient that it decreases the central temperature below the threshold required for H-burning. For a solar composition, the value of MMHB is $\sim 0.075M_\odot$, and it increases with decreasing stellar metallicity. Less massive stars are not able to reach the temperature for H-burning, and so after a burning phase of light elements such as Li and D, they become fainter and fainter. These are the so called *brown dwarfs*.

These objects will evolve cooling down in the same way as white dwarfs (see Section 7.4). The difference is that brown dwarfs have an homogeneous chemical composition – due to the fact that they are fully convective – where hydrogen is the dominant species (hence crystallization is not attained) and the degenerate electrons never become relativistic. The non-degenerate envelope determines the rate of cooling of brown dwarfs, as in the case of white dwarfs.

5.7 The mass–luminosity relation

Stellar evolution models predict that stars spend a sizeable fraction of their core H-burning evolutionary lifetime close to their ZAMS location. This theoretical result is strongly supported by empirical evidence in galactic stellar clusters.

One of the most important empirical properties of stars on or near the ZAMS is that they display a tight relationship between their total mass and the surface luminosity, the so-called 'mass–luminosity' relation. The existence of this relationship can be proved, on qualitative grounds, by a simple analysis of the stellar structure equations.

Let us assume that the energy transport is purely radiative; if in the derivative in the left-hand side of the radiative transport equation we consider the differences between the centre and the surface of the star, and use in the right-hand side the central temperature T, surface radius R and denote with M the total mass of the star, the radiative transport equation can be written as

$$L \propto \frac{R^4}{M} T^4$$

where L is the surface luminosity, R the total radius and T the central temperature. Following the same approach for the hydrostatic equilibrium equation, one derives

$$P \propto \frac{M^2}{R^4}$$

Assuming now that the stellar matter behaves as an ideal gas ($P \propto \rho T$) and remembering that $\rho \propto M/R^3$, one obtains

$$T \propto \frac{M}{R}$$

By substituting this dependence into the first equation, one finds

$$L \propto M^3$$

Due to the several approximations made in this sketchy derivation, one has to expect that the relation $L \propto M^3$ is only roughly followed by 'real' stars. Empirical data for stars of approximately solar chemical composition provide $L \propto M^{3.6}$ for masses between $\sim$2 and $20M_\odot$, $L \propto M^{4.5}$ in the range between $\sim$2 and $0.5M_\odot$, and $L \propto M^{2.6}$ in the range between $\sim$0.5 and $0.2M_\odot$. Figure 5.11 shows a theoretical mass–luminosity relation for ZAMS stars in the mass range $0.1M_\odot$–$10M_\odot$ provided by detailed evolutionary computations, that displays trends with mass roughly consistent with the observations.

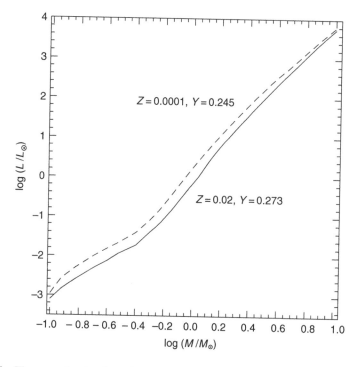

Figure 5.11 The mass–luminosity relation for ZAMS stars in the mass interval $0.1M_\odot$–$10M_\odot$, for two initial chemical compositions

For more massive stars $(M > 10 M_\odot)$, the mass–luminosity relationship becomes less steep, and one derives that $L \propto M$. It can be shown that this behaviour is due to the increasing contribution of radiation pressure to the total pressure in massive stars.

5.8 The Schönberg–Chandrasekhar limit

Regardless of the H-burning mechanism at work during the MS phase, at the exhaustion of central H the star is left with an He core surrounded by an H rich envelope. At the base of this envelope the temperature is high enough for H-burning which continues in a shell. Given that there is no nuclear burning inside the He core and the temperature gradient is radiative, its thermal stratification is isothermal (see Equation (3.15)). In 1942, Schönberg and Chandrasekhar investigated, with a pure analytical approach, the hydrostatic equilibrium conditions for an isothermal He core with an ideal gas EOS. They found a fixed limiting value for the ratio between the core mass and the total stellar mass (M_{core}/M_{tot}) the so-called *Schönberg–Chandrasekhar limit*. If this mass ratio is larger then the Schönberg–Chandrasekhar limit, the core must contract on Kelvin–Helmholtz timescales (virial theorem) because the isothermal core cannot support the pressure exerted by the overlying envelope layers.

The existence of an upper limit to the ratio M_{core}/M_{tot} can be qualitatively understood as follows. We first apply to the isothermal core (with mass M_{core}, radius R_c and temperature T_c) the virial theorem in case of non-vanishing surface pressure P_0 (Equation (3.46)) and solve for P_0. We obtain

$$P_0 = K_1 \frac{M_{core} T_c}{R_c^3} - K_2 \frac{M_{core}^2}{R_c^4}$$

where the first term in the right-hand side comes from the total internal energy of the core (this internal energy is $\sim (3/2)(K_B/\mu m_H) T_c M$) and the second term comes from the gravitational potential Ω; K_1 and K_2 are constants. For a given value of M_{core} the pressure P_0 attains a maximum value $P_{0,m} = K_3 T_c^4 / M_{core}^2$ when the radius is equal to $K_4 M_{core}/T_c$ (K_3 and K_4 are constants). It is important to notice that the maximum pressure $P_{0,m}$ decreases for increasing M_{core}. For the star to be in equilibrium $P_{0,m}$ must be larger than, or at least equal to, the pressure P_e exerted by the non-degenerate envelope on the interface with the core. If we now apply the dimensional analysis performed to obtain the mass–luminosity relationship for MS stars – and assume that the functional dependences found for the central values also hold for any other point within the star – we can roughly approximate $P_e \propto M_{tot}^2 / R^4$ and $T_c \propto M_{tot}/R$, where M_{tot} and R are the total mass and the total radius of the star; hence at the interface with the core $P_e \propto T_c^4 / M_{tot}^2$. Therefore, the condition $P_e \leq P_{0,m}$ dictates the existence of an upper limit to the ratio M_{core}/M_{tot}.

The exact value of the Schönberg–Chandrasekhar limit depends on the ratio between the mean molecular weight in the envelope and in the isothermal core

$$\left(\frac{M_{core}}{M_{tot}}\right)_{SC} = 0.37 \left(\frac{\mu_{env}}{\mu_{core}}\right)^2$$

At the end of the MS phase of a solar chemical composition object $\mu_{env} \sim 0.6$ and $\mu_{core} \sim 1.3$ (the core is essentially made of pure helium); the Schönberg–Chandrasekhar limit is therefore equal to $(M_{core}/M_{tot})_{SC} \sim 0.08$.

This means that if the He core mass is larger than or equal to about 10 per cent of the total mass, it must contract. This value is lower than the relative He core mass of stars with total mass larger than ~ 2.5–$3M_{\odot}$ at the central H exhaustion (the exact value depending on the initial chemical composition) but larger than the He core mass of stars belonging to the Lower Main Sequence.

5.9 Post-MS evolution

After exhaustion of H in the core, there remains an He core that does not provide any contribution to the energy production (the temperature being too low to allow He-burning). This is surrounded by an H-rich envelope with a shell burning at its base. The next question to address is how this configuration evolves and what the evolutionary and structural changes are for the stars. Although all evolutionary computations provide very similar results, we still lack a definitive explanation of the precise physical reason(s) that drive the evolution of stars just after the end of central H-burning to configurations with very large radii and low effective temperatures (see the fundamental work by [160]).

In the following, the Post-MS evolution of stars is discussed, and we will show how it strongly depends on the value of the total stellar mass. Stars that are going to ignite He in an electron degenerate He core are called 'low-mass' stars ($M \leq 2.3M_{\odot}$). As discussed in Chapter 7, stars with larger mass are called intermediate-mass or massive stars, depending on the thermal properties of the CO core at the end of the central He-burning.

5.9.1 Intermediate-mass and massive stars

In this mass range the He core mass at the end of the central H-burning phase is typically larger than the Schönberg–Chandrasekhar limit.[3] Numerical computations show that H-burning continues in an initially rather broad shell around the He core,

---

[3] In stars between $\sim 2.3M_{\odot}$ and $\sim 3.0M_{\odot}$ this limit is reached later, thanks to the increase of the He core mass caused by the fresh He produced in the overlying H-burning shell.

which later narrows in mass size. As a general rule, we notice that when H-burning occurs in a shell, the main burning mechanism is always the CNO cycle.

During this phase the core contracts very slowly and the envelope expands: the gravitational energy release changes sign at the point where the H-burning efficiency is largest. Due to the expansion, the outer layers cool and envelope opacity increases drastically. As a consequence energy trapping in the envelope becomes increasingly efficient, and supports the expansion of the outer layers. During this phase the structure moves in the HRD from the blue to the red side at almost constant surface luminosity. This is called *Sub-Giant Branch* (SGB). The evolutionary rate during this phase is roughly the Kelvin–Helmholtz timescale, being of the order of ~12 Myr for a $3M_\odot$ and ~1 Myr for a $6M_\odot$. This evolutionary phase is so short for stars of high and intermediate mass that the chance of observing objects in this phase is very small. This leads to the presence of the so called *Hertzsprung gap* (a lack of stars along the SGB) in the HRD of stellar systems populated by intermediate-mass and massive stars during the post-MS phases.

The most important structural change in these stars is that the stellar envelope becomes convective, as the outer layers cool due to their expansion. The appearance of an outer convection zone is very important for the final fate of the star: convection is a very efficient energy transport mechanism and acts to slow down the expansion of the star, and thus prevent its complete dissolution. This evolutionary phase corresponds to point D in Figure 5.7 and it marks the beginning of the Red Giant (RG) configuration. From now on, any further expansion occurs at almost constant effective temperature, while the luminosity increases.

For high- and intermediate-mass stars, the central density at the beginning of the RG phase is low enough to prevent the onset of electron degeneracy. Core contraction therefore causes an increase of the temperature in the interior of the star. This allows the structure to reach the thermal conditions ($T \approx 10^8$ K) required for efficient He-burning. When this occurs (point E in Figure 5.7) the core stops contracting and the star is fully supported by nuclear burning. This marks the end of the RG phase, which is very short for these stars, and the beginning of the core He-burning stage which is the longest phase after central H-burning.

For increasing values of the total mass the core contracts much faster and the temperature required for He-burning is reached sooner. As a consequence the RG lifetime is significantly shorter, and the RG phase may even disappear as stars start to burn He near the MS. This behaviour is also characteristic – but for different reasons – of less massive stars but with extremely low metallicity, as will be discussed in Section 5.11.

5.9.2 Low-mass stars

Contrary to the case of more massive stars, for low-mass stars the transition from core to shell H-burning is not so fast. The fractional mass of the He core at the end of the MS is below the Schönberg–Chandrasekhar limit, and even when it grows (due to the production of He from the H-burning shell) above this limit, the electron degeneracy of the He core provides the pressure necessary to support the overlying envelope. In fact, due to the larger central density at the end of the MS phase – with respect to more massive stars – the electron gas in the He core becomes electron degenerate (see Figure 5.12).

Low-mass stars approaching central H-exhaustion have a radiative core (or only small convective cores), and as the central H abundance drops below a critical value, the maximum in the nuclear energy release ceases to be located at the centre as it begins to move outward. At the turn off, the maximum in ϵ_{nucl} occurs at $M_r \approx 0.1 M_\odot$, and ~90 per cent of the nuclear energy is generated in a thick shell of about $0.2 M_\odot$. During the evolution from the turn off to the RG phase (the Sub-Giant phase), due to the large dependence of the CNO cycle efficiency on temperature, the shell becomes increasingly thinner as H is rapidly depleted in the inner portion of the shell and as the temperature drops in the envelope. When the star arrives at the base of the Red Giant Branch (RGB) – the HRD region populated by stars evolving through their RG phase – (see Figure 5.13) the thickness of the shell is only ~$0.001 M_\odot$ and it will

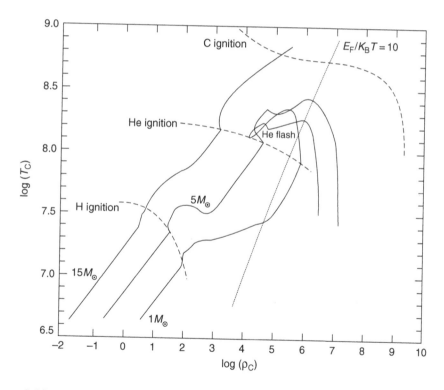

Figure 5.12 The evolution of the thermal properties of the stellar core of stars of various masses. The dashed lines show the location of the H-, He- and C-ignition loci. The dotted line – corresponding to the locus where the Fermi energy (E_F) is equal to $10 K_B T$ – qualitatively marks the separation between the electron degenerate zone and non-degenerate region (above the line)

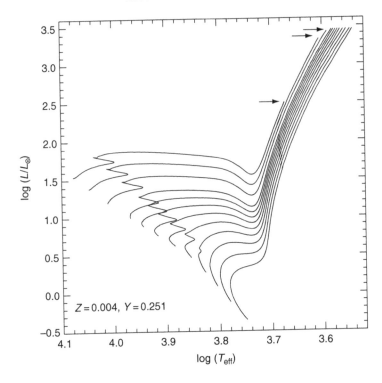

Figure 5.13 The HRD for both the core and shell H-burning phases of low-mass stars for the labelled chemical composition. The RG phase begins when the stars start to evolve at almost constant T_{eff} and increasing luminosity. The various evolutionary tracks correspond to the following stellar masses: $M/M_\odot$ =0.8, 0.9, 1.0, 1.1, 1.2, 1.3, 1.4, 1.5, 1.6, 1.8, 2.0, 2.2. The arrows mark the location of the tip of the RGB for the $2.2M_\odot$ and $2M_\odot$ models, and for those less massive (that has an approximately constant luminosity)

continue to decrease, to about $0.0001M_\odot$ at the tip of the RGB. During the SGB the evolution – in a similar way to the case of more massive stars – the luminosity is approximately constant and, due to the expansion and cooling of the outer layers, the convection already present in the stellar envelope penetrates deeper into the star.

An important property of low-mass stars during the RGB evolution is the tight correlation between the surface luminosity and the mass of the electron degenerate He core: the so-called *He core mass–luminosity* ($M_{cHe}-L$) relation whereby an increase of the He core mass causes an increase of the surface luminosity. The physical reason for this is that the surface luminosity is almost fully provided by the H-burning shell and the thermal properties of this shell and, in turn, the nuclear burning rate are only determined by the mass M_{cHe} and radius R_{cHe} of the He core (see [115]); the initial chemical composition also has some effect, but it is of secondary importance. The thermal properties of the envelope have no effect on the H-burning shell because the mean density inside the He core is very high, whereas that of the envelope is very low. The pressure gradient just above the border of the He core is so large that the pressure decreases by several orders of magnitude as one moves outwards from the H-burning shell. Therefore, the H-burning shell cannot 'feel' the presence of the expanded envelope. This is also the reason why mass loss from the stellar surface – which is efficient along the RGB – does not affect the nuclear burning, M_{cHe} and the stellar luminosity. What is affected by mass loss is the star radius; in brief, the radius (hence T_{eff}, given that the luminosity is unchanged) of mass-losing RGB stars tends to readjust to dimensions appropriate to the actual value of their total mass. As we will see later, when evolving at constant M, lower-mass RGB stars at a given luminosity have larger radii and lower T_{eff}; therefore mass loss tends to shift the star T_{eff} towards progressively lower values, without affecting its evolutionary timescale, luminosity and M_{cHe}.

Figure 5.14 shows the behaviour with time of the base of the convective envelope, and of the boundary of the He core, which roughly coincides with the H-burning shell. As convection penetrates deeper into the star, some of the He produced during the central H-burning phase is mixed in. Therefore, the surface abundance of He monotonically increases until the convection reaches its maximum penetration: this phase is called the *first dredge up*. Other chemical species involved in the H-burning are also mixed in, such as ^{3}He and the CNO elements. The major results are the

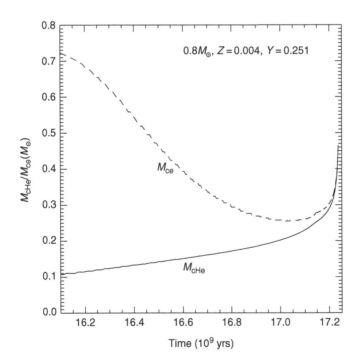

Figure 5.14 The behaviour with time of the base of the outer convective envelope and of the He core mass for a $0.8M_\odot$ star

following: a doubling (roughly) of the surface ^{14}N abundance, a reduction in the surface ^{12}C abundance by approximately 30 per cent, the formation of a surface ^{12}C/^{13}C ratio of about 20–30, a reduction of the envelope Li and Be abundances by several order of magnitudes and a very slight change of the abundance of ^{16}O. Any changes to the parameters which affect the size of the convective envelope, i.e. the total mass, metallicity, He content, efficiency of the superadiabatic convection, can produce sizeable differences in the surface abundances of these elements.

Along the RGB (for both low-mass stars and more massive objects) the H-burning shell moves steadily towards more external mass layers that contain fresh hydrogen, and the lower boundary of the envelope convection recedes towards the surface without overlapping with the shell. As the convective boundary recedes, a chemical discontinuity is left at the layer of maximum inward penetration of the envelope convection zone (see Figure 5.15). When the H-burning shell encounters this discontinuity, the rate at which the star climbs the RGB temporarily drops and even reverses for a while (see Figure 5.16). This behaviour is due to the change in the H-burning efficiency caused by the increase of H abundance, hence decrease of the mean molecular weight, because $L_H \propto \mu^7$. After the shell source has crossed the discontinuity, the mean molecular weight remains at a fixed value (the envelope is now fully mixed by

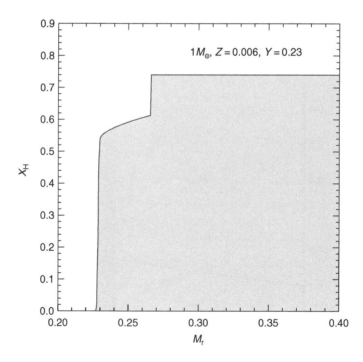

Figure 5.15 Hydrogen abundance profile within a $0.8M_{\odot}$ star, after the first dredge up. The bottom end of the convective envelope at its maximum extension corresponds to the abundance discontinuity at $M_r \sim 0.26$

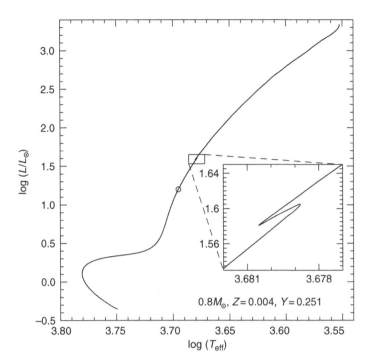

Figure 5.16 The HRD of the same $0.8M_\odot$ star of Figure 5.14. The inset shows the trend of surface luminosity with the effective temperature when the H-burning shell crosses the chemical discontinuity produced by the first dredge up (see Figure 5.15). The point along the evolutionary track corresponding to the first dredge up is marked with an open circle

the convection) and the surface luminosity grows monotonically with increasing core mass. As a consequence, the star will cross three times the same luminosity interval (see the inset in Figure 5.16), and hence one can predict an increase of the star counts in this luminosity interval, i.e. a peak in the luminosity function (see Section 5.10.2 for a discussion on this subject): the so called *bump* of the RGB Luminosity Function. The time spent during the RGB Bump phase is a significant fraction (~ 20 per cent) of the total RGB lifetime.

Let us now consider what occurs inside the He core during the RGB evolution. Since the mass of the He core continuously increases as H is converted into He in the H-burning shell, the density in the He core also increases. The gravitational energy generation, ϵ_g, is then positive within the core. However when neutrino energy losses, ϵ_ν, become relevant, ϵ_ν may exceed ϵ_g at the centre. This usually happens for $M_r \leq 0.3M_{tot}$, and as $\epsilon_\nu + \epsilon_g < 0$ in this region, dL_r/dM_r becomes negative near the centre, i.e. a temperature inversion develops in the central portion of the He core and the temperature maximum is no longer at the centre, but in a shell.

An important point concerning the evolution of low-mass stars along the RGB is related to the electron degeneracy. During the RGB evolution the He core mass increases, its central density increases, and also the degree of electron degeneracy becomes increasingly stronger, causing a large decrease of the conductive opacity in the He core. This makes the conductive energy transport very efficient and, in turn, increases the cooling of the central portion of the He core. In spite of the energy losses from the core associated with the neutrino flux and electron conduction, the maximum temperature in the core increases monotonically. This is because with growing M_{cHe} the core contracts and the associated gravitational energy release heats up the layers below the H-burning shell, where the transition from degenerate to non-degenerate matter takes place and, indeed, the whole He core. When the maximum temperature reaches a value $\approx 10^8$ K, He-burning is ignited. This occurs when the mass of the core is equal to $M_{cHe} \sim (0.48-0.50)M_{\odot}$, almost regardless of the initial total mass – a consequence of the fact that stars in this mass range develop very similar electron degeneracy levels.

Due to the highly degenerate state, the nuclear burning in the He core is unstable and this causes a thermal runaway at the tip of the RGB: the so-called *He flash*. This marks the end of RGB evolutionary phase.

5.9.3 The helium flash

At the moment of He ignition in the core, the central density and temperature are typically of the order of 10^6 g cm^{-3} and 8×10^7 K respectively. In the core, He is under conditions of very strong, partially relativistic electron degeneracy. We have already shown that under such thermal conditions, the gas pressure is not sensitive to temperature changes. When a nuclear burning ignites in a non-degenerate core, the new energy source causes a temperature increase and correspondingly, a pressure increase; to maintain hydrostatic equilibrium the core expands and cools, and this expansion prevents the continuous increase of the energy generation. The same does not hold in the case of an electron degenerate core; the initial temperature rise is not followed by an immediate expansion so the rate at which the He-burning reactions (triple-α reactions) occur increases dramatically, with a continuous increase of the temperature, a so-called *thermal runaway*.

During this runaway, there is a huge production of nuclear energy; in a few seconds an amount of energy of the order of $10^{10}L_{\odot}$ is produced (see Figure 5.17). However, almost none of this energy reaches the stellar surface, as it is absorbed by the surrounding non-degenerate layers. The expansion of the layers just outside the shell where He-burning ignites is the first factor which contributes to the moderation of the He flash strength. The second factor is that convection sets in, a consequence of the large energy flux, and this dilutes the produced energy over progressively larger mass. Canonical stellar models have clearly shown that even if, during the He flash, convection reaches layers very close – both in radius and mass – to the lower boundary of the convective envelope, the huge jump in pressure and entropy existing between the two convective zones, due to the presence of the H-burning shell, prevents the possibility of any mixing. However, this process could occur in stars igniting the He flash under conditions of extreme electron degeneracy as those stars

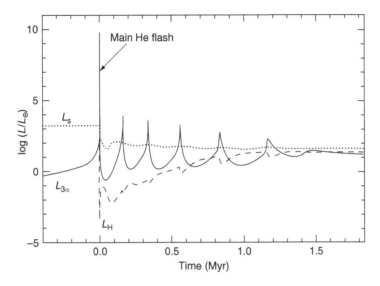

Figure 5.17 The evolution with time of the surface luminosity (L_s) the luminosity produced by 3α-reactions ($L_{3\alpha}$) and that from the H-burning (L_H) during the onset of He-burning at the tip of the RGB in a low-mass star. The time has been offset so that zero corresponds to the start of the main He flash

which experience this event while cooling along the white dwarf cooling sequence ([22],[45]) as well as in extremely metal-poor stars ([196]).

The initial temperature rise at constant density helps to remove the electron degeneracy at the point where the He ignites. However, the region of the core interior to this flash site remains degenerate following the main flash, as there is insufficient time during the main flash for heat to diffuse inward into this region. The degeneracy of this inner region is subsequently removed through a series of much lower strength secondary flashes, until the He burning eventually reaches the star centre. As a consequence of the nuclear burning occurring during the He flash, about 5 per cent of He in the core is converted into carbon. The whole evolutionary phase from the start of the main He flash to the beginning of the core He-burning phase lasts $\sim 10^6$ years.

5.10 Dependence of the main RGB features on physical and chemical parameters

In this section, we will discuss the effect of the initial He content, metallicity and efficiency of superadiabatic convection, on the main features of the RGB such as its location in the HRD, the RGB bump luminosity and the luminosity of the tip of the RGB (the point of He ignition).

5.10.1 The location of the RGB in the H–R diagram

The location of the RGB in the HRD strongly depends on any structural and/or physical parameter which affects the size of the convective envelope, and is similar to the case of the Hayashi track, discussed in Section 4.3.

Numerical computations show that for each fixed chemical composition, the RGB becomes cooler with decreasing stellar mass.

The dependence of the RGB location on the chemical composition is due to changes induced in the radiative, low-temperature opacity by any change in the He content and metallicity. An increase in the He content at fixed Z causes a decrease of the envelope opacity, as the He opacity is lower than the H opacity. This occurrence causes a reduction in the mass extension of the envelope convection zone and, in turn, a hotter RGB.

The abundance of heavy elements is the parameter which most affects the RGB morphology; any increase of Z produces a larger envelope opacity and, in turn, a more extended envelope convection zone and a cooler RGB. The strong dependence of the RGB effective temperature on the metallicity makes the RGB one of the most important metallicity indicators for stellar systems such as galaxies and galactic star clusters. An important issue is the dependence of the shape and location of the RGB on the distribution of the metals; different heavy elements have different ionization potentials, and contribute differently to the opacity of the envelope. The abundance of low ionization potential elements such as Mg, Si, S, Ca, Ti and Fe strongly influences the RGB effective temperature, through their direct contribution to the opacity due to the formation of molecules such as TiO which strongly affects the stellar spectra at effective temperatures lower than 5000–6000 K, and through the electrons released when ionization occurs, which affect the envelope opacity via the formation of the H^- ion – one of the most important opacity sources in RGB structures. As an example, a change of the heavy elements mixture from a scaled solar one to an α-element enhanced distribution with the same iron content produces a larger envelope opacity and the RGB becomes cooler and less steep. The change in the slope is due to the increasing contribution of molecules to the envelope opacity when the stellar effective temperature decreases along the RGB.

The RGB morphology depends also on the efficiency of the convection in the outer stellar layers. Due to the low density which characterizes these layers, a significative fraction of the envelope shows a temperature gradient which is strongly superadiabatic. Therefore, any change in the efficiency of convection in these layers has a strong impact on the RGB effective temperature. If the value of the mixing length is increased, which corresponds to an assumption of more efficient convection, the thermal gradient decreases and in turn the effective temperatures along the RGB increase. Due to the larger extension of the superadiabatic region in RGB stars, the dependence of the RGB temperature on the adopted mixing length value is much stronger than that for MS stars as is clearly shown in Figure 5.9. The same figure also indicates that when the mixing length parameter is set to zero, the star will stop climbing up the RGB, but continues to expand steadily until its complete disruption; this is an indirect proof that if the energy transport mechanism associated with convection did not exist, we would live in a Universe without RG stars.

5.10.2 The RGB bump luminosity

The physical reasons for the occurrence of the bump in the RGB Luminosity Function (LF) have been previously discussed, and it is clear that the luminosity of the bump must strongly depend on any parameters which can change the location of the H abundance discontinuity left over by the envelope convection after the first dredge up. The bump luminosity decreases as the location of the discontinuity moves deeper into the star, given that the H-burning shell will encounter it at earlier times, i.e. at lower surface luminosity.

This occurs with decreasing stellar mass, as the base of the envelope convection is located deeper in the structure during the first dredge up.

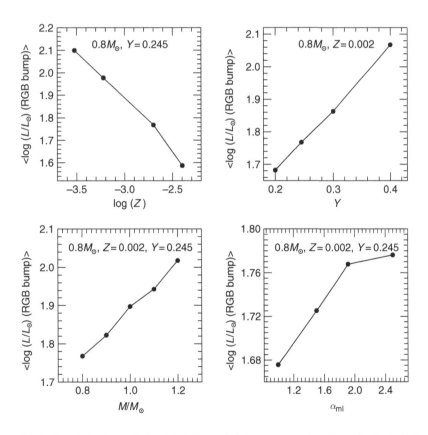

Figure 5.18 The behaviour of the RGB bump brightness as a function of selected physical parameters: stellar mass, mixing length parameter, metallicity and helium content (see also data in Table 5.4)

A decrease in He abundance, or alternatively an increase in the heavy element abundance, due to the corresponding changes – previously discussed – in the opacity of the envelope, moves the location of the chemical discontinuity deeper into the star and, in turn, reduces the bump luminosity.

The efficiency of superadiabatic convection clearly affects the thermal stratification of the whole convective envelope. A larger efficiency, i.e. a larger adopted value for the mixing length, decreases the mass extension of the outer convection zone at the first dredge up. As a result, the chemical discontinuity is located in more external layers and the bump luminosity is increased.

Figure 5.18 shows the dependence of the RGB bump luminosity on the discussed parameters, while in Table 5.4 we report the RGB bump luminosity values for different choices about these parameters.

5.10.3 The luminosity of the tip of the RGB

The tip of the RGB marks the evolutionary phase corresponding to He-burning ignition through the He flash, and this (see the discussion in the previous section) occurs when the He core mass has reached a well-defined value. This means that the luminosity of the tip is in general a function of the He core mass at the He flash. Any parameter affecting the size of the He core at the He-burning ignition affects the luminosity of the RGB tip: if the He core mass at the flash increases, the brightness of the RGB tip also increases. In Table 5.5 are listed the values of the surface luminosity and of the He core mass at the tip of the RGB for different initial chemical compositions and a mass of the RGB star equal to $\sim 1 M_{\odot}$.

The value of M_{cHe} does not depend strongly on the stellar mass for masses lower than $\sim 1.8 M_{\odot}$ at solar chemical composition (this limit depends strongly on the adopted chemical composition, significantly decreasing with decreasing metallicity and/or increasing He content). In fact, all stars with a mass below this limit develop very similar levels of electron degeneracy within the He core, and the mass of the He core has to reach almost the same value before He-burning ignites. When the stellar mass value exceeds this limit, the electron degeneracy in the core is at a lower level as a consequence of the larger temperatures and lower densities. This has the effect that the luminosity of the tip of the RGB is almost constant for masses below the quoted limit, and strongly decreases at larger masses. This behaviour is shown in Figure 5.19. After a minimum has been attained, the value of M_{cHe} starts to increase again with increasing total mass (the electron degeneracy is by now removed from the core) as a consequence of the increasing mass of the convective core during the core H-burning phase.

The significant changes in the properties of the RGB, which are expected in stellar systems as a consequence of the differences of the age, and in turn of the masses evolving along the RGB, is called *RGB phase transition* after [211].

An increase in the initial He content increases the interior temperatures of a star through the increase in the mean molecular weight; this, in turn, decreases the electron degeneracy level in the He core during the RGB evolutionary phase. As a

Table 5.4 The mean luminosity of the bump along the RGB as a function of various parameters

$Z = 0.002, Y = 0.245, \alpha_{\mathrm{ml}} = 1.91$	
$M(M_\odot)$	$\log(L/L_\odot)$
0.8	1.768
0.9	1.822
1.0	1.897
1.1	1.943
1.2	2.017

$M = 0.8M_\odot, Z = 0.002, Y = 0.245$	
α_{ml}	$\log(L/L_\odot)$
1.00	1.676
1.50	1.725
1.91	1.768
2.50	1.776

$M = 0.8M_\odot, Y = 0.245, \alpha_{\mathrm{ml}} = 1.91$	
Z	$\log(L/L_\odot)$
0.0003	2.099
0.0006	1.977
0.002	1.768
0.004	1.587

$M = 0.8M_\odot, Z = 0.002, \alpha_{\mathrm{ml}} = 1.91$	
Y	$\log(L/L_\odot)$
0.200	1.682
0.245	1.768
0.300	1.863
0.400	2.067

result the star ignites He at a lower He core mass and the luminosity of the tip of the RGB decreases.

With increasing metallicity, for a fixed He content and stellar mass, the He core mass at the He flash decreases. This is because at a higher metallicity the H-burning in the shell is more efficient; as a consequence the growth and, in turn, heating of the He core is faster. In this way the thermal conditions for igniting He are reached at a lower He core mass. This is shown in Figure 5.20, where the behaviour of the maximum temperature in the He core as a function of the He core mass is plotted for structures with same mass and initial He content, but different values of the

Table 5.5 Bolometric luminosity, He core mass and surface helium abundance at the tip of the RGB for a low-mass ($\sim 1 M_\odot$) star for various choices about the initial chemical compositions. The other two columns report the bolometric luminosity and mass of the star with an effective temperature equal to $\log(T_{eff}) = 3.85$ belonging to the Zero Age Horizontal Branch locus (see Chapter 6) whose RGB progenitor has the evolutionary properties listed in the previous columns

Z, Y	$\log(L_{tip}/L_\odot)$	$M_{cHe}(M_\odot)$	Y_{surf}	$\log(L_{3.85}/L_\odot)$	$M_{3.85}(M_\odot)$
0.0001, 0.245	3.332	0.511	0.253	1.780	0.821
0.0003, 0.245	3.369	0.505	0.256	1.732	0.721
0.001, 0.246	3.405	0.498	0.260	1.687	0.650
0.002, 0.248	3.422	0.495	0.262	1.653	0.619
0.004, 0.251	3.445	0.491	0.269	1.614	0.594
0.008, 0.256	3.463	0.487	0.277	1.561	0.572
0.01, 0.259	3.465	0.485	0.280	1.540	0.565
0.02, 0.273	3.475	0.478	0.296	1.489	0.543

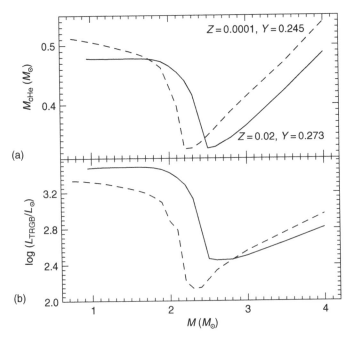

Figure 5.19 (a) The behaviour of the mass of the He core and (b) of the surface luminosity at the onset of He-burning as a function of the stellar mass for two selected initial chemical compositions

metallicity. Note that, with increasing metallicity – regardless of the decrease of the He core mass at the tip of the RGB – the stellar luminosity at He ignition increases, at odds with the previous general rule. This is because the M_{cHe}–L relation is partially affected by the initial chemical composition; an increase of the metallicity increase

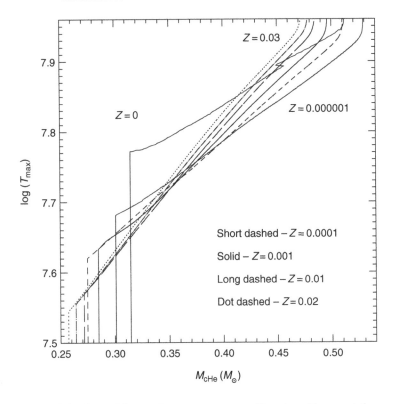

Figure 5.20 The behaviour of the maximum temperature off-centre with respect the mass of the He core, during the RGB phase of a $0.8M_\odot$ star and for various values of the metallicity

the H-burning rate in the shell and, in turn, the surface luminosity at each fixed value of the He core mass.

The luminosity of the tip of the RGB is not affected by changes in the efficiency of superadiabatic convection. This is because the thermal stratification within the He core does not depend on the thermodynamical properties of the outer envelope.

5.11 Evolutionary properties of very metal-poor stars

According to the standard cosmological model, primeval matter emerged from the Big Bang as a mixture of H and He with negligible quantities of elements heavier than ^{7}Li. Since then, stellar nucleosynthesis has been at work, progressively increasing the abundance of metals in the Universe. In this context, the empirical evidence that disk stars – Population I objects – are more metal rich than halo stars – Population II objects – has long been considered a proof of the chemical evolution in our own galaxy. However, when comparing the predictions of Big Bang nucleosynthesis with the spectroscopic estimates of the metallicity of Population II stars, one has to face the evidence that the old halo Population II should have been preceded by an earlier stellar population, responsible for enhancing the amount of heavy elements from the cosmological value – $Z \sim 10^{-12}-10^{-10}$ if not zero – to the typical values of Population II stars ($Z \sim 10^{-3}-10^{-4}$). This primordial stellar population, characterized by a negligible, if not vanishing, metallicity, is called *Population III*.

The major difference between Population III stars and those with normal metal content lies in the efficiency of the H-burning mechanism. Owing to the lack of CNO elements, even high-mass stars are forced to contract until the central density and temperature are high enough so that the p–p chain is able to provide all the energy required by the equilibrium conditions. Because of the small dependence of p–p chain efficiency on temperature compared with the CNO cycle, very high temperatures have to be attained in the core. In low-mass stars, the p–p chain can provide the amount of energy necessary for the star to be in equilibrium, whereas at larger masses this configuration is achieved only at much higher temperatures.

Once the central temperature reaches $T \sim 10^8$ K, He-burning reactions become efficient. This occurrence marks a fundamental event: a small amount of carbon is self-produced by the star. As a consequence, the produced ^{12}C leads to the activation

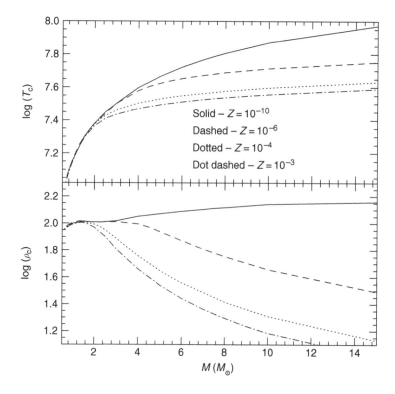

Figure 5.21 The behaviour of temperature and density at the centre of the star on the ZAMS, as a function of mass, for extremely metal-poor stellar populations. For the sake of comparison, the same trends for metallicities ($Z = 10^{-4} - 10^{-3}$) typical of Population II stars are also shown

of the CNO cycle, which then starts providing nuclear burning energy in competition with p–p chain. Detailed evolutionary computations ([40]) have shown that the threshold abundance of C which allows the CNO cycle to begin is of the order of $X_C \sim 10^{-10}$–10^{-9}. The self production of ^{12}C occurs at earlier stages with increasing stellar mass; in the less massive structures the production of C occurs towards the end of the central H-burning phase, and it does not occur at all in stars less massive than $\sim 0.8 M_\odot$. For massive stars $(M > 15$–$20 M_\odot)$ the He-burning reactions occur even before the star has burned a significant fraction of its H. In these objects the central H-burning is then controlled by the CNO cycle, and their evolutionary behaviour is not very different from that of more metal-rich stars. In order to illustrate this issue more clearly, Figure 5.21 shows the trend of the core temperature and density as a function of stellar mass at the ZAMS.

In Figure 5.22 and 5.23 we display the evolutionary path in the HRD for a $1 M_\odot$ and $5 M_\odot$ model with initial metallicity $Z = 10^{-10}$. For comparison, the evolutionary tracks of models with the same mass but different metallicities are also shown.

As soon as the CNO cycle is activated, the stellar core reacts by expanding due to the additional energy input, with a consequent decrease of the central density. The activation of the CNO cycle produces sizeable effects on the development and

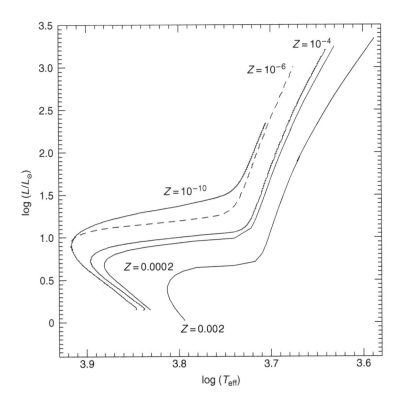

Figure 5.22 The HRD for a $1 M_\odot$ star for different values of the stellar metallicity

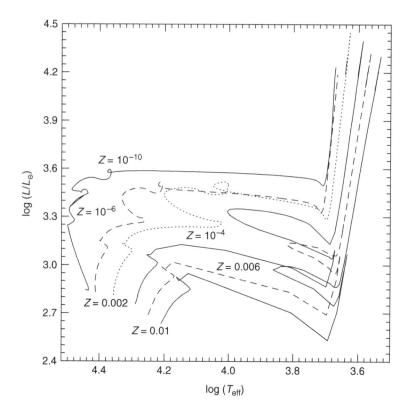

Figure 5.23   The HRD for a $5M_\odot$ star with different values of the stellar metallicity, from the ZAMS until more advanced phases following the end of central He-burning

evolution of the convective core during the central H-burning phase. For stars in the mass range $2–5M_\odot$, after the temporary occurrence of the convective core associated with the ^{3}He equilibrium, H-burning occurs in a radiative region but as soon as the CNO cycle becomes the main H-burning process, a convective core appears. In more massive stars, however, the transition of the core from convective to radiative and back again does not occur (see Figure 5.24).

The most significant difference during the post-MS evolutionary phase with respect to normal metallicity stars is that, due to the larger central temperatures of low-mass stars, the level of electron degeneracy is lower, making easier He-ignition. For this reason, Population III low-mass stars experience the He flash at a lower He core mass and, in turn, at a lower luminosity of the tip of the RGB compared with Population II stars. Therefore, at very low metallicity ($Z < 10^{-6}$) the trend of He core mass at the tip of the RGB with the metallicity is reversed, as it is shown in Figure 5.20. In addition, the maximum stellar mass experiencing He ignition through the He flash is much lower; equivalently the RGB phase transition occurs at a lower mass, i.e. at a greater age, compared with more metal-rich stellar populations.

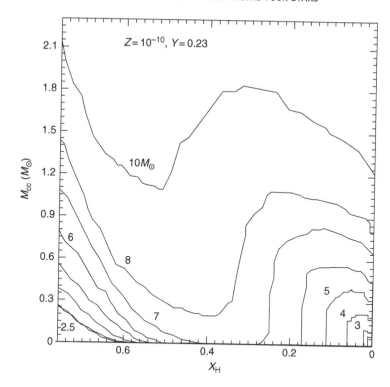

Figure 5.24 The behaviour of the mass of the convective core as a function of the central abundance of H for various stellar masses and for a very low metallicity ($Z = 10^{-10}$)

The increase of core temperatures in Population III stars is so great that intermediate-mass stars do not experience the RGB evolutionary phase; they are able to ignite He-burning quietly soon after the exhaustion of H in the core, on the blue side of the HRD (see Figure 5.23).

6 The Helium Burning Phase

6.1 Introduction

In the previous chapter the structural and evolutionary properties of stars during the main H-burning phase were discussed, as well as their dependence on the most important parameters such as total mass and initial chemical composition. Here we will describe the evolution of stars during the next evolutionary phase, the He-burning phase. It has already been mentioned that all stars with initial total mass larger than $\sim 0.5 M_{\odot}$ are able to attain, in their interiors, the thermal conditions required for the onset of He-burning.

Unlike during the H-burning, the physical processes at work in stars of different masses during the main core He-burning stage are quite similar. Nevertheless, since the morphology of the evolutionary tracks during this phase is strongly dependent on the total stellar mass, we again prefer to discuss separately the low-, intermediate-, and high-mass stars.

6.2 The nuclear reactions

The first and fundamental reaction in the He-burning process is the production of ^{12}C from the fusion of three ^{4}He nuclei. This reaction is called the *triple alpha (3α) reaction*. A triple encounter being of very low probability, this reaction usually occurs in two separate steps:

$$^4\text{He} + {}^4\text{He} \rightarrow {}^8\text{Be}$$

$$^8\text{Be} + {}^4\text{He} \rightarrow {}^{12}\text{C} + \gamma$$

The first reaction is endothermic by about 91.8 keV, meaning that in a short time $(\sim 10^{-16}\,\text{s})$ ^{8}Be decays back into two α particles. The possibility of the second reaction

---

Evolution of Stars and Stellar Populations   Maurizio Salaris and Santi Cassisi
© 2005 John Wiley & Sons, Ltd

is therefore extremely low. In this context, ^{8}Be behaves as a secondary element, being involved simultaneously in a destruction and a generation process. However, when the interior temperature rises, the probability of the second reaction increases, due to these combined effects:

- the number of $\alpha + \alpha$ reactions increases and, the ^{8}Be decay lifetime remaining constant, the concentration of ^{8}Be significantly increases,

- the nuclear cross section of the reaction ^{8}Be $+ \alpha$ strongly increases.

Both effects increase the probability that a nucleus of carbon is produced before ^{8}Be decay; a temperature of the order of $\sim 1.2 \times 10^8$ K is necessary before 3α reactions produce a sizeable amount of energy. In addition, any increase in the density strongly favours this nuclear reaction, $\epsilon_{3\alpha}$ being proportional to the square of the density.

The evaluation of the nuclear cross sections for both reactions is complicated by the presence of several nuclear resonances. The amount of energy which is released for any ^{12}C nucleus produced is equal to ≈ 7.27 MeV, which corresponds to ~ 0.6 MeV per nucleon. This is more than one order of magnitude smaller than the amount of energy per nucleon released during the H-burning via the CNO cycle. This explains why, for a fixed stellar mass, the core He-burning lifetime is about a factor of 100 shorter than the core H-burning lifetime.

The temperature sensitivity of the 3α reaction, is quite strong: $\epsilon_{3\alpha} \propto T^{40}$ for $T \sim 10^8$ K and $\epsilon_{3\alpha} \propto T^{20}$ for $T \sim 2 \times 10^8$ K. For the same physical reasons discussed for the CNO cycle, one can foresee that during the core He-burning stage, the stars have extended convective cores.

The other nuclear reactions involved in the He-burning process are:

$$^{12}C + \alpha \rightarrow {}^{16}O + \gamma$$

$$^{16}O + \alpha \rightarrow {}^{20}Ne + \gamma$$

$$^{20}Ne + \alpha \rightarrow {}^{24}Mg + \gamma$$

$$^{24}Mg + \alpha \rightarrow {}^{28}Si + \gamma$$

Only the first two reactions, together with the 3α reaction, are really important. A relevant consequence of the He-burning process is to transform He into a mixture of ^{12}C and ^{16}O with traces of ^{20}Ne.

It is worth emphasizing that the ^{12}C$(\alpha, \gamma)^{16}$O reaction is one of the most important in stellar evolutionary computations for the following reasons.

- The value of the corresponding nuclear cross section strongly affects the C/O ratio in the core of carbon–oxygen (CO) white dwarfs and, in turn, their cooling times.

- More importantly, when the abundance of He inside the convective core, during the core He-burning stage, is significantly reduced, the ^{12}C$(\alpha, \gamma)^{16}$O reaction becomes strongly competitive with the 3α reactions (which need three α particles) in contributing to the nuclear energy budget (the two processes produce a similar amount of energy per reaction). This means that the cross section of this reaction has a strong influence on the core He-burning phase lifetime.

Unfortunately, this reaction has a resonance and a very low cross section ($\sim 10^{-17}$ barn) at low energies, and so the nuclear parameters are difficult to measure experimentally or to calculate by theoretical analysis. According to [51], an uncertainty of the order of a factor of two is reasonable for this nuclear reaction rate (but see also the more recent analysis by [121]).

6.3 The zero age horizontal branch (ZAHB)

About one million years after the He-ignition in the core of low-mass stars, electron degeneracy is fully removed. During this period, a series of recurrent flashes occurs in the core, each one removing the degeneracy closer and closer to the centre. In the meantime, the surface luminosity drastically decreases – by about one order of magnitude – with respect to the luminosity at the RGB tip, because of the huge expansion of the He core occurring after the main He flash, which cools down the H-burning shell.

The probability of observing stars during this evolutionary stage is very small, if not negligible, as the corresponding evolutionary lifetime is very short. This is the main reason[1] why many authors initiate their core He-burning evolutionary sequences from an equilibrium model, neglecting the computation of the evolution from the RGB to the beginning of the core He-burning stage (see the discussion in [225]). In any case, in the computation of this initial model, the effect of the He-burning occurring during the He flash, is accounted for by considering that a mass fraction of ~ 5 per cent of carbon is produced.

Equilibrium models that burn He in a chemically homogeneous core, and H in a shell with chemical stratification similar to the one at onset of the He flash (in some sense this is equivalent to saying that the CNO elements are at their equilibrium in the H-burning shell) are called *Zero Age Horizontal Branch* (ZAHB) models.

The structural and evolutionary properties of a ZAHB star are fixed by four parameters: the He core mass M_{cHe}, the total mass M, the abundance of He and the metallicity in the envelope. It is clear that the value of M_{cHe} depends on the initial chemical composition as well as on the total mass, although for low-mass stars ($M \leq 1.4 M_{\odot}$), i.e. those with an age (at the RGB tip) larger than 4–5 Gyr, M_{cHe} is only weakly dependent on the stellar mass. This means that for low-mass RGB progenitor, the parameters which characterize the corresponding ZAHB structure are reduced to the He and the metal abundances in the envelope and the total mass. The

---

[1] One also has to notice that the computation of an evolutionary sequence through the He flash is an extremely time-consuming procedure.

He abundance in the envelope of ZAHB stars is slightly enriched compared to the initial value ($\Delta Y \sim 0.02$–0.04) because of the first dredge up.

From a structural point of view, a ZAHB star is characterized by two nuclear burning sources: the He-burning located in the interior of the He core that causes the presence of a convective core, and the H-burning in a shell surrounding the He core. For each fixed mass of the He core, the efficiency of the H-burning shell is modulated by the mass of the envelope – the larger the mass of the envelope, the hotter the H-burning shell, and thus the more efficient the H-burning process is.

We now consider the location on the HRD of different ZAHB models characterized by the same He core mass and envelope composition but of different total masses, i.e. different envelope masses. Detailed numerical computations show that the models define an almost horizontal locus on the HRD: those with the lowest mass envelopes are located in the hottest part of the branch, the location moving to cooler values with increasing envelope mass (see Figure 6.1). This is the reason why this phase is called the *Horizontal Branch* (HB) phase ([48]). This theoretical prediction is confirmed by observations of old stellar clusters.

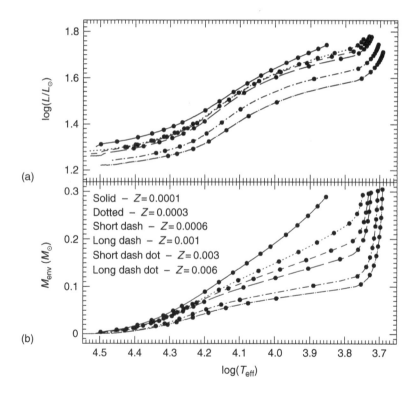

Figure 6.1 (a) The HRD of the ZAHB for different values of the metallicity but the same RGB star progenitor, with $M = 0.8M_\odot$. (b) The corresponding trend of the envelope mass as a function of their effective temperature. Solid dots in both (a) and (b) mark the location of selected ZAHB models

A close inspection of the theoretical predictions clearly shows that, due to the dependence of the H-burning efficiency on the envelope mass, the ZAHB locus is not perfectly horizontal, but becomes slightly brighter with increasing total mass. One can see that the effective temperature extension of the ZAHB can be quite large, going from $\sim$35 000 K for the stars with a negligible envelope mass ($M_{env} \sim 10^{-4} M_\odot$) to $\sim$4000 K for more massive envelopes ($M_{env} \sim 0.4 M_\odot$).

The reason for the mass spread among ZAHB structures in a real stellar system, can easily be accomplished when accounting for the evidence that RGB stars lose a significant ($\sim 0.3 M_\odot$) portion of their total mass, the amount varying from star to star as a consequence of the fact that mass loss is an intrinsically stochastic process.

It appears that, for a fixed envelope chemical composition, the luminosity of a ZAHB star is predominantly fixed by the mass of its He core, and secondly by the mass of the envelope. Its effective temperature, for a fixed He core mass, depends only on the envelope mass. The He core mass is almost constant for stars of ages (at the RGB tip) larger than 4–5 Gyr, i.e. initial mass lower than $\sim 1.4 M_\odot$. Thus, the fact that the ZAHB luminosity is predominantly fixed by the He core mass is the main reason why the ZAHB and, more generally, the HB brightness, is one of the most important standard candles for Population II stellar systems (see Section 9.2.6).

6.3.1 The dependence of the ZAHB on various physical parameters

In view of the possible use of the ZAHB brightness as a distance indicator, it is important to know how the observational properties of the ZAHB react to any change in the most relevant parameters such as helium content, metal abundance and any additional (non-canonical) process which can change the mass size of the He core at He-ignition.

An analysis of how the properties of a ZAHB model change when changing either the initial He abundance or metallicity or the value of M_{cHe}, keeping everything else fixed, provides

$$\left(\frac{d\log(L_{ZAHB}^{3.85})}{dM_{cHe}} \right)_{Y,Z} \sim 3.04$$

$$\left(\frac{d\log(L_{ZAHB}^{3.85})}{dY} \right)_{M_{cHe},Z} \sim 2.07$$

$$\left(\frac{d\log(L_{ZAHB}^{3.85})}{d\log(Z)} \right)_{M_{cHe},Y} \sim -0.04$$

where each derivative is computed at $\log(T_{eff}) = 3.85$, typical of RR Lyrae variable stars (see Section 6.6.1).

One has to bear in mind that these results serve only as approximate guidelines, because a change of the abundances of He and metals also cause a change of the He core mass at the RGB tip (see Section 5.10). In the following discussion about the chemical composition effects on the ZAHB location we also include the induced change to M_{cHe}.

- *The He content*: for a fixed metallicity, an increase in the initial helium content, Y, causes a decrease in M_{cHe}. This should cause a decrease in the ZAHB brightness. However, for ZAHB stars with massive enough envelopes and so with efficient H-burning in the shell, the increase of the envelope He abundance causes a large increase in the H-burning efficiency, counterbalancing the decrease of the ZAHB brightness due to the change in M_{cHe}. As a final result, with increasing He abundance, the blue part of the ZAHB (populated by stars with low-mass envelopes) becomes fainter, and the red part brighter as shown in Figure 6.2. For fixed total mass, the stars become slightly hotter.

- *The metal abundance*: any increase in the metal abundance, at fixed He abundance, makes the ZAHB fainter. This occurrence is due to the combination of two effects: (1) the decrease of the value of M_{cHe} at the He flash, (2) the increase of the envelope opacity. For fixed total mass, the increase of the envelope metallicity makes the ZAHB location cooler.

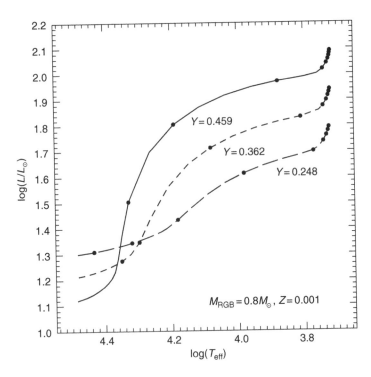

Figure 6.2 HRD of three ZAHBs computed with different choices of the initial He abundance but the same metallicity and progenitor mass. Solid dots mark the location of selected models

- *Non-canonical processes*: there are many physical processes such as rotation and non-canonical mixing processes, commonly not accounted for in standard evolutionary computations, which can produced sizeable effects on the ZAHB properties. Regardless of the detailed physical description of the process, we can easily predict, at least qualitatively, the effect on the ZAHB by simply considering the effect on the parameters characterizing a ZAHB star, i.e. the He core mass and the total mass. Any physical process able to delay the ignition of the He flash at the RGB tip, such as rotation[2], will cause an increase of the He core mass and, on average, a decrease in the total mass. This is because mass loss along the RGB then has more time to work. As a consequence, the ZAHB location will be brighter and hotter.

Table 5.5 lists the values of the ZAHB luminosity and the ZAHB total mass at a fixed effective temperature ($\log(T_{\text{eff}}) = 3.85$) for different initial chemical compositions.

6.4 The core He-burning phase in low-mass stars

The rule governing the morphology of the evolutionary tracks of HB stars during their main He-burning phase is, in some sense, the inverse of that described for the ZAHB. The main characteristic of the HB evolution is that the H-burning shell efficiency monotonically decreases as the efficiency of the central He-burning steadily increases. As long as the luminosity produced by the H-burning shell is larger than that produced by 3α reactions, the star evolves towards larger effective temperatures. When the central He-burning process becomes dominant in the stellar energy budget, the evolutionary path reverses towards the red side of the HRD. As a consequence, the stars perform a loop on the HRD, and the effective temperature extension of this loop strongly depends on the parameters affecting the H-burning shell efficiency, mainly the envelope mass and the envelope He abundance (see Figure 6.3).

It has already been emphasized that, because of the large dependence on temperature of the He-burning efficiency, HB stars always burn helium inside a convective core. This produces further evolutionary properties worth discussing in some detail.

6.4.1 Mixing processes

The physical processes occurring within the convective core of He-burning low-mass stars are very important, as their efficiency determines not only the time spent by the star during the main core He-burning phase, but also its path in the HRD

---

[2] Rotation makes the stellar interiors cooler compared with non-rotating models with the same initial mass and chemical composition. This causes a stronger electron degeneracy in the He core during the RGB evolution.

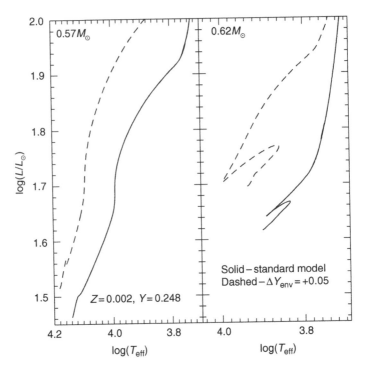

Figure 6.3 The evolutionary tracks of HB stars of various masses for different values of the He abundance in the envelope

and its lifetime during the subsequent evolutionary phase, i.e. the shell He-burning stage.

Inside the convective core, the transformation of helium into carbon strongly increases the free–free opacity, being $\kappa_{ff} \propto X_i Z_i^2$ (where X_i and Z_i represent here the abundance by mass and the atomic charge respectively, of the element i). Because of this opacity increase due to transformation of helium into carbon, the effect of chemical evolution on the radiative temperature gradient ∇_{rad} overcomes those due to the changes of other physical quantities within the convective core. As a result, ∇_{rad} increases monotonically with time in the whole convective region, an occurrence that prevents any decrease of the mass size of this convective zone. One has also to notice that convection occurs on a timescale much shorter than the nuclear burning timescale – which means that nuclear burning has no time to modify the chemical abundances in the convective core before mixing occurs.

When one neglects convective overshoot at the border of the convective core the evolution with time of ∇_{rad} within the star is shown in Figure 6.4. One immediately notices the growing discontinuity of the radiative gradient at the boundary of the convective core, due to the changing chemical composition and, in turn, the radiative opacity, caused by the He-burning process.

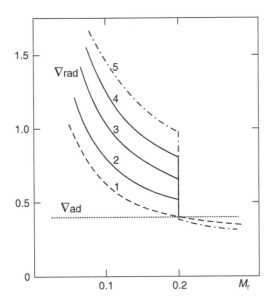

Figure 6.4 The growing discontinuity of the radiative temperature gradient at the boundary of the convective core for a core He-burning low-mass star. The numbers mark a time sequence (time is increasing going from stage 1 to stage 5)

In a realistic case, one can assume that a certain amount of overshoot (albeit possibly even very small) at the Schwarzschild border of the convective core has to occur. Due to the presence of the chemical discontinuity previously discussed and the increasing opacity inside the convective core (because of the conversion of He into C) the mixing of a radiative shell – surrounding the convective core – caused by the convective overshoot, causes a local increase of the opacity and, in turn, a convective boundary instability. As a result, we can expect that a *self-driving mechanism* for the extension of the convective core occurs in the star: any radiative shell which is mixed as a consequence of the convective core overshoot, will definitely become part of the convective core. This physical behaviour is fully supported by detailed computations of HB stellar models. Due to this process, at the boundary of the resulting enlarged convective core, the radiative gradient becomes equal to the adiabatic one (see Figure 6.5(b)).

After this early phase, during which the extension of the convective core closely follows the increase of ∇_{rad}, the radiative gradient profile shows a minimum (see profile 3 in Figure 6.6). This is a consequence of the progressive shift outwards – within an He-rich region – of the convective boundary, due to the self-driving mechanism. The occurrence of this minimum depends on the complex behaviour of the physical quantities involved in the definition of ∇_{rad}, such as opacity, pressure, temperature and local luminosity. All numerical simulations show that when this minimum appears, there is a close coupling between the physical and chemical

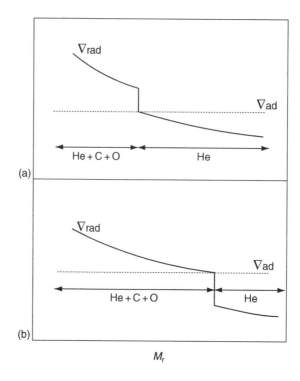

Figure 6.5 Qualitative behaviour of the radiative temperature gradient near the boundary of the convective core during the core He-burning phase: (a) with an increasing chemical discontinuity (see also Figure 6.4); (b) in the case where convective overshoot is allowed to occur

evolution of the convective zone outside the minimum and the mass of the surrounding radiative layers which are engulfed.

The mixing of a radiative shell produces a general decrease of the radiative gradient in the whole convective core (see profile 4 in Figure 6.6). This effect is mainly due to the combination of the mixing of He-rich matter and of the change in the physical properties of the mixed shell. The radiative gradient will eventually decrease to the value of the adiabatic gradient at the location of the minimum (profile 5 in Figure 6.6).

The main problem in the computation of HB evolutionary models is related to the treatment of the intermediate convective zone, located between the minimum of the radiative gradient and the outer radiative zone. A full mixing between the convective core located inside the minimum and the external convective shell cannot occur because at the minumum the radiative gradient is equal to the adiabatic one. Therefore, the convective shell outside the minimum is no longer mixed with the core.

The result of this decoupling between the convective core and the convective shell outside the minimum, is the formation of an extended, partially mixed region – a *semi-convective* region – around the fully mixed core, between the minimum of ∇_{rad} and the outer radiative zone. The chemical composition at each point within

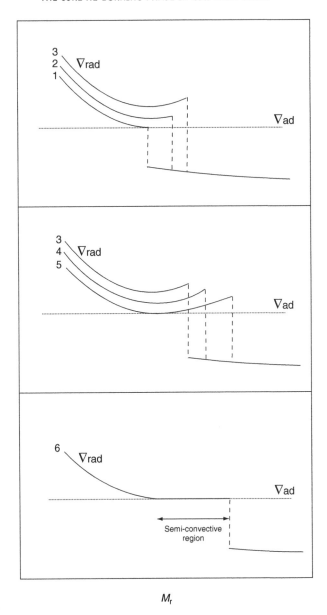

M_r

Figure 6.6 Qualitative behaviour of the radiative temperature gradient near the boundary of the convective core during the core He-burning phase, showing the time sequence of the events which result in the appearance of a semi-convective region. The numbers mark a time sequence (time is increasing going from stage 1 to stage 6)

this semi-convective region is that required to fulfill the equality $\nabla_{rad} = \nabla_{ad}$ (see profile 6 in Figure 6.6). Therefore, HB models that account for the presence of a semi-convective region are computed following these prescriptions: $\nabla_{rad} \simeq \nabla_{ad}$ at the border of the convective core and $\nabla_{rad} = \nabla_{ad}$ in the whole semi-convective region ([49],[50]).

This situation is similar to that for massive stars during the main sequence evolutionary phase. However, in massive stars near the end of the core H-burning stage, an outer zone unstable against convection exists around the convective core, and this zone tends to be stabilized by mixing. In HB structures, the outer region is by itself stable against convection and it is the convective core that *induces* a partial mixing in the surrounding regions. For this reason, the partial mixing phenomenon that occurs in HB stars is often called *induced semi-convection* ([49]).

The mass location of the minimum of the radiative gradient changes (increases) with time, because of the evolution of the chemical abundances caused by nuclear burning. As a final result, the region being enriched by the carbon and oxygen produced by He-burning, increases outwards up to a maximum extension.

The effects of semi-convection on the evolution of HB models are: (1) the evolutionary tracks perform more extended loops on the HRD, (2) the central He-burning phase lasts longer, since the star has a larger amount of fuel to burn, (3) the mass size of the He-depleted core at the He-exhaustion is larger.

When the core abundance of He has been lowered by nuclear burning to about $Y \sim 0.10$, a convective instability can affect the convective core. In fact, when the core He abundance is lower than this limit, α-captures by ^{12}C nuclei tend to overcome ^{12}C production by 3α reactions, thus He-burning becomes mainly a $^{12}C + \alpha$ production of oxygen, whose opacity is even larger than that of ^{12}C. This causes – as described above – an increase of the size of the semi-convective region and, in turn, fresh helium is transferred into the core, which is now nearly He-depleted. As shown by [210], even a small amount of He driven into the core becomes very important in comparison with the vanishing amount of He otherwise present in the core. This enrichment of the He abundance in the core succeeds in enhancing the rate of energy production by He-burning, and thus the luminosity increases, driving an increase in the radiative gradient. As a consequence, a phase of enlarged convection zone is started – the so-called *breathing pulse*. After a pulse, the star readjusts itself to burn steadily in the core the fresh He driven there by the convection. Detailed numerical simulations show that three major breathing pulses are expected before the complete exhaustion of He in the core. The evolutionary effects of the breathing pulses are: (1) the star performs a loop on the HRD at each pulse (see Figure 6.7), (2) the He-burning lifetime is slightly increased, (3) the mass of the CO-core at the He exhaustion is increased.

The ratio of evolutionary times between the subsequent Asymptotic Giant Branch phase and the HB is very sensitive to the occurrence of breathing pulses. From the observational point of view, the ratio of the star counts along these two phases – the so-called $R2$ parameter ([36]) – is a measure of the corresponding lifetime ratio (see Chapter 9). Models without breathing pulses predict $R2 \sim 0.12$–0.15 whereas inclusion of breathing pulses gives $R2 \sim 0.08$. The value $R2 = 0.14 \pm 0.05$ observed

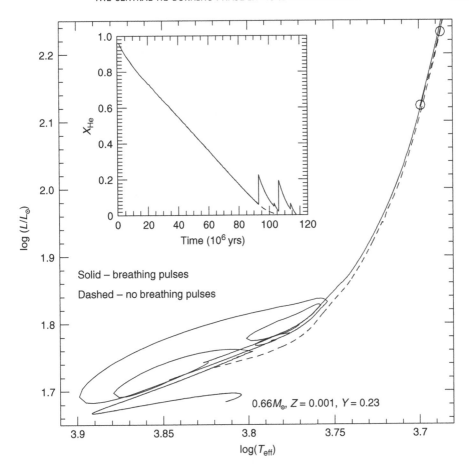

Figure 6.7 The evolutionary tracks of HB stars computed alternatively neglecting or accounting for the occurrence of breathing pulses at the end of the core He-burning phase. The inset shows the behaviour of the central abundance of He as a function of time during this evolutionary phase

in a sample of Galactic globular clusters suggests that the efficiency of the breathing pulse phenomenon is very low, if not zero ([36], [44]). The appearance of breathing pulses in theoretical stellar models may be related to the approximation of instantaneous mixing when convection sets in, that could possibly break down in this late HB evolutionary stage.

6.5 The central He-burning phase in more massive stars

Section 5.9.1 discussed the evolutionary and structural properties of stars more massive than $\sim 2.3 M_\odot$, from the central exhaustion of hydrogen until the ignition of 3α reactions in the He core, at the tip of the RGB.

Owing to the absence of electron degeneracy within the He core, the onset of He-burning is not characterized by the He flash, as in low-mass stars. Helium is ignited when the central temperature reaches $\sim 10^8$ K and central density $\approx 10^4$ g cm^{-3}; this occurrence terminates, and reverses the upward climb along the RGB.

The star now begins an extended phase of He-burning in a steadily growing convective core (we note that the previous discussion on mixing processes in the convective core of He-burning low-mass stars also applies to intermediate-mass stars, although now the semi-convective region is a very small fraction of the whole convective zone). H-burning in a thin shell continues to provide the bulk of the surface luminosity, and so the mass size of the He core continues to grow during this evolutionary phase. The core He-burning lifetime is of the order of ~ 20 per cent of the core H-burning lifetime, and it is ~ 22 Myr for a $5M_\odot$ star and ~ 4 Myr for a $10M_\odot$ star of solar chemical composition. The duration of the core He-burning phase is fairly large when one considers that the star is approximately two orders of magnitude brighter than during the MS phase, and that the specific gain of energy (per unit of mass of burnt fuel) is one tenth of that for H-burning. This is due to the very large contribution to the energy budget provided by the H-burning shell.

Figure 6.8 shows the evolutionary track of a $5M_\odot$ model on the HRD. Looking at the part of the track corresponding to the core He-burning phase (points E to I) one can easily note that after point F, the star moves from the RGB to the blue side of the HRD. The *bluest point*, G, in a $5M_\odot$ model, is reached when the central abundance of He is $Y \sim 0.50$. During this period before point G, the energy release in the H-burning shell has been steadily increasing. The maximum efficiency is achieved at point G, where the fraction of energy produced by the 3α reactions is of the order of ~ 20 per cent. After this point, the fraction of energy produced via H-burning decreases steadily. The model then comes back toward the Hayashi line, so performing a loop in the HRD, the so called *blue loop* (sometimes the evolution is still more complicated by the occurrence of secondary loops). At point H, the core abundance of He is $Y \sim 0.02$, while $L_{3\alpha}$ accounts for ~ 33 per cent of the energy produced by nuclear burning. At point K, the fraction of energy produced by the core He-burning process is ~ 60 per cent.

The importance of the blue loop stems from the fact that it occurs during a long-lasting evolutionary phase, in which the star has a large probability of being observed (as it really occurs in galactic young stellar clusters and young and intermediate-age clusters in the Magellanic Clouds). Further details on the blue loop will be given in the next section.

We close this section with a brief discussion on the core He-burning phase in massive stars, $M > 8$–$10M_\odot$. The evolutionary properties of these stars before the ignition of the He-burning are quite similar to those of intermediate-mass stars (see the discussion in Section 5.9.1). However, in massive stars, He is ignited in the core before the star reaches the RGB configuration, and the star continues to evolve monotonically to the red on the HRD, while He burns in a growing convective core. H-burning in a shell continues to supply most of the surface luminosity. Almost immediately following the exhaustion of He in the core, it attains temperatures and

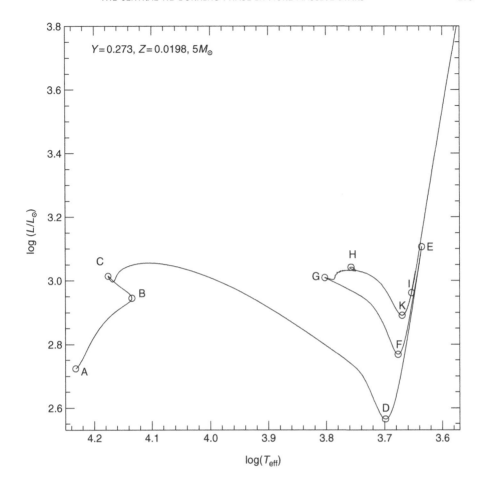

Figure 6.8 The evolutionary track of a $5M_\odot$ model during the H- and He-burning phases. The different evolutionary stages discussed in the text are marked A–K

densities high enough to ignite carbon in non-degenerate conditions (more will be discussed in Chapter 7).

6.5.1 The dependence of the blue loop on various physical parameters

For a long time, the physical reasons for the blue loops challenged our understanding of stellar evolution, and it is still not possible to predict easily the response of an intermediate-mass star, during this phase, to changes in the physical parameters and/or the physical assumptions adopted in the evolutionary computations. This is in contrast to low-mass stars, for which the induced changes to their HRD locations can be easily predicted. The reason for this difference is that in intermediate-mass stars, the contribution to the stellar energy budget provided by the H-burning shell is significantly larger than in low-mass stars; in addition, the relative energy contributions of the He- and H-burning change significantly during the core He-burning phase (see Figure 6.9). As a consequence, any variation of the physical inputs which can modify the H-burning efficiency can either trigger, or inhibit, the loop.

A detailed analysis of the role of several factors known to affect the blue loop was performed by [160]. Only the problem of the dependence of the blue loop on the most important physical parameters, such as the stellar mass and chemical composition, and on the mixing processes accounted for in stellar model computations are addressed below. It is emphasized that both the morphology, and the actual occurrence itself, of blue loops has a highly non-linear dependence on the physical inputs and assumptions made in the evolutionary computations. What is discussed in the following should be considered only as a rule of thumb for understanding the general behaviour of the evolutionary tracks during the core He-burning stage.

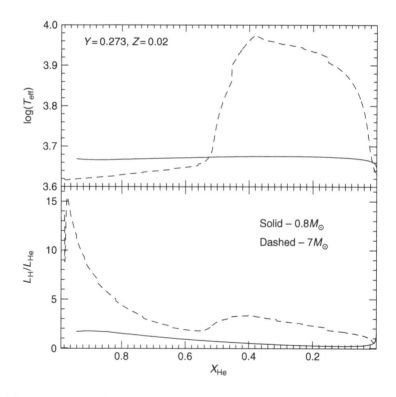

Figure 6.9 The evolution of the effective temperature and of the ratio between the energy released by the H-burning shell and that produced by central He-burning process, as a function of the central helium mass fraction, for a $0.8M_\odot$ and a $7M_\odot$ model during the core He-burning phase

- *Stellar mass*: for stars less massive than about $10-12M_\odot$, the extension of the blue loop generally increases with mass (see Figure 6.10). In more massive stars, the blue loop disappears as the star is able to ignite He before reaching its Hayashi track.

- *Chemical composition*: the general rule is that if the initial He abundance is increased, the extension of the blue loop during the core He-burning phase is larger. The opposite is true when increasing the heavy element abundance. However, as already stated, this behaviour is sometimes contradicted by evolutionary computations which, for example, clearly show that the behaviour of the blue loop with both helium content and metallicity is non-linear (see Figures 6.11 and 6.12).

- *Mixing processes*: by increasing the efficiency of the convective core overshoot during the central H-burning phase, the extension of the blue loops is strongly reduced (see Figure 5.7). The role played by convective envelope overshoot has been investigated by several authors (see, e.g. [2], [206] and [160]) although its effect is still somewhat uncertain. Evolutionary computations suggest that a blue

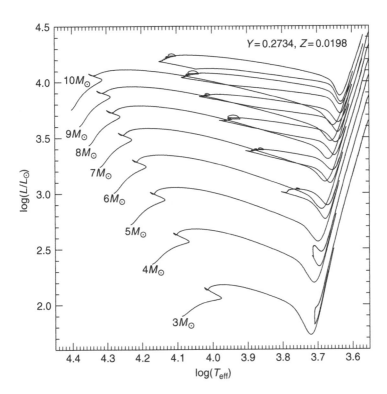

Figure 6.10 The HRD of evolutionary tracks for different intermediate-mass stars with solar chemical composition

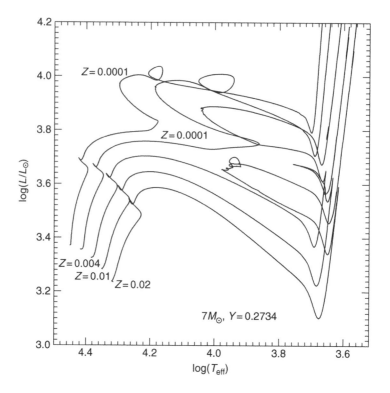

Figure 6.11 The HRD of evolutionary tracks of a $7M_\odot$ star for various values of the metallicity but with the same value for the initial He content

loop is apparently favoured by a sharper H profile in the chemical stratification of the envelope, as a consequence of the change in the H-burning efficiency when the shell encounters the discontinuity. This means that envelope overshoot, which can enhance the sharp discontinuity during the first dredge up phase (a deeper convective envelope reaches regions with more He produced during the MS phase) may be able to trigger a loop that otherwise would not occur. However, the picture is more complicated as it also depends on when the H-burning shell encounters this discontinuity – if this occurs before core He-ignition, envelope overshoot has no effect at all on the development of the blue loop.

Detailed numerical computations have shown that the effect of semi-convection on the morphology of the blue loops is negligible. It would be reasonable to suppose that the presence of a semi-convective region at the edge of the convective core could have a significant effect on the occurrence of the blue loops by changing the rate at which He is mixed into the core. One would suppose that the larger the semi-convective region, the lower the chance for the occurrence of blue loops. The fact that semi-convection has a negligible effect is because, in intermediate-mass stars,

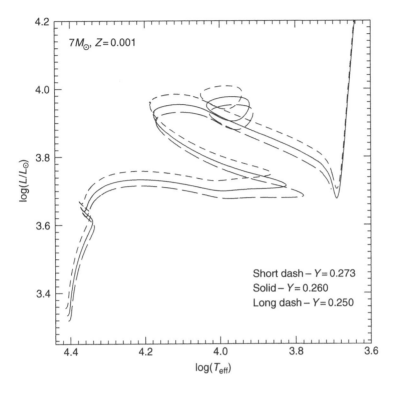

Figure 6.12 The HRD of evolutionary tracks of a $7M_{\odot}$ star for various values of the initial He abundance but with the same metallicity ($Z = 0.001$)

the semi-convective zone is only a small fraction of the whole convective core (≈ 10 per cent for a $4M_{\odot}$ model and ≈ 1 per cent for a $7M_{\odot}$ model).

6.6 Pulsational properties of core He-burning stars

When stars cross a number of well-defined regions in the HRD, they are affected by stable radial pulsations. As a consequence, they change their brightness periodically and become *variable stars*. These peculiar regions of the HRD are called *instability strips*. For a detailed discussion of the different instability strips as well as the physical reasons for which a star pulsates, the reader is referred to the book by [61] and the review papers by [79] and [80]. In this section the *Cepheids instability strip*, populated by stars in the core He-burning stage, is discussed. This instability strip is related to the two most important classes of radially pulsating stars, the *RR Lyrae* stars and the *Classical Cepheid*. These represent probably the most important standard candles for Population II and Population I stellar systems, respectively.

Before discussing the general properties of these two families of radial variables, it is useful to provide some information about stellar pulsations. In his 1879 pivotal work, Ritter demonstrated that the pulsation period P of a homogeneous sphere experiencing adiabatic radial pulsations is related to its surface gravity and radius through the relation

$$P \propto \sqrt{R/g}$$

Since $g \propto M/R^2$, we rewrite this relation as

$$P \propto \sqrt{R^3/M}$$

by using the relation between the mean stellar density, mass and radius, we obtain

$$P\sqrt{\rho} = Q$$

where Q is the pulsational constant, which depends slightly on the mass of the variable. More detailed investigations have shown that this relation is, in fact, roughly valid for real stars. The existence of this relation, connecting an empirical quantity such as the period of pulsation, with a structural property such as the mean stellar density, highlights the importance of stellar pulsations to stellar evolution theory; the analysis of radial pulsations provides a formidable tool for investigating stars, allowing a quantitative test of the reliability of theoretical evolution predictions.

The basic theory of pulsating stars was developed by Eddington ([68],[69]) who showed the physical reasons why some stars pulsate while others do not. In principle, any star experiencing a transient perturbation in its interior or in the external regions may experience radial pulsations. However, as demonstrated by Eddington, these induced pulsations would be damped out very quickly, on a timescale of few thousand years. For a given stellar mass, the radius of a star is roughly fixed by the energy flow through the star. In order to have stable radial pulsations, this energy flux and, in turn, the radius has to vary in a periodic way.

It was recognized early on that the necessary modulation of the outward energy flux would be achieved if the opacity at some suitable level in the stellar envelope were to increase during the phase of compression (during which the envelope becomes hotter) and decrease during the expansion phase, so releasing the energy absorbed during compression ('κ mechanism'). Detailed analysis showed that the He- and H-ionization zones located near the surface (the total mass above the base of the He-ionization zone is only $\approx 10^{-7} M_\odot$ in a RR Lyrae variable) are responsible for driving the pulsations, during which the stellar luminosity for an RR Lyrae star can vary by a factor of two and the radius can change by ≈ 20 per cent.

These ionization zones have to contain a sufficient fraction of the mass of the star in order to drive the pulsations efficiently, therefore they cannot be too close to the low-density surface layers; this explain the existence of the hot boundary of the instability strip, given that increasing the stellar T_{eff} moves the ionization regions towards the surface. On the other hand, when the stellar T_{eff} becomes sufficiently low, envelope convection sets in and the κ mechanism is no longer able to drive the pulsation efficiently; this is the reason for the existence of the cool boundary of the instability strip.

We close this brief introduction pointing out that these radial pulsations involve only the external layers of the stars crossing the instability strip, without affecting their interiors and the energy generation efficiency.

6.6.1 The RR Lyrae variables

These variables have periods in the range 0.2 to 0.9 days and light curve (e.g. the trend of the luminosity with time) amplitudes between 0.2 and 1.6 mag in the *B* photometric band (see Chapter 8). They are classified into two groups: the *ab* type variables, characterized by asymmetric light curves of large amplitude, and *c* type variables with nearly sinusoidal light curves and small amplitude. The *ab* type RR Lyrae pulsate in the fundamental mode, and *c* type variables in the first overtone mode. There exists also a third group, the *d* type RR Lyrae, pulsating simultaneously in both modes. RR Lyrae are low-mass stars, which are crossing the instability strip during their core He-burning phase.

The topology of the RR Lyrae instability strip is shown in Figure 6.13. One can note that the hottest, *blue edge*, of the instability strip is located at an effective temperature of the order of $\sim 7200\,K$ at the ZAHB luminosity level, a value which slightly decreases with increasing stellar luminosity. The reason for the existence of a blue edge (and of its change in effective temperature when increasing the luminosity) has to do with the fact that, for a given mass and luminosity, as the surface temperature increases, there is progressively less mass above the ionization zones and, in turn, their contribution to the pulsational driving decreases. Moving from the blue edge to the red side of the HRD, there is a small region within which only the first-overtone mode is stable (the so-called *FO zone*) and on the right of this on the HRD, both modes can be stable (the *OR region*). Here a star can pulsate as fundamental, or first-overtone, or double-mode variable. Moving further to the red, a region appears where only the fundamental mode is stable (the *F zone*). The red side of the instability strip is limited by the so-called *red edge*, located at $T_{eff} \sim 5900\,K$, whose existence is related to the presence of envelope convection in the stars; being an efficient energy transport mechanism, it acts to quench the process of pulsation. It is also worth noting that both the width and the topology of the RR Lyrae instability strip is marginally dependent on the stellar metallicity, at least in the heavy element abundance range appropriate for the old galactic stellar systems which host RR Lyrae variables.

For a long time ([224]) it has been known that the period of an *ab* RR Lyrae variable is connected to the main structural parameters of the star such as the luminosity,

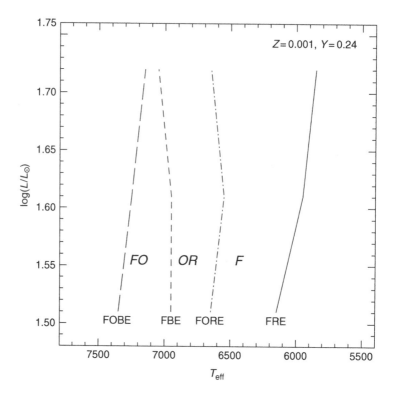

Figure 6.13 The location of the RR Lyrae instability strip in the HRD. The labels denote the location of the zones where the various pulsational modes are unstable. The blue and red edges of the various zones are also labelled (BE and RE)

the effective temperature and the mass. The most up-to-date relation between these quantities is ([16]):

$$\log(P) = 11.627 + 0.823 \log(L/L_{\odot}) - 0.582 \log(M/M_{\odot}) - 3.506 \log(T_{\text{eff}})$$

which can be also written in the form:

$$\log(P) = 11.627 + 0.823 A - 3.506 \log(T_{\text{eff}})$$

where A is equal to $\log(L/L_{\odot}) - 0.707 \log(M/M_{\odot})$, and depends on the mass-to-luminosity ratio of the star. A similar relation can be also derived for first-overtone variables. These analytical relations provide a useful guide for predicting the trend of the period of pulsation with the luminosity, mass and effective temperature of the variable stars.

By comparing theoretical results with observations, these relations supply a useful tool for testing the consistency and reliability of the stellar evolution framework. In passing, we notice that since the luminosity of an HB star is strongly dependent on the He content (see the discussion in Section 6.3.1) the relation between the period of pulsation and the parameter A represents a useful tool for measuring the initial He content of RR Lyrae stars ([35], see also Section 9.2.4).

It is important to note that RR Lyrae stars provide two other important tools for checking evolution theory: the *Bailey's* and the *Petersen's diagram*. From the analysis of empirical light curves one derives the pulsational period and also the amplitude. The same information can be obtained by detailed numerical computations of pulsational models adopting different assumptions about the mass and luminosity of the pulsators. Therefore, the comparison between theoretical predictions and empirical estimates for the behaviour of amplitude with period (the *Bailey's diagram*) allows one to test the stellar pulsational and evolutionary models (see Figure 6.14).

The Petersen diagram is a very useful diagnostic for double-mode RR Lyrae variables, being based on the analysis of the trend of the ratio between the two pulsational periods with the fundamental period. It has been shown by [149] that this diagram, reproduced in Figure 6.15, can be used to derive direct information about the mass of double-mode RR Lyrae variables.

6.6.2 The classical Cepheid variables

Classical Cepheid variables are important because of their famous period–luminosity ($P–L$) relation, which is used to establish the basic distance scale of the universe. It is therefore worthwhile to devote some attention to the main pulsational properties of these variable stars.

Classical Cepheids are very bright objects, with intrinsic luminosity in the range from $\sim 300 L_\odot$ to $\sim 25\,000 L_\odot$. Their pulsational periods are mainly confined to the range 1–50 days, with a few extreme examples of up to 250 days. The shape of their light curves varies quite smoothly when moving from short-period variable to long-period ones: the shorter-period ones show steep, narrow maxima, and the longer-period Cepheids have broader maxima. From the point of view of stellar evolution Cepheids are intermediate-mass stars going through the stage of core He-burning. The evolutionary tracks of intermediate-mass stars can cross the instability strip up to three times. All stars cross the Cepheid instability strip during their expansion toward the Red Giant configuration, at the point of exhaustion of H in the core, but this crossing is so rapid that there is almost no chance to observe a star pulsating as a Cepheid variable in this stage. So the Cepheid behaviour can be observed only for those stars that experience a blue loop, extended enough to cross the pulsational instability strip.

On general grounds, the physical reasons determining why an intermediate mass star is affected by stable radial pulsations, and the morphology of the instability strip for this class of variables, are identical to the case of RR Lyrae stars. This notwithstanding, accurate and realistic pulsational models for these variables ([18]) have shown that the topology of their instability strip shows significant differences to that of RR Lyrae stars. It appears clear from data shown in Figure 6.16 and

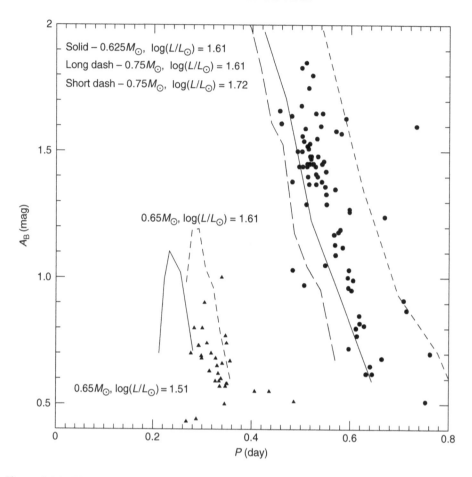

Figure 6.14 The Bailey diagram: the pulsational amplitude in the B photometric band as a function of the period, for various values of the mass and luminosity (in solar unit) of the RR Lyrae variable. The solid points correspond to the empirical data for the sample of RR Lyrae variables in the old galactic cluster M3. The data plotted on the lower left corner refer to first-overtone variables

a comparison with Figure 6.13 that Cepheids present a 'wedge-shaped' instability strip rather than the 'rectangular-shaped' strip of RR Lyrae stars. In addition, both the shape and location of the Cepheid instability strip are significantly affected by metallicity – the larger the metallicity, the cooler and steeper is the location of the instability strip on the HRD.

For Cepheid variables, it is possible to derive pulsational relations connecting the period to the stellar mass, luminosity and effective temperature. For a solar chemical composition and for fundamental pulsators, the relation derived by [18] is the following:

$$\log(P) = 0.987 - 3.108 \log(T_{\text{eff}}) - 0.767 \log(M/M_{\odot}) + 0.942 \log(L/L_{\odot})$$

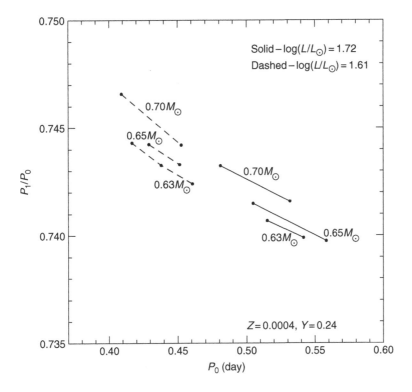

Figure 6.15   The Petersen diagram: the ratio of fundamental to first-overtone period as a function of the fundamental period for double-mode RR Lyrae stars. The adopted masses, luminosity levels and chemical composition are labelled

All the coefficients of this relation show a non-negligible dependence on the stellar metallicity (but the sign of each numeric coefficient is preserved). At a fixed luminosity and mass, decreasing the effective temperature produces (according to this relation) an increase in the pulsational period. Since an increase of the metallicity has the effect of shifting the Cepheid instability strip towards a lower effective temperature, and therefore longer periods, one should expect that the period–luminosity relation for Cepheid variables depends on metallicity.

The Cepheid period–luminosity relation has long played a pivotal role in constraining Galactic and extra-galactic distances. The existence of this correlation can easily be understood by considering the dependence of the pulsational period on stellar luminosity, mass and effective temperature, and the prediction of evolution theory that there is, for a set chemical composition, a tight relation between the mass and surface luminosity of stars at the beginning of the core He-burning phase (luminosity increases for increasing mass) – the so-called *mass – luminosity relation* (in analogy with ZAMS stars). As a consequence, for any set chemical composition, the brightness of a Cepheid variable is a function of its period and effective temperature. Moving into the observational plane (see Chapter 8), this means that a relation

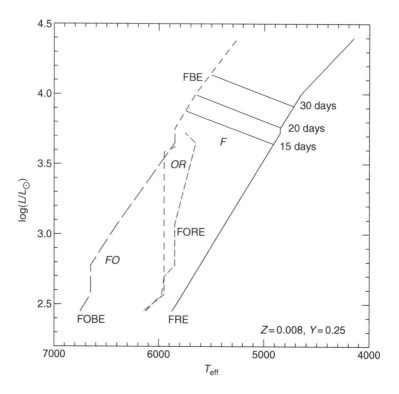

Figure 6.16 The location on the HRD of the boundaries of the classical Cepheid instability strip. The labels indicate the location of the zones where the various pulsational modes are unstable. The lines of constant pulsational period for some selected values are also shown

of the kind $< M_A > = a + b \log P + c(CI)$ has to hold, where $< M_A >$ is the mean absolute magnitude in a generic A bandpass and (CI) is the colour index. We refer the reader to Section 9.3.3, for a detailed discussion about the use of the Cepheid period–luminosity relation for distance estimates.

7 The Advanced Evolutionary Phases

7.1 Introduction

This chapter will discuss the evolution of stars during the evolutionary phases after the exhaustion of helium in the core. We will first outline the evolution of those stars which develop an electron degenerate carbon–oxygen (CO) core and thus evolve along the Asymptotic Giant Branch (AGB) and their later evolution as white dwarfs. The evolution of stars massive enough to ignite the burning of elements heavier than helium and, at the end, to explode as Type II supernovae is discussed in Section 7.5, and finally the problem of the evolution of Type Ia supernovae progenitors is addressed.

7.2 The asymptotic giant branch (AGB)

When the abundance of He becomes low enough, the stellar tracks, regardless of the value of the initial mass, move on the HRD towards a lower effective temperature and larger luminosity. This is the *Asymptotic Giant Branch* (AGB) which corresponds to the He-shell burning phase and shows a close similarity with the previous RGB phase. The designation *asymptotic* comes from the fact that in low-mass AGB stars, i.e. less than $\sim 2.5 M_\odot$, the effective temperature–luminosity relationship is very similar, albeit slightly hotter, to that of low-mass RGB stars. For more massive stars, the term *asymptotic* has no morphological significance.

After He-exhaustion, He-burning shifts to a shell around the CO core, whose mass-size increases as a consequence of the conversion of He to carbon and oxygen in the He-burning shell. The overlying H-burning shell, which has burnt outwards for some time, extinguishes due to the expansion and consequent drop of its temperature, caused by the onset of He-burning in the shell. In low-mass stars, the onset of the He-shell burning induces a temporary drop of the surface luminosity and the star

Evolution of Stars and Stellar Populations Maurizio Salaris and Santi Cassisi
© 2005 John Wiley & Sons, Ltd

crosses the same region of the HRD three times (as during the RGB bump phase). As a consequence, there is a good probability of observing an AGB star during this phase, which is called the *AGB clump*; this is indeed the case in well-populated old galactic stellar systems.

During the first part of the AGB phase (see Figure 7.1) – usually called the *early AGB* – while He-burning is progressively moving outwards inside the He core, and the mass of the CO core is increasing, an important convective episode occurs in stars more massive than $3-5\,M_\odot$, the precise limit being a function of the initial composition. In these stars, the large energy flux produced by the He-burning shell causes the base of the H-rich envelope to expand and cool, so that H-burning in the shell is immediately switched off. When this occurs, the outer convection zone penetrates inwards, into the H-depleted zone. This process is known as the *second dredge up*. For lower mass objects, H-burning in the shell remains quite efficient, and prevents the outer convection from penetrating deeper into the star (therefore the second dredge up does not occur).

In the dredged-up material, which can be as much as $\sim 1\,M_\odot$ for the most massive AGB stars, hydrogen has been completely converted into helium, and both ^{12}C and ^{16}O have been converted almost completely into ^{14}N. In addition, the second dredge up has the effect of reducing the mass-size of the H-exhausted region, thus preventing the later formation of very massive white dwarfs.

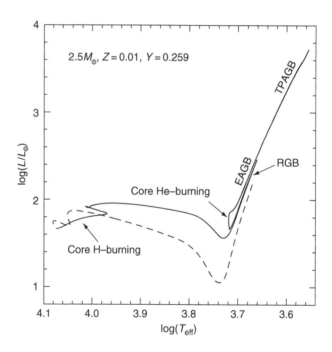

Figure 7.1 Evolutionary track of a $2.5\,M_\odot$ star from the PMS (dashed line) to the advanced TPAGB phase

In stars less massive than 8–$10\,M_\odot$, core He-burning produces a compact CO core. Soon after core He-exhaustion, the central density rapidly increases, reaching values ranging from $10^5\,\mathrm{g\,cm^{-3}}$ up to $10^8\,\mathrm{g\,cm^{-3}}$, depending on the total stellar mass. As a consequence, electron degeneracy attains a high value and a huge energy loss by plasma neutrinos occurs. As neutrino energy losses are only partially balanced by the gravitational energy release associated with core contraction, the thermal content of the core is used to reduce the energy deficit. As this cooling is larger in the central regions where the density is larger, the maximum value of the temperature is located out of the stellar centre and moves progressively outwards.

There is a very important limiting mass value for any stellar population, the so-called M^{up} limit, which corresponds to the largest mass at which the electron degeneracy of the CO core is high enough to prevent carbon ignition. The exact value of M^{up} depends strongly on the initial chemical composition, and is of the order of $\sim 8M_\odot$ at a solar metallicity and for extremely metal-poor populations, and has a minimum at $\sim 4M_\odot$ for a metallicity of $Z \sim 0.001$. Stars more massive than M^{up} will ignite (quietly or through a violent flash, depending on the mass) carbon in the core, while stars less massive than M^{up} enter the thermally pulsing AGB phase (detailed in the following section). Stars that ignite He in a non-electron degenerate He-core but that develop an electron degenerate CO core at the end of central He-burning are defined as 'intermediate-mass' stars. Stars with masses above M^{up} are so-called 'massive' stars.

7.2.1 The thermally pulsing phase

During the early AGB phase, the He-burning shell moves outward in the He-rich envelope whose mass is not significantly increased since the H-burning shell is essentially inactive. When the He-burning shell approaches the H/He discontinuity it dies down and, after a rapid contraction, the H-burning shell becomes fully efficient to supply the energy necessary for the star's needs. This temporary stop of the He-burning shell marks the beginning of the *thermally pulsating AGB phase* (TPAGB).

As hydrogen burns, the helium ashes which are accreted above the degenerate CO core, are compressed and heated. When the mass of these ashes reaches a critical value, of the order of $10^{-3}M_\odot$ for a CO core mass of $0.8\,M_\odot$ (the precise mass limit depends on the CO core mass – roughly speaking it increases by one order of magnitude when decreasing the CO mass by $\sim 0.2M_\odot$) He ignites and a thermonuclear runaway occurs. Thermonuclear runaway means that the shell reacts to the new energy input with an increase of temperature that, in turn, increases the energy generation even more. This runaway occurs because of the small geometrical thickness of the shell and, in spite of the low degree of electron degeneracy (one would expect a runaway in case of strong degeneracy, as in the case of central He ignition in low-mass stars), in the He layer on top of the CO core. It can be shown (e.g. [115], [155]) that in a thin shell of thickness s, located at a radial coordinate r, pressure and density changes are related through:

$$\frac{dP}{P} = 4\frac{s}{r}\frac{d\rho}{\rho}$$

From the equation of state one obtains

$$\frac{dP}{P} = \alpha\frac{d\rho}{\rho} + \beta\frac{dT}{T}$$

with α and β positive constants. Using this latter result we get

$$\frac{d\rho}{\rho}\left(4\frac{s}{r} - \alpha\right) = \beta\frac{dT}{T}$$

A stable burning requires that the extra energy input induces an expansion of the layers (decrease of density) and a decrease of temperature in the shell. Given that β (and α) is positive, this implies that

$$4\frac{s}{r} > \alpha$$

When the geometrical thickness s of the shell is sufficiently small, as in the case of the thermally pulsing phase, this condition is not satisfied and the expansion of the shell induces an increase of the temperature (notice that in the case of the He flash the runaway proceeds at constant density).

At the peak of the flash, the rate of nuclear energy release can reach values as high as $L_{He} \sim 10^7 - 10^8 L_\odot$. Most of this energy goes into heating up the nuclear burning layers causing them, and the layers above them, to expand against gravity. As a consequence of being pushed out to very low temperatures and densities, the hydrogen burning in the shell is switched off.

Stellar models predict the formation of a convective zone extending from the He-burning shell up to the H/He discontinuity, as a consequence of the huge energy release during the thermonuclear runaway, as shown in Figure 7.2. As the material in the convective shell continues to expand outwards, the shell source is now widely expanded and the condition for the onset of a runaway discussed above is no longer applicable. The shell starts to cool and the rate of He-burning drops precipitously. In AGB stars of large enough core mass, say $M_{CO} \geq 0.7 M_\odot$, the decrease in L_{He} causes the convective shell to disappear. At lower masses, the base of the convective shell reaches into the zone where incomplete He-burning occurs, and so the products of He-burning (mainly carbon) are mixed into the whole region located in between the He-burning shell and the H/He discontinuity. Eventually, a steady state is established between nuclear energy production and energy flow outwards.

The AGB star then continues through a quiescent He-burning phase, which lasts until the total amount of material which is processed by 3α reactions equals the

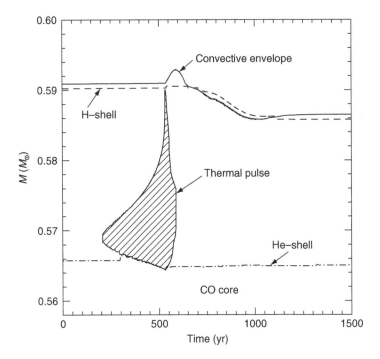

Figure 7.2 Evolution with time of part of the internal structure of a $2\,M_\odot$ star with $Z = 0.015$ and $Y = 0.275$, during the tenth thermal pulse; the locations of the H-burning shell, the He-burning shell and the base of the convective envelope are shown. The origin of the time coordinate has been arbitrarily shifted. The dashed zone shows the development of the convective region during the thermal pulse. Notice the occurrence of the *third dredge up* about 200 yr after the onset of the thermal pulse (courtesy of O. Straniero)

amount of material which was processed by the H-burning prior to the flash. At this point, hydrogen near the H/He discontinuity is reignited and the star embarks on another long phase of quiescent H-burning. When the mass of the He-rich layers reaches the critical value previously specified, another thermal pulse is initiated and this cycle is repeated many times.

The time dependence of L_{He}, L_{H}, the surface luminosity and the effective temperature during a series of thermal pulses is shown in Figure 7.3. The pulse amplitude grows with each succeeding pulse, rapidly for the first five or 10 pulses, then more slowly for the next ~ 10 pulses, until an asymptotic regime is approached.

During the inter-pulse phases, the H-burning shell, whose efficiency is determined by the mass-size of the H-exhausted core, is able to provide all the energy necessary to compensate for the surface energy losses. Therefore, as for during the previous RGB phase, there exists a direct correlation between the surface luminosity and the mass of the H-exhausted core (i.e. an $M_{\mathrm{cHe}}-L$ relation).

Due to their importance for explaining the chemical anomalies observed in AGB stars, it is worth discussing in some detail the convective phenomena that could occur

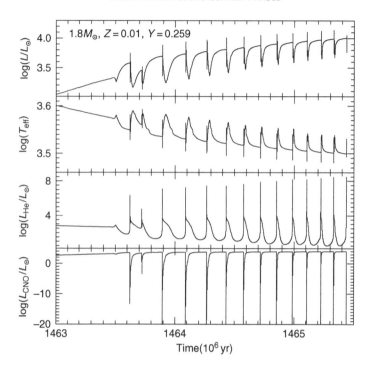

Figure 7.3 Evolution with time of the effective temperature, surface luminosity, luminosity associated with the He-burning in the shell, and the luminosity produced through the CNO cycle in the H-burning shell, during the first series of thermal pulses of a $1.8 M_\odot$ star

during the TPAGB phase. The occurrence of an inter-shell convective episode has been discussed previously; one has to note that, once the H-burning has been switched off after the He-shell burning reignition, the convection envelope can move inward in mass, crossing the H/He discontinuity. If this is the case, a new dredge up occurs – the so-called *third dredge up* (see Figure 7.2). If this dredge up is able to penetrate deep enough into the inter-shell region mixed during the previous thermal pulse, then helium, products of He-burning (essentially carbon) and heavy s-elements (see below) are thereupon dredged up into the envelope and brought to the surface where they can be observed. As a consequence of recurrent third dredge up episodes, a carbon-rich star (C stars are those structures with a photospheric ratio O/C < 1) can be produced.

The third dredge up is driven by the expansion and subsequent cooling of the envelope which occurs during the thermal pulse. It penetrates deeper when the strength of the pulse is stronger. The strength of a pulse is fixed by the thermal conditions at the bottom of the He-rich layers, which are strongly dependent on the rate at which He is accreted by the H-burning shell: as a general rule, the slower the H-burning is, the higher the density at the bottom of the He-shell and, in turn, the stronger the thermal pulse. Therefore, the strength of a thermal pulse is essentially regulated by the H-burning rate; any parameter, such as initial chemical composition, envelope mass and mass of the H-exhausted region, which affects the pace at which H is burnt, affects the thermal pulse strength.

The amount of material dredged up in each single episode initially increases, because the core mass increases. After attaining a maximum value, it starts decreasing because the envelope mass decreases, due to the combined effects of H-burning in the shell and of mass loss which is extremely effective during this phase.

For each fixed core and envelope mass, the third dredge up can penetrate deeper into the star as the metallicity decreases. The dependence of the third dredge up efficiency on the metallicity is very important in explaining the differences in the surface chemical abundances between Galactic AGB stars and those belonging to more metal-poor systems.

Evolutionary computations show that there is a minimum envelope mass for the occurrence of the third dredge up, the value of which is of the order of $0.4 M_\odot$. This has the consequence that stars of initial mass below a given limit (of the order of 1.2–$1.5 M_\odot$) cannot experience the third dredge up since at the beginning of the TPAGB phase they already have a residual envelope mass which is too small. As a consequence, we do not expect to observe C-rich stars in very old stellar populations.

In more massive AGB stars, with $M > 6$–$7 M_\odot$, a peculiar physical process, first recognized by [209], can occur at the base of the convective envelope: the temperature can be so large ($T \sim 8 \times 10^7$ K) that significant burning can occur at the bottom of the convective zone, the so-called *hot-bottom burning*. The consequences of the coupling between nuclear burning and convective mixing are many: (1) the surface luminosity increases significantly, breaking the core mass–luminosity relation; (2) the surface chemical composition is strongly affected – carbon is converted into nitrogen, preventing the formation of C-rich stars and an enhancement of lithium can occur through the so-called Cameron–Fowler mechanism ([33]). The general idea is that when the envelope is mixed down to temperatures of the order of 10^8 K, ^{3}He is transformed very efficiently into ^{7}Be, according to the reaction

$$^3\text{He} + {}^4\text{He} \rightarrow {}^7\text{Be} + \gamma$$

The produced ^{7}Be is then transformed into ^{7}Li through

$$^7\text{Be} + \text{e}^- \rightarrow {}^7\text{Li} + \nu_e$$

The only lithium that can survive, however, is the one that stays below its burning temperature of $\sim 2.5 \times 10^6$ K. The amount of surface lithium preserved after the envelope retreats towards the surface is therefore determined by the balance between the ^{7}Be transported out towards the surface by the convective mixing, and the ^{7}Li produced at low enough temperatures that it is then transported back inward by the convection.

This mechanism may explain the empirical evidence ([231]) that almost all the brightest Long Period Variables (commonly identified as bright TPAGB stars experiencing large amplitude radial pulsations) are super-lithium rich objects, their surface lithium abundance being about three orders of magnitude larger than normally expected.

7.2.2 On the production of s-elements

Spectroscopical observations show that the surface of AGB stars is strongly enriched in s-process elements, i.e. those elements beyond the iron peak, such as Sr, Y, Zr, Ba, La, Ce, Pr and Nd, which are produced through *slow* neutron captures (slow compared with beta decay). The spectroscopic detection of unstable ^{99}Tc, whose half-life is of the order of 2×10^5 yr, provides clear evidence that this enrichment cannot be accounted for by pollution of the primeval stellar material from which these stars formed, but is rather due to the production of these elements in the stellar interiors.

The key ingredient in activating the s-process reactions is the neutron source. In AGB stars, Cameron ([30], [32]) first recognized two major neutron sources: those from the reactions ^{22}Ne$(\alpha, n)^{25}$Mg and ^{13}C$(\alpha, n)^{16}$O.

During the early phase of the convective episode occurring in the inter-shell region, the material is fully mixed and ^{14}N, which is present in this region as a consequence of the previous H-burning process (whose abundance is almost equal to the sum of the initial abundances, by number, of the CNO elements), is converted into ^{22}Ne by means of the nuclear reaction branch

$$^{14}\text{N}(\alpha, \gamma)^{18}\text{F}(\beta^+, \nu)^{18}\text{O}(\alpha, \gamma)^{22}\text{Ne}$$

If, during the thermonuclear runaway, the temperature at the base of the inter-shell region attains a high enough value ($T \sim 3.5 \times 10^8$ K) the reaction ^{22}Ne$(\alpha, n)^{25}$Mg can occur, so providing a sizeable flux of neutrons. It has been demonstrated ([103]) that this is the case for intermediate-mass AGB stars. In AGB stars with $M < 3M_\odot$ the base of the inter-shell region does not attain a high enough temperature – at maximum it is $T \sim 3 \times 10^8$ K – to provide enough neutrons in this way.

In low-mass AGB stars, an alternative source of neutrons is provided by the reaction ^{13}C$(\alpha, n)^{16}$O. In order to be effective, this reaction requires a temperature of $\sim 9 \times 10^7$ K, which can easily be attained in the inter-shell region. Model computations predict that between two consecutive thermal pulses, some ^{13}C is created at the top of the inter-shell region due to the reactions in the H-burning shell. Nevertheless, the burning of this ^{13}C is not an efficient neutron source: in a region which has been processed by H-burning ^{14}N is more abundant than ^{13}C, and so any neutrons released by ^{13}C would be immediately captured by ^{14}N. An alternative source of ^{13}C is necessary in a region where ^{14}N has already been exhausted. A possible way out is as follows. In the inter-shell region, after the convective shell episode, there is a huge amount of ^{12}C, and ^{14}N is missing, having been converted into ^{22}Ne. ^{13}C is then produced via the nuclear chain ^{12}C$(p, \gamma)^{13}$N$(\beta^+, \nu)^{13}$C. This appears to be the most promising approach for having ^{13}C at the right place. However, there is a problem with this solution. How is it possible to have sufficient protons in the He-rich inter-shell region after the convective shell episode?

Numerical simulations ([104]) have shown that a more promising phase for the formation of a ^{13}C pocket within the inter-shell, is the one corresponding to third dredge up and during the so-called *post-flash dip*, which is the period that immediately follows the third dredge up. At the time of the third dredge up, a sharp discontinuity between the H-rich envelope and the He- and C-rich inter-shell is produced, and this situation holds until H-burning is ignited again, i.e. for a time of the order of $\sim 10^4$ yr. If, during this time, a few protons can diffuse into the underlying layers, at the reignition of the H-burning shell the upper layers of the He-rich inter-shell region heat up and a ^{13}C pocket can be formed. However, it is important to avoid an excess of protons, because in this case the production of ^{13}C is followed by its destruction via ^{13}C(p, γ)^{14}N reaction. It has been shown by [208] that the ^{13}C produced according to this scenario is fully burned via the ^{13}C(α, n)^{16}O reaction in a radiative region during the time between two consecutive thermal pulses, when the temperature attains a value of $\sim 9 \times 10^7$ K. As a consequence of this process, the neutron density would be of the order of 10^7 cm^{-3}.

Many different processes have been proposed as the physical mechanisms producing the ingestion of the right amount of protons at the right place: atomic diffusion, mechanical overshoot, rotational mixing, gravity waves and so on. However, we still lack a firm explanation of the process and, for this reason, current evolutionary computations treat it as an ad hoc mechanism with free parameters to be calibrated on observations.

7.2.3 The termination of the AGB evolutionary phase

Once the star settles into the thermally pulsating phase, in principle it can experience a huge number of thermal pulses, the number being limited only by the mass of the H-rich envelope or by the mass of the CO core. If the CO core mass exceeds the value $\sim 1.4 M_\odot$ (the so-called *Chandrasekhar mass limit*, see Section 7.3) the ignition of carbon burning can occur and this marks the end of the AGB phase. In reality, due to mass loss during the TPAGB, we do not expect that any star will reach the Chandrasekhar limit – as a consequence of the increase of the mass of the CO core during the TPAGB evolution – and ignite carbon. The H-burning shell source cannot burn further when it reaches a few $10^{-3} M_\odot$ below the surface; when this minimum mass limit for the envelope is achieved, additional thermal pulses cannot occur and the star leaves the AGB, moving towards hotter effective temperatures.

The mechanism(s) driving mass loss during the AGB phase and the occurrence of a 'super-wind' ([157]) – with extremely high mass-loss rates – that terminates the AGB phase, are topics which have been addressed for a long time and for which we do not yet have a full explanation. What we know is that stars climbing along the AGB can experience large amplitude pulsations (the so called *Mira* variable stars) and that it has to be expected that a strong correlation between pulsations and mass-loss efficiency exists. The gas compression due to the pulsation and, in turn, the density increase, makes the formation of molecules and dust grains easier. These easily trap the outgoing radiation flux, so driving a strong wind (this is the radiation-driven wind scenario). Even if empirical evidence shows a correlation between mass-loss rate and pulsational period, it has to be considered that there is a quite large spread in the data at each given period, and that these empirical data refer to stars with a large range in core mass, total mass and composition. Thus, any detailed conclusions achieved by adopting a single period–mass-loss rate relation could be very uncertain. Observational measurements do, however, provide a clear indication of the real mass-loss rates of AGB stars; a range from a few $10^{-8} M_\odot \mathrm{yr}^{-1}$ to a few $10^{-4} M_\odot \mathrm{yr}^{-1}$. In addition, it is clear that the brightest AGB stars in different systems, such as the Magellanic Clouds, show essentially the same luminosity. This evidence can be explained only by assuming that the mass-loss rate strongly increases to a much higher value ($\sim 10^{-4} M_\odot \mathrm{yr}^{-1}$) than the average rate towards the end of the AGB phase (the super-wind?). Some commonly used relationships for the mass loss along the AGB phase are ([134]):

$$\log \left(\frac{dM}{dt} \right) = -11.4 + 0.0123 \log(P) \quad P \le 500 \tag{7.1}$$

$$\frac{dM}{dt} = 6.07023 \times 10^{-3} \frac{L}{cv_e} \quad P > 500 \tag{7.2}$$

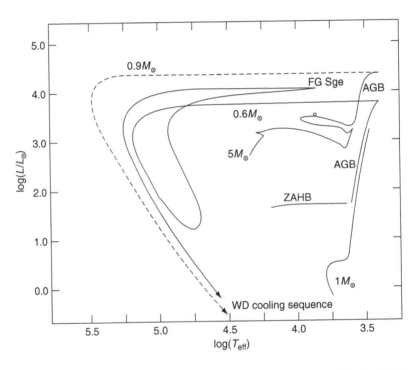

Figure 7.4 Evolutionary tracks from the MS to the WD cooling sequence. The dashed line is the evolutionary track of the $0.9\,M_\odot$ object left at the end of the AGB, whose initial MS mass was equal to $5\,M_\odot$; the solid line labelled $0.6\,M_\odot$ is the evolutionary track of an object whose initial mass was equal to $1\,M_\odot$. This star experiences a final He flash in the shell before reaching the WD cooling sequence. The location of the observational counterpart represented by the star FG Sagittae is also shown

where the mass-loss rate is given in solar masses per year, and

$$v_e = -13.5 + 0.056P \tag{7.3}$$

$$\log(P) = -2.07 + 1.94\log(R) - 0.9\log(M) \tag{7.4}$$

where P is the pulsation period in days, the stellar radius R and mass M are in solar units, and v_e denotes the terminal velocity of the stellar wind in km s^{-1}.

It is now worth discussing briefly the evolution of low-mass stars from the end of the TPAGB phase to the beginning of the following evolutionary phase, i.e. the white dwarf (WD) stage (the corresponding evolutionary track on HRD is shown in Figure 7.4). As a consequence of radial pulsations (or any other mechanisms) the star loses a lot of its mass during the brightest part of the AGB. Once the mass of the H-rich envelope decreases below a critical value, the remnant star evolves rapidly towards hotter effective temperatures. It continues to burn hydrogen in a thin shell until the bluest point along its evolutionary track at almost constant luminosity. At this point, H-burning is switched off and both the H-rich envelope and He-rich layer contract rapidly. Depending on thermal conditions of the envelope, different evolutionary channels can be followed.

1. Nuclear burning definitively dies out and the remnant cools as a white dwarf.

2. Heating of the He-rich layers leads to an He-burning thermonuclear runaway in the He shell, the consequence of which is to carry the star back to the AGB. This is the so called *born-again AGB scenario*, the most remarkable empirical evidence for which is provided by the object FG Sagittae (see Figure 7.4). During the subsequent quiescent He-burning phase, the star evolves to the blue side of the HRD on a timescale approximately three times longer than that of the previous shell H-burning phase. At the end of this, the star becomes a white dwarf.

3. If heating of the H-rich envelope induces an H-burning thermonuclear runaway, the stellar remnant becomes a *self-induced nova*. To outline the final fate of a self-induced nova is not an easy task, due to the difficulties encountered in the numerical simulations. This notwithstanding, it is possible to make some predictions: (a) if the process is very violent and dynamic, this can produce the ejection of the whole H-rich envelope and so the star becomes a white dwarf with no trace of H on its surface (the so called *DB* white dwarf or *non-DA* white dwarf) (b) if the process is only mildly dynamical and only a small portion of the H-rich envelope is lost, the mass of this envelope will be decreased by nuclear burning. After a phase of quiescent H-burning, the star will start to cool down the white dwarf cooling sequence. It is possible that another H flash event occurs leading to a further reduction through nuclear burning of the H-rich envelope mass. This process could, in principle, be reiterated many times until there is too little

hydrogen left to burn. During the subsequent WD evolution, even a very modest wind could cause the loss of this layer, so the WD would appear as a non-DA white dwarf.

Before closing this section, it is worth mentioning that the evolutionary path followed by a star from the end of the AGB to the beginning of the cooling phase along the WD sequence is the one corresponding to the formation of Planetary Nebulae. After the super-wind phase has finished, the material ejected during the mass-loss process – whose composition is that of the envelope near the termination of the TPAGB phase – keeps expanding, while the remnant star continues its evolution towards higher temperatures. When the effective temperature reaches a large enough value, $T_{eff} \sim 30\,000$ K, the ejected material is ionized by the photons from the central star remnant, and it assumes the characteristics (and the shape) of a Planetary Nebula.

7.3 The Chandrasekhar limit and the evolution of stars with large CO cores

At the beginning of this chapter we introduced the mass limit M^{up}. This corresponds to the highest mass where, after the central He-burning stage, the electron degeneracy level in the CO core is strong enough to prevent the non-explosive ignition of carbon burning. Now we discuss the final fate of stars with mass around the M^{up} limit, and define the Chandrasekhar mass limit.

Chandrasekhar developed a fundamental theory of white dwarf stars and discovered that there exists an upper mass limit for a fully relativistic electron degenerate core in hydrostatic equilibrium ([54]). This value is equal to

$$M_{ch} = \left(\frac{2}{\mu_e}\right)^2 1.459 M_\odot$$

or the *Chandrasekhar mass* (see Section 7.4 for an approximate derivation of this limit). For any given chemical composition, this relation allows the computation of the maximum mass for a fully relativistic electron degenerate structure. In the case of a CO core, $\mu_e = 2$, so the Chandrasekhar limit is equal to $M_{ch} = 1.46 M_\odot$. This limit is only strictly valid for the ideal case of a fully relativistic electron degenerate core. Nevertheless, it is quite important and its physical meaning is that one cannot expect to observe stars with degenerate CO cores more massive than $\sim 1.4 M_\odot$. Indeed, so far, no WD with mass around this limit has ever been discovered. The Chandrasekhar limit is also very important when analysing the evolutionary channels which could produce explosive outcomes, resembling Type Ia supernova events.

The evolutionary behaviour of a degenerate CO core closely resembles that of a degenerate He-core during the RGB stage for low-mass stars. The physical properties of the core are almost completely unaffected by the envelope – at least until the mass of the envelope is so small that it does not allow further nuclear burning, at the end of the AGB phase.

While the CO core mass increases because of the He-burning in the shell, it also contracts and becomes denser. The release of gravitational energy should heat the core, but the increasing densities in the core strongly favour neutrino energy losses, and the central portion of the core cools. However, with increasing CO core mass, the nuclear burning of carbon is activated – although at this stage ϵ_C is less than ϵ_ν. When, and if (see below), the CO core mass approaches the Chandrasekhar limit, ϵ_C becomes comparable to ϵ_ν. Ultimately, the carbon burning energy release will overcome the neutrino energy losses, and a thermonuclear runaway occurs.

If this thermonuclear runaway occurs at the centre of the star, we expect an explosive process that would cause its complete destruction (the so-called Type I 1/2 supernova event). In the case where the carbon burning starts in a shell far enough from the centre, after one – or more – flash(es) carbon burning quietly settles in the centre of the star. The final fate of these stars is eventually to experience thermal pulses at a very high luminosity and finally to become white dwarfs with a degenerate O–Ne core.

Whether a star can develop an electron degenerate CO core of the order of the Chandrasekhar mass during its evolution along the AGB, depends not only on its initial mass but also on the efficiency of mass loss along the previous evolutionary phases. In the absence of mass loss all stars more massive than about $1.4M_\odot$ and less massive than the M^{up} value, should be able to ignite C-burning under degenerate physical conditions. In the real world, mass loss significantly reduces the stellar mass so that very few, if any, stars with mass around M^{up} (or slightly larger) are able to reach the conditions for the ignition of carbon.

All stars more massive than about M^{up} do not develop electron degeneracy in the CO core at the end of the He-burning phase. They are able to start non-explosive C-burning in their cores and will experience all the subsequent nuclear burning phases until an iron core is created.

7.4 Carbon–oxygen white dwarfs

As discussed in the previous sections, stars with initial masses up to $M \sim 6$–$8M_\odot$ (the precise value depends on the initial chemical composition and the extension of the overshoot from the convective cores during the previous evolutionary phases) lose all their envelope during the thermal pulses and continue the evolution with an electron degenerate core made almost exclusively of carbon and oxygen (whose stratification is determined by the nuclear transformations of the previous evolutionary phase) plus a residual non-degenerate envelope, with thickness of about $10^{-2}M_{tot}$, where M_{tot} is the total mass of the white dwarf (WD). No nuclear energy processes are active, since temperatures are not high enough, and the stars evolve at approximately constant radius and decreasing luminosity.

The faintest WDs detected to date have $L \approx 10^{-4.5} L_\odot$, the observed WD mean mass is ~ 0.55–$0.60 M_\odot$, in agreement with the CO masses of WD progenitors at the beginning of the thermal-pulse phase. Due to our imperfect knowledge of the mass-loss processes during the AGB phase (and, more generally, along all the various evolutionary phases) we cannot exactly predict the final WD mass produced by a progenitor with a given initial total mass M. In general, the CO core mass at the beginning of the TPAGB phase is a reasonable approximation since one does not expect a large increase of the core during the thermal pulses. The location of WD tracks in the HRD and a relationship between initial mass on the MS and final WD mass (for an initial solar chemical composition) are shown in Figure 7.5.

The large range of WD progenitor masses encompasses the vast majority of stars formed in the Galaxy, and thus WD stars represent the most common end-point of stellar evolution. It is probable that more than 95 per cent of the stars in the Galaxy will eventually end up as white dwarfs.

As mentioned above, the typical mass of these objects is of the order of $0.6 M_\odot$, while their size is more similar to that of a planet. Their compact nature gives rise to large average densities, large surface gravities and low luminosities. The main properties of WD stars obtained from the full system of equations of stellar structure, can be highlighted using some simplified physics, as follows.

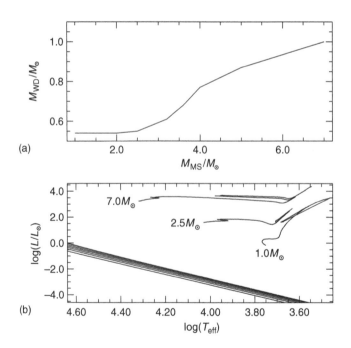

Figure 7.5 (a) shows an approximate relationship between the initial stellar mass (M_{MS}) and the final WD mass (M_{WD}) for a range of low- and intermediate-mass stars. (b) displays the HRD location of WD evolutionary tracks for different values of M_{WD} (between 0.54 and 1 $M_\odot$) compared with previous evolutionary phases

The equations of continuity of mass and hydrostatic equilibrium can be rewritten in an approximate form as

$$\frac{M}{R^3} \propto \rho$$

$$P_c \propto G \frac{M^2}{R^4}$$

Here we have considered the differentiation between two points, i.e. the centre of the star where $m = 0$, $P = P_c$ and $r = 0$ and the surface, where $m = M$, $P \sim 0$, $r = R$, and used the total mass, radius and an average density in the right-hand side of the equations.

When we approximate the gas pressure with the degenerate electron pressure in the non-relativistic case and request that this pressure must balance the gravitational potential, we get from the second of the previous equations:

$$\frac{M^{5/3}}{R^5 \mu_e^{5/3}} \propto \frac{GM^2}{R^4}$$

This equation provides the mass–radius relation:

$$R \propto \frac{1}{G \mu_e^{5/3}} M^{-1/3} \qquad (7.5)$$

Hence, WDs with non-relativistic degeneracy have a hydrostatic equilibrium radius inversely proportional to their mass. Also, for a given mass, the radius tends to decrease for increasing μ_e (i.e. for WDs made of heavier nuclei). If a WD of mass M_1 has a radius R_1, increasing its mass up to a value M_2 causes an increase of the gravitational energy that is balanced by the pressure forces for a decrease of R (the pressure increases with decreasing R faster than the gravitational effects) hence its equilibrium radius is smaller. Typical radii of WD stars are of the order of $\sim 0.01\, R_\odot$, and densities are of the order of $10^6\, \mathrm{g\, cm^{-3}}$. Figure 7.6 shows a comparison between the mass–radius relationship from theoretical models ([176]) and observations ([156]).

Given that increasing mass implies a decrease of R, the density of the WD will increase with M_{WD}. When the density is high enough the electron degeneracy becomes relativistic and the hydrostatic equilibrium requires that

$$\frac{M^{4/3}}{R^4 \mu_e^{4/3}} \propto \frac{GM^2}{R^4}$$

This equation provides

$$M \propto \frac{1}{G^{3/2} \mu_e^2}$$

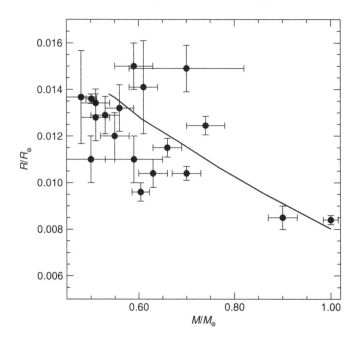

Figure 7.6 Comparison of the theoretical mass–radius relationship predicted by theoretical models ([176]) with observations ([156])

independent of R. The meaning of this relationship for the case of relativistic degeneracy is the following. If the equilibrium configuration is independent of R, hydrostatic equilibrium can be found only by adjusting the mass to a specific value M that we denote with M_{ch}. If $M > M_{ch}$ the pressure forces due to gravitation increase and they cannot be balanced by the decrease of the radius as in the non-relativistic case. If $M < M_{ch}$ the gas pressure makes the star expand, the EOS becomes less relativistic, and the star recovers a mass–radius relationship, hence a suitable equilibrium radius. This means that there exists a limiting mass M_{ch} – called the Chandrasekhar mass – for a WD to be in hydrostatic equilibrium. A rough numerical estimate of M_{ch} may be obtained along the same lines of the previous derivation, after including the constants entering the relationships between P and ρ and considering an average density $\rho = (3/4\pi)(M/R^3)$ instead of simply $\rho \sim (M/R^3)$. The condition of equilibrium between gravitational and pressure forces becomes

$$\frac{GM^2}{R^4}\frac{1}{4\pi} \sim K_0 \frac{3^{4/3}M^{4/3}}{(4\pi)^{4/3}R^4\mu_e^{4/3}}$$

where

$$K_0 = \left(\frac{3}{\pi}\right)^{1/3}\frac{hc}{8m_H^{4/3}} = 1.2435 \times 10^{15} \ (\text{cgs})$$

This relationship gives

$$M_{ch} \equiv M \sim 9 \left(\frac{K_0}{G} \right)^{3/2} \left(\frac{1}{4\pi} \right)^{1/2} \frac{1}{\mu_e^2} \sim \frac{2.54}{\mu_e^2} 1.27 M_\odot$$

The numerical constants that multiply the factor $(K_0/G)^{3/2}$ are very approximate, due to the simplifications made in dealing with the structure equations. A more precise value for the Chandrasekhar mass is

$$M_{ch} = \left(\frac{2}{\mu_e} \right)^2 1.459 M_\odot \tag{7.6}$$

Mass loss during the AGB evolution is expected to prevent the CO cores of WD progenitors to reach M_{ch}.

These general properties of WDs are also valid when one takes into account the existence of the thin non-degenerate envelope. Different thicknesses of the envelope change the mass–radius relationship slightly at a given WD mass, but the effect is largely a second-order effect. Instead, when studying the evolution with time of the WD luminosities, the envelope plays a fundamental role, as we will see below.

According to the virial theorem, a star with a degenerate electron component and ideal ions radiates the thermal energy of the ions and lowers its temperature. The evolution of a WD is therefore often labelled as a cooling process, and we are now going to obtain an approximate relationship for the evolution of the WD luminosity with time. We consider first the non-degenerate envelope and suppose it is radiative. If we divide the equation of hydrostatic equilibrium (with r as independent variable) by the equation of radiative transport we obtain

$$P dP = \frac{4ac}{3} \frac{4\pi G M K_B}{\kappa_0 L \mu m_H} T^{7.5} dT$$

having used the EOS of a perfect gas, $\kappa = \kappa_0 \rho T^{-3.5}$, and assumed that $m_r \sim M$ (M being the total mass) in the envelope. This equation can easily be integrated from the surface to a generic point within the envelope using as surface boundary conditions $P \sim 0$ and $T \sim 0$, and assuming the luminosity L is constant throughout the envelope and equal to the surface value. The integration provides P as a function of T plus constants. After using the ideal gas EOS to replace P with ρ we finally obtain

$$\rho = \left(\frac{32\pi ac G M \mu m_H}{8.5 \ 3 \kappa_0 L K_B} \right)^{1/2} T^{3.25} \tag{7.7}$$

with μ being the molecular weight of the matter in the envelope. We consider the base of the envelope, i.e. the transition point to the electron degenerate interior, to be the point where the degenerate electron pressure equals the pressure of the ideal electron gas in the envelope (ions are an ideal gas at both sides of this transition point). This means that at the interface (the values of density and temperature at this point are denoted with the subscript b, μ_e is the electron molecular weight of the degenerate core)

$$\frac{\rho_b K_B T_b}{\mu_e m_H} = 1.0036 \times 10^{13} \left(\frac{\rho_b}{\mu_e}\right)^{5/3} \text{(cgs)} \tag{7.8}$$

Equating the densities given by Equation (7.7) (evaluated at the base of the envelope) and (7.8) provides

$$\frac{L}{L_\odot} = 6.4 \times 10^{-3} \frac{\mu}{\mu_e^2} \frac{M}{M_\odot} \frac{1}{\kappa_0} T_b^{3.5} \tag{7.9}$$

This temperature T_b at the transition from degenerate core to envelope is approximately equal to the temperature at the centre. In fact, due to the very low opacity of degenerate electrons the core is practically isothermal. Therefore Equation (7.9) is actually a relationship between the surface luminosity and the central temperature T_c of the WD. As one can easily see, this relationship depends on the chemical composition (through the value of μ) and opacity of the envelope. The typical luminosities of faint WDs ($L \approx (10^{-3}-10^{-4})L_\odot$) imply cold interiors, with $T < 10^7$ K. More detailed L–T_c relationships determined from the solution of the full system of stellar evolution equations ([177]) are displayed in Figure 7.7.

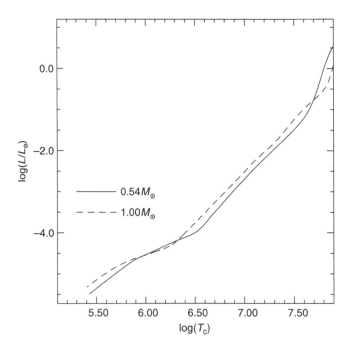

Figure 7.7 Detailed L–T_c relationships for WDs

The size of the envelope can easily be derived as follows. If we insert in the equation of radiative transport the Kramers' opacity $\kappa = \kappa_0 \rho T^{-3.5}$ we obtain

$$\frac{dT}{dr} = -\frac{3\kappa_0 \rho^2}{4acT^{6.5}} \frac{L}{4\pi r^2}$$

Now using Equation (7.7) for the density ρ we obtain the simple differential equation

$$\frac{dr}{r^2} = 4.25 \frac{K_B}{\mu G m_H} \frac{1}{M} dT$$

which can easily be integrated from the surface (where $r = R$) to the base of the envelope (where $r = r_b$). The solution (considering that at the surface $T \ll T_b$) is

$$T_b = \frac{1}{4.25} \frac{\mu G m_H}{K} \frac{M}{R} \left(\frac{R}{r_b} - 1 \right)$$

Temperatures T_b of the order of 10^6–10^7 K imply a negligible radial extension of the envelope, of the order of ~ 1 per cent of the total WD radius.

These simple properties of WD stars allow one to determine a good approximation of their cooling law easily, i.e. how the WD luminosity changes with time. The virial theorem applied to degenerate objects tells us that the energy radiated away by a WD is equal to the rate of decrease of the thermal energy of its ions (we neglect here any contribution from the non-degenerate envelope). Considering a perfect monatomic ion gas, the thermal energy per ion in the electron degenerate core is given by $c_V T = (3/2) K_B T$, where T is the temperature of the isothermal core. The total thermal energy is

$$E_i = \frac{3}{2} K_B T \frac{M}{\mu_i m_H}$$

where the term $M/(\mu_i m_H)$ gives the number of ions contained in the core (μ_i is the ionic molecular weight in the core). For a temperature $T \sim 10^7$ K E is $\sim 10^{48}$ erg.

We now rewrite Equation (7.9) in the form $L = \alpha M T^{7/2}$, where α is a constant; recalling that the WD luminosity is governed by

$$L = -\frac{dE_i}{dt}$$

we substitute in the left-hand side the expression for L obtained from Equation (7.9) which yields

$$\alpha T^{7/2} = -\frac{dT}{dt} \left(\frac{3K_B}{2\mu_i m_H} \right)$$

The integration of this equation between an initial time t_0 (when the WD stage starts) and a generic time t provides

$$(t - t_0) = \frac{3K_B}{5\alpha\mu_i m_H}(T^{-5/2} - T_0^{-5/2})$$

where T_0 is the temperature of the core at t_0 and T the temperature at time t. When the WD has cooled down significantly, $T_0 \gg T$ and therefore

$$\Delta t \equiv (t - t_0) \sim \frac{3K_B}{5\alpha\mu_i m_H}(T^{-5/2})$$

This simple relationship relates the WD cooling time to its core temperature. A relationship between t and the luminosity L can be obtained using $L = \alpha M T^{7/2}$ (from the envelope integration) to write T in terms of L, which provides

$$\Delta t \propto \left(\frac{L}{M}\right)^{-5/7} \approx \frac{4.5 \ 10^7}{\mu_i}\left(\frac{LM_\odot}{ML_\odot}\right)^{-5/7} \tag{7.10}$$

where Δt is in years. This approximated cooling law is called Mestel law ([137]) and shows for example that higher WD masses evolve more slowly; this is easy to understand in terms of more ionic thermal energy stored because of the higher mass. Also, increasing the ionic molecular weight at a given M decreases the evolutionary times, since there are less ions in the star. This simple law predicts cooling ages (i.e. ages from the beginning of the WD phase) of the order of 10^9 years when $L \approx 0.001 L_\odot$.

Although the Mestel law is a good zero-order approximation of the real WD cooling law, complete models are more complicated, and in the following we will discuss physical effects not included in Equation (7.10).

7.4.1 Crystallization

The main difference between the complete WD cooling law and the Mestel law (apart from corrections to the degenerate electron EOS accounting for the fact that T is not zero) is due to the treatment of the EOS for the ions. To derive the Mestel law, ions have been approximated with an ideal gas with $c_V = (3/2)K_B$ per particle. The specific heat c_V is obviously a crucial quantity for the determination of the cooling time, and we will now see that the simple approximation of the ideal gas is not good when the core cools down. Due to the steady decrease of the temperature, the ions in the core tend to move less freely because the Coulomb interactions play an increasingly major role in determining their thermal properties. During the cooling the core gets cooler whereas the density is almost constant, hence Γ increases; when Γ becomes larger than 1 the ions tend to behave like a liquid, and with still decreasing temperatures, approaching $\Gamma \sim 180$, they form a periodic lattice structure that minimizes their total energy. This latter phenomenon is called crystallization, and when $\Gamma > 180$ ions behave like a solid.

The Coulomb interactions obviously affect the free energy of the ions, hence their EOS and the derived value of c_V. In general, the total free energy F will be the sum of the ideal part plus a Coulomb correction part F_C. In the case of a single chemical species, F_C is given ([202]) by

$$F_C = NK_BT(-0.89752\,\Gamma + 3.78176\,\Gamma^{1/4} - 0.71816\,\Gamma^{-1/4} + 2.19951\,\ln\Gamma - 3.30108)$$

when $1 < \Gamma < 180$ (liquid phase) and

$$F_C = NK_BT\left(-4.29076 + 4.5\;\ln\Gamma - \frac{1490}{\Gamma^2}\right)$$

when $\Gamma \geq 180$ (solid phase) with N being the number of particles per unit mass.

During the solid phase there is an additional term to the total free energy F, due to the oscillations of the ions around their equilibrium positions in the lattice configuration; at increasingly lower temperatures somewhat uncertain ionic quantum corrections have to be added to F. The quantum corrections become important when the dimensionless parameter $\theta = \theta_D/T$ is larger than 1, where $\theta_D \approx 4 \times 10^3 \rho^{1/2}$ K is the so called Debye temperature.

The influence of these non-ideal effects on the value of c_V for the ions can be summarized as follows. When $\Gamma < 1$ the specific heat per ion is well approximated by $c_V = (3/2)K_B$. Above this threshold a lattice starts progressively to form and c_V increases with increasing Γ reaching a maximum value $c_V \sim 3K_B$ at about crystallization ($\Gamma = 180$). Upon further cooling the temperature reaches θ_D and from this moment on c_V starts to decrease according to $c_V \propto T^3$.

This variation of c_V with Γ (hence temperature) is the first modification of the Mestel cooling law due to a realistic treatment of the ion EOS. There are, however, two other major effects. The first one is the release of latent heat during the crystallization process. When the WD cools down, since the density is higher in the centre than at the boundary of the core and T is almost uniform, $\Gamma = 180$ will first be reached in the centre and then, with decreasing T, this crystallization front will move progressively towards the core boundary. At $\Gamma = 180$ a phase transition occurs, like water solidifying into ice. As a consequence, the solid usually has a different density and latent heat q is released, that is the difference in entropy between the two phases. Typically $q \sim K_BT$ per ion. This means that during the core crystallization extra energy is released, first at the centre and then at progressively more external core layers. This latent heat (that in a full stellar evolution computation is included as an additional term to the energy generation coefficient ϵ) slows down the cooling of the WD with respect to the Mestel law, given that it is an extra energy contribution to the ionic energy budget. Crystallization starts in the centre at $\log(L/L_\odot) \sim -2.7$ and reaches the boundary of the degenerate core at $\log(L/L_\odot) \sim -4.2$ for a WD of $1M_\odot$. Less massive WDs start crystallizing at lower luminosities, hence lower core temperatures, because they are less dense and reach $\Gamma = 180$ when their cores are cooler. As an example, a WD of $0.6M_\odot$ starts to crystallize at the centre when $\log(L/L_\odot) \sim -3.6$, and the whole core is in the solid phase when $\log(L/L_\odot) \sim -4.5$. It is important to notice that the earlier the crystallization, the shorter the delay Δt of the cooling (with respect to the Mestel law) caused by the extra energy input, since $\Delta t \sim \Delta E/L$, where ΔE is the energy injected during the crystallization and L the WD luminosity; earlier crystallization means higher L hence shorter Δt for a given ΔE.

There is an additional energy release during the crystallization process, due to the phenomenon of chemical separation ([77], [107]). The physical principle behind this phenomenon is that, given a CO binary mixture with a given abundance ratio X_C/X_O in the gas and liquid phase, at crystallization the same abundance ratio cannot be maintained, because the two elements are not fully miscible in the solid phase. As an example, consider a WD core made of a mixture of two elements with mass fractions X_1 and X_2 in the liquid phase, uniform throughout the core. The abundance of the lighter element is X_1, and $X_1 < X_2$; this situation reflects the abundance ratios in the central parts of the WD core, where oxygen is more abundant than carbon. A representative phase diagram for this binary mixture is displayed in Figure 7.8; the shape of this phase diagram (there are many possible shapes, depending on the element mixture) is similar to the shape of the diagram for a CO mixture, so that the qualitative results of the discussion that follows also hold for a realistic WD composition.

In general, the phase diagram tells us how the abundances of the two elements have to be changed during the phase transition at a given point within the WD, because

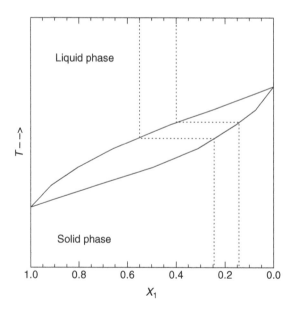

Figure 7.8 Representative phase diagram for the mixture of two elements with mass fractions X_1 and X_2. The graphical method to determine the abundances in the solid phase is also illustrated

of the non-miscibility of their mixture. Assume that $X_1 = 0.4$ when the mixture starts to crystallize at the centre. To determine graphically the chemical composition in the solid phase, one has to draw a vertical line with horizontal coordinate equal to 0.4, that runs through the region belonging to the liquid regime until it intersects the upper line describing the phase diagram, as shown in Figure 7.8. The vertical coordinate of the intersection point corresponds to the crystallization temperature of the WD centre. From this point one draws an horizontal line until the point where it intersects the lower line of the phase diagram. The horizontal coordinate of this latter intersection point gives the mass fraction of element $X_1 \sim 0.14$ in the solid phase, and of course $X_2 = 1 - X_1$. Since X_1 in the now crystallized centre is lower than the initial value, X_1 in the liquid phase at the crystallization boundary is necessarily increased with respect to the original value, due to conservation of mass. This means that right above the crystallized boundary the molecular weight is lower than in the overlying layers still in the liquid phase (where the ratio X_2/X_1 is higher). According to Equation (3.4) an increase of molecular weight for increasing distance from the centre causes a convective instability to develop. The convective mixing rehomogenizes the liquid phase very fast, overall enhancing the average X_1 value. This mixing region extends outwards in mass as long as the new rehomogenized average X_1 abundance is higher than the abundance of the following unperturbed layer.

Let us suppose the new value of X_1 at the boundary of the solid core is equal to 0.55. When this layer crystallizes, at a lower temperature than the core because of lower density, the abundances in the solid phase can be derived in the same way as before, and it is equal $X_1 = 0.25$ with our hypothetical phase diagram. This implies that right outside the centre the abundance of X_2 is lower than at the centre. The convective instability in the liquid phase ensues again (for the same reasons explained before) and the cycle is repeated until the whole degenerate core is crystallized. The final profile of X_1 and X_2 will not be homogeneous any longer, but X_2 will decrease from the centre outwards, and the opposite will be true for X_1.

The local change of chemical composition – hence molecular weight – due to this process of chemical separation at crystallization causes a release of energy (in addition to the latent heat discussed before) by virtue of the term $(dU/d\mu)_{T,v}(d\mu/dt)$ in the ϵ_g coefficient, that we have until now discarded because it is negligible when nuclear reactions are efficient. In the case of WDs, when there are no active nuclear reactions, this term is important and causes an increase of the cooling times by ~ 10 per cent.

In realistic WD computations the CO profile is not flat and the process of rehomogenization of the liquid phase during crystallization is slightly more complicated, but the main idea and the qualitative effect are the same as discussed before. Figure 7.9 shows the oxygen profile within the core of WDs of different masses and solar initial chemical composition (the carbon mass fraction X_C is essentially equal to $1 - X_O$). Different initial metallicities change the CO ratio but not too much. The initial chemical profile at the end of the AGB phase displays a constant oxygen mass fraction in the inner core, out to a point that marks the maximum extension of the central convective region during the core helium burning phase. The bump in the abundance just above this point (Figure 7.10) is produced when the He burning shell crosses

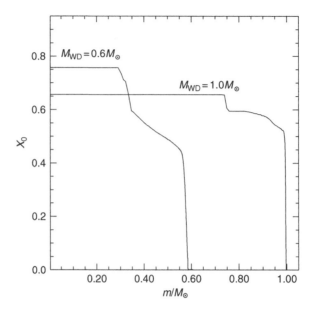

Figure 7.9 Oxygen profile within the core of two WD models

the semi-convective region partly enriched in carbon and oxygen, and carbon is converted into oxygen through the $^{12}C + \alpha \rightarrow ^{16}O + \gamma$ reaction. Beyond this region, the oxygen profile is built when the helium burning shell is slowly advancing towards more external layers. A convective instability at the beginning of the WD phase (the bump causes an increase of molecular weight for increasing distance from the centre) mixes all the inner region of the WD, producing a flat central profile (see Figure 7.10) that is maintained until the onset of crystallization. At the end of crystallization the oxygen profile is modified as shown in Figure 7.10.

7.4.2 The envelope

The generally accepted ideas about WD envelopes envisage the existence of a helium layer with mass $M_{He} \sim 10^{-2} M_{tot}$ (M_{tot} being the total WD mass) on top of the CO core, surrounded by an external hydrogen envelope of mass $M_H \sim 10^{-4} M_{tot}$. The uncertainty is due to our poor knowledge of the mass-loss processes that drive the evolution of stars to the WD stage.

Observations show that the external WD layers are either made of pure hydrogen (DA white dwarfs) or helium (non-DA) sometime with traces of metals, but typically $Z \sim 0$ at the WD surface. The explanation for the lack of metals at the surface is the efficiency of atomic diffusion. The typical observed ratio of DA to non-DA WDs is 4:1, but there is strong observational evidence for an evolution of the surface chemical composition among WDs, i.e. some of the DA WDs stars become non-DA objects,

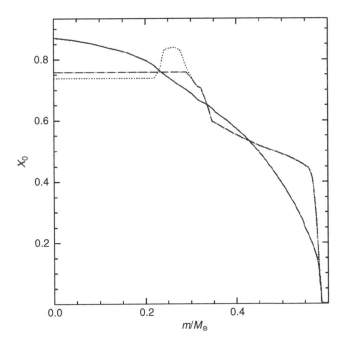

Figure 7.10 Oxygen profile within the core of a $0.6\,M_\odot$ WD at the end of the AGB phase (dotted line) after the convective instability at the beginning of the WD cooling (dash-dotted line) and after crystallization is completed (solid line)

and vice versa, during the cooling sequence; in fact the ratio of DA to non-DA white dwarfs changes as a function of effective temperature along the cooling sequence. Moreover, there exists a T_{eff} interval from ~45 000 K to about 30 000 K in which no WD with helium atmospheres has been found, while at $T_{\mathrm{eff}} > 70\,000$ K no DA objects are observed, in spite of the fact that helium rich WDs are observed well over $T_{\mathrm{eff}} >$ 100 000 K. Well-established explanations for these phenomena have not been found yet, although it seems reasonable to assume that at least some of the WD stars change their surface chemical composition from helium to hydrogen and back to helium again as evolution proceeds. It is suspected that a complicated interplay between diffusion, radiative levitation and convective mixing is responsible for this. Another possibility is that these two broad classes of WDs are produced at birth, i.e. the exact phase between the thermal pulses when a star leaves the AGB due to mass loss, determines whether or not it retains its hydrogen envelope. This idea, however, encounters difficulties to explain the existence of T_{eff} regions devoid of either DA or non-DA WDs.

In the practical case of computing extended sets of WD evolution models for various masses (and eventually different initial chemical compositions of their progenitors) it is usually assumed that DA WDs have a pure helium layer of $M_{\mathrm{He}} \sim 10^{-2} M_{\mathrm{tot}}$ on top of the CO core, and an external pure hydrogen envelope with $M_{\mathrm{H}} \sim 10^{-4} M_{\mathrm{tot}}$. In the case of non-DA WDs it is considered that there is only a pure helium layer with $M_{\mathrm{He}} \sim 10^{-2}$–$10^{-3} M_{\mathrm{tot}}$.

This non-degenerate envelope contributes to the energy budget by slowly contracting according to the virial theorem. In addition, since they are extremely opaque to radiation with respect to the highly conductive isothermal core, they essentially regulate the energy outflow from the star (we have already seen in the simplified cooling law described previously, that it is the structure of the envelope that provides the link between the progressive decrease with time of the core ionic thermal energy and the surface luminosity). Hydrogen envelopes are more opaque than helium ones, therefore the existence of an H layer on top of the helium envelope in DA WDs produces longer cooling times at fixed mass. This can easily be seen from Equation (7.9) which shows how an increase in opacity (the constant κ_0) induces higher central temperatures at a given luminosity.

It is important to notice that if the external hydrogen layer (whose thickness cannot be precisely predicted by stellar evolution theory) is above $\sim 10^{-4} M_\odot$, WDs with mass $\sim 1 M_\odot$ or above can experience some H-burning through the proton-proton chain at the hotter and denser bottom of the hydrogen envelope, when the WD is bright. This burning would produce an extra energy input and reduce the initial thickness of the hydrogen layer (see Section 7.2.3).

During the cooling of a WD at constant radius, its luminosity decreases hence its $T_{\rm eff}$ also decreases, and convection develops in the envelope, so that our simplified assumption of radiative transport is no longer valid. When convection is established, the boundary conditions and the superadiabatic gradient calibration will play an important role in determining the envelope structure and extension, and the cooling time. When $T_{\rm eff} < 6000\,{\rm K}$ boundary conditions based on grey $T(\tau)$ relationships are no longer good approximations, and the results from full non-grey model atmospheres have to be used ([91], [177]).

7.4.3 Detailed WD cooling laws

In this section we compare the cooling law (time spent along the WD cooling sequence as a function of the luminosity) obtained from full stellar evolution models of DA WDs, with the Mestel law. At high luminosities ($\log(L/L_\odot) > -1.0$) and high central temperatures, energy is lost from the star not only as radiation from the surface but also through neutrino emission from the degenerate core. Plasma neutrino emission, in particular, is very efficient during the hot WD phase, and the cooling is very fast, faster than the results given by the Mestel law (Equation (7.10)) because there is an additional energy loss. When the luminosity drops below $\log(L/L_\odot) \sim -1.0$ neutrino losses are negligible and the cooling slows down. After this stage cooling times become longer than the result from Equation (7.10) due to the energy gained from the slow contraction of the non-degenerate envelope, and the increase of c_V during the liquid phase. With the onset of crystallization latent heat and chemical separation provide extra energy sources with respect to the Mestel law (hence even longer cooling times) whereas at lower luminosities and temperatures, when $\theta_{\rm D}$ is

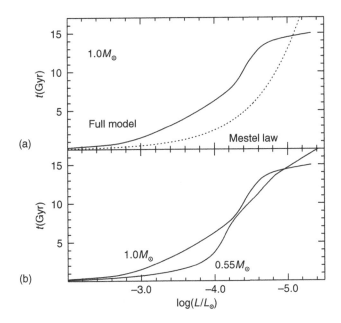

Figure 7.11 (a) This shows a comparison between the cooling law of a $1 M_\odot$ DA WD obtained from a full model computation (solid line) and the Mestel law (dotted line); (b) this compares the cooling laws obtained from a full model computation of two DA WDs with different masses

attained in the core, c_V starts to decrease, cooling times tend to approach the Mestel law again and eventually become shorter.

Figure 7.11 shows a detailed cooling law for two WDs of different masses, and also a comparison between the complete cooling law and the Mestel approximation (we assumed in this latter case an average homogeneous value of μ_i for the whole core) for the more massive object. The zero point of the cooling time t_{cool} is set by the moment the WD starts on its cooling sequence. If one wants to know the total age of the WD star, one has to include the evolutionary time of the progenitor until the WD formation.

The more massive WD has typically longer cooling times as predicted by the Mestel law; however, the core of the more massive WD reaches its Debye temperature θ_D at higher luminosities (because it is denser and θ_D increases with ρ) and, due to the decreasing c_V, the cooling times get increasingly fast, so that t_{cool} at low luminosities becomes shorter than the Mestel law and also shorter than the value for the lower mass, whose core will reach θ_D at lower L.

7.4.4 WDs with other chemical stratifications

Oxygen–neon WDs are the end-point of the evolution of stars with initial masses between $\sim 8 M_\odot$ and $\sim 11 M_\odot$. At the end of the central carbon burning (see Section 7.5.1) the

ONe core becomes electron degenerate; due to mass loss, the envelope mass is strongly reduced and the ONe core, surrounded by a thin non-degenerate envelope, settles on its cooling sequence. The masses of ONe WDs are between ~ 1.0 and $\sim 1.2 M_\odot$ ([78]) again the efficient mass loss preventing the core from reaching M_{ch}; typical chemical abundances in the core are about 64 per cent of oxygen (by mass) and 25 per cent of neon, and minor fractions of other elements, including some unburnt carbon.

The cooling of ONe WDs is similar to the CO case, but with shorter timescales (for a fixed envelope chemical composition) because of the reduced thermal content of the core. In fact, the atomic weight of a mixture of oxygen and neon is larger than the case of carbon and oxygen mixtures, and the number of ions in the core – hence the thermal content of the core at a fixed temperature – is lower, even accounting for their slightly higher masses. Moreover, the average charge of the ONe mixture is larger than the CO case and the onset of crystallization happens at higher temperatures.

Helium core WDs correspond to the electron degenerate He core of RGB stars that have lost their envelope through mass transfer in an interacting binary system or extremely high mass loss due to stellar winds or dynamical interactions in dense stellar systems. The expected masses of this class of WDs range from $0.15 M_\odot$ up to about $0.5 M_\odot$. Above the electron degenerate helium core there is a thin hydrogen envelope with mass thickness of the order of $10^{-4} M_\odot$. The qualitative properties of helium WDs are similar to the CO counterpart. The limiting mass M_{ch} is marginally larger because the electron molecular weight in the degenerate core is slightly smaller; radii are larger than for the CO counterpart, due to the fact that helium WDs have smaller masses, but also because μ_e in the core is smaller.

Their cooling is conceptually similar to the case of CO WDs. The main difference being that the atomic weight of He is lower than C and O by a factor of three and four, respectively, and therefore helium WDs contain more heat (at a given temperature) than carbon or oxygen cores, even accounting for their lower masses. Thus the helium core WDs are brighter at a fixed age than the CO counterpart of the same mass ([91]). Due to the lower charge of He with respect to the CO mixture, helium WDs do not crystallize within cosmological ages.

7.5 The advanced evolutionary stages of massive stars

In the previous chapters we have discussed the evolution of low- and intermediate-mass stars during the H- and He-burning phases. The evolution of stars more massive than M^{up} during these early stages (see Figure 7.12) is quite similar, the only exception being the sensitivity of the location on the HRD to the mass loss – which can be huge for these stars – as well as to the treatment of convection (the use of the classical Schwarzschild criterion versus the Ledoux one, see Section 5.4). In particular, the choice of the convection criterion affects the ratio between the time spent on the blue and the red side of the HRD by these stars. In more detail, stellar models based on the Schwarzschild criterion spend more time on the blue side of the HRD compared with models based on the Ledoux criterion.

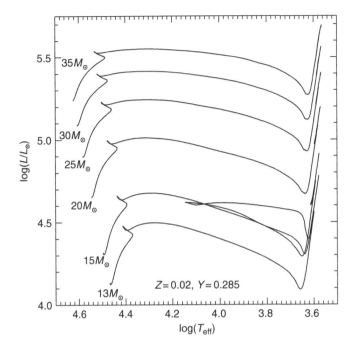

Figure 7.12   Evolutionary tracks of massive stars with different masses (courtesy of M. Limongi)

An additional peculiarity is that while, as in less massive stars, the mass of the He core is fixed both by the mass-size of the convective core during the central H-burning phase and by the amount of hydrogen processed by the H-burning shell during the core He-burning stage, the mass of the CO core is essentially fixed by the size of the convective core during the central He-burning phase. This is because the evolutionary phases subsequent to the core H-burning stage, are so fast that the He-burning shell has no time to advance significantly outwards. In addition, the chemical profiles of carbon and oxygen in the core are flat, being the result of the fully efficient mixing into the core during the central He-burning phase.

As has been known for some time ([5]) the core abundance of carbon is a fundamental quantity that strongly affects all the successive evolutionary stages, since it essentially determines the amount of carbon available for the core and shell C-burning. Needless to say, the amount of carbon is strongly sensitive to the nuclear cross section for the $^{12}C(\alpha, \gamma)^{16}O$ reaction – this is the reason for which the rate of this nuclear reaction adopted in stellar computations strongly affects the predicted final nucleosynthesis – and to the efficiency of all mixing processes which can occur inside the convective core.

In this section, the late evolutionary stages which precede the core collapse and the explosion of these stars as *Type II supernovae* will be discussed briefly. It is worth emphasizing that the evolution of massive stars in the advanced pre-supernova stages is quite a hot topic in stellar astrophysics and so many changes and improvements are expected in the next few years. For this reason, only the most important and well-established properties of these stars are discussed (see [234] for a detailed review).

In Table 7.1 are listed the main structural and evolutionary properties of stars with masses 15, 20 and $25M_\odot$ during the evolutionary phases from the H-burning stage to

Table 7.1 Selected quantities for $15M_\odot$, $20M_\odot$ and $25M_\odot$ stars with solar chemical composition ([130]). The evolutionary lifetimes refer to the core burning phases and M_{cc} corresponds to the maximum mass of the convective core, M_{Hec} and M_{CO} are the mass of the He core and CO core at the central exhaustion of H and He, respectively. M_{Fe} is the mass of the iron core of the last model computed by [130]

Quantity	$15M_\odot$	$20M_\odot$	$25M_\odot$
H-burning			
t_H (Myr)	10.70	7.48	5.93
$M_{cc}(M_\odot)$	6.11	9.30	13.77
$M_{Hec}(M_\odot)$	4.10	5.94	8.01
He-burning			
t_{He} (Myr)	1.40	0.93	0.68
$M_{cc}(M_\odot)$	2.33	3.63	5.23
$M_{CO}(M_\odot)$	2.39	3.44	4.90
C-burning			
t_C (10^3 yr)	2.60	1.45	0.97
$M_{cc}(M_\odot)$	0.41		
Ne-burning			
t_{Ne} (yr)	2.00	1.46	0.77
$M_{cc}(M_\odot)$	0.66	0.50	0.50
O-burning			
t_O (yr)	2.47	0.72	0.33
$M_{cc}(M_\odot)$	0.94	1.12	1.15
Si-burning: radiative core			
t_{Si-rad} (10^{-2} yr)	29.00	2.80	1.94
$\log T_c$	9.420	9.443	9.434
$\log \rho_c$	8.092	7.818	7.798
Si-burning: convective core			
$t_{Si-conv}$ (10^{-3} yr)	20.00	3.50	3.41
$M_{cc}(M_\odot)$	1.14	1.11	1.12
$M_{Fe}(M_\odot)$	1.43	1.55	1.53

the silicon burning stage as given by [130]. From these data, one can easily see how fast the advanced evolutionary stages are with respect the H- and He-burning lifetimes.

The timescale needed for these stars to move from the red side of the H–R to the blue side and vice versa, which is of the same order of magnitude as the Kelvin–Helmholtz timescale for the envelope, is about 10^2–10^3 years, the exact value depending on the stellar mass and on its location on the HRD. This has the important consequence that as the advanced evolutionary phases – successive to the core He-burning stage – are extremely fast, the surface evolution is, in some sense, frozen out, and so the surface luminosity and effective temperature do not change significantly until the star explodes. For this reason, as recognized by [5], the advanced evolutionary phases of massive stars can be described as the *neutrino-mediated Kelvin–Helmholtz contraction* of the CO core.

7.5.1 The carbon-burning stage

At the exhaustion of central helium, the CO core immediately starts to contract so that gravitation can supply the energy necessary to support the star. During this phase, the core temperature of these massive stars exceeds the value $\sim 5 \times 10^8$ K, entering into a region of the density–temperature (ρ – T) plane (see Figure 7.13) where

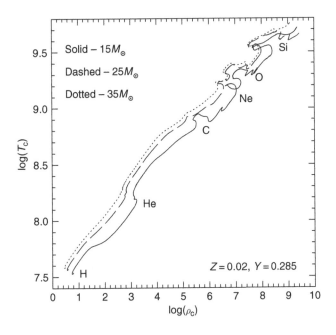

Figure 7.13 Evolution of the core of selected massive stars in the central temperature–central density plane. The loci corresponding to the onset of the various burning stages are marked (courtesy of M. Limongi)

neutrino losses from pair annihilation start to dominate the stellar energy budget. Radiative energy transport and convection are still important in determining the stellar structure, but it is the neutrino luminosity that mainly balances the energy generated by the gravitational contraction and by nuclear burning. To elaborate this point, in Figure 7.14 we show the evolution with time of the surface luminosity, neutrino luminosity and of the nuclear luminosity for two massive stars. From this figure, it is evident that up to C-ignition, the main energy losses are from the surface but later neutrino energy losses dominate. Since the energy produced by nuclear burning has to supply the energy lost from the stellar surface, the nuclear luminosity initially follows the behaviour of the surface luminosity, but from the ignition of carbon it closely

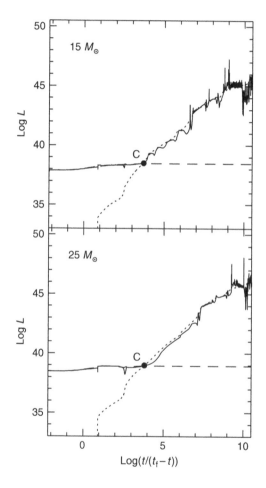

Figure 7.14 Behaviour with time of various luminosities (in erg s^{-1}) for two massive stars: the dashed line corresponds to the surface luminosity, the solid one to the energy produced by the nuclear burning, while the dotted lines refers to the neutrinos' energy. The solid dot marks the ignition of the core C-burning (courtesy of M. Limongi)

follows the neutrino luminosity. One should also note that for the $25M_\odot$ model, from the ignition of C-burning until the start of Ne-burning, the neutrino luminosity exceeds the nuclear luminosity. This means that in these stages, nuclear burning is not able to provide all the energy required by the star. As a consequence, the star rapidly contracts to produce the missing energy through the work of gravitation.

For stars with mass in the range 15–$30M_\odot$, C-burning ignites when the core temperature attains a value in the range $(0.3$–$1.2) \times 10^9$ K. The principal nuclear reaction is the fusion of two ^{12}C nuclei to produce a compound ^{24}Mg nucleus in a highly-excited nuclear state, which then decays through three channels:

$$^{12}C + {}^{12}C \longrightarrow {}^{24}Mg \longrightarrow {}^{23}Mg + n$$
$$\longrightarrow {}^{20}Ne + \alpha$$
$$\longrightarrow {}^{23}Na + p$$

The properties of the C-burning phase depend strongly on the amount of carbon in the CO core produced during the central He-burning phase. The abundance of C has a large influence on the size of the convective core during this phase. In fact, a convective core forms only if a positive energy generation rate (nuclear rate minus neutrino rate) exists, and a larger fraction of C implies more efficient nuclear burning. The interplay between the nuclear energy generation rate and the neutrino loss rate is the key to understanding the decrease of the mass of the convective core with increasing stellar mass. With increasing stellar mass, the core temperature increases and does the neutrino emission rate. As a consequence, at large enough stellar masses, the convective core during this stage will disappear. This occurrence can also break the correlation existing between the mass of the convective core and central abundance of C.

Once carbon is exhausted in the core, the burning shifts quietly to a shell. C-burning in the shell is characterized by the occurrence of one (or three, depending on the total mass and on details of the numerical simulations) convective episode(s), e.g. a convective shell appears. During the C-burning stage, particularly during the shell burning, a significant fraction of s-elements are also produced the most efficient neutron source being – during shell C-burning – the reaction ^{22}Ne$(\alpha, n)^{25}$Mg.

7.5.2 The neon-burning stage

Following carbon burning, the stellar core contains mainly ^{16}O (~ 0.7 by mass fraction) ^{20}Ne (~ 0.2–0.3 by mass fraction) and ^{24}Mg. It is worth noting that oxygen has the smallest Coulomb barrier, but before the temperature necessary for O-burning is achieved, the reaction ^{20}Ne$(\gamma, \alpha)^{16}$O (which is endothermic) is allowed thanks to the presence of high-energy photons. The α-particle separation energy of ^{16}O is almost a factor of two larger than that requested for ^{20}Ne, so ^{20}Ne is the more fragile nucleus. Even if the ^{20}Ne$(\gamma, \alpha)^{16}$O reaction is endothermic, the whole set of exothermic nuclear reactions occurring during this evolutionary phase makes the Ne-burning an exothermic process. The net effect is to produce mainly ^{16}O and ^{24}Mg at the expense of ^{20}Ne.

For stars in the same mass range considered previously for C-burning, core Ne-burning occurs in the temperature range $(1.2–1.9) \times 10^9$ K. A characteristic of this burning phase is that a convective core is always formed, regardless of the stellar mass, but it exists only for the short time during which the nuclear burning rate overcomes neutrino energy losses. Thus core Ne-burning occurs partially in a radiative region and it lasts only for a short time.

As the central abundance of neon is vanishing, Ne-burning shifts into a shell whose evolution is limited to the short time intervals between two consecutive core burning stages. In fact, the various central burning stages are so fast (see Table 7.1) that the neon shell has no time to significantly advance through the star.

7.5.3 The oxygen-burning stage

At the end of core Ne-burning, the chemical composition of the core is mainly ^{16}O, ^{24}Mg and ^{28}Si. In the mass range $15–30M_\odot$, O-burning ignites when the core temperature attains a value in the range $(1.5–2.6) \times 10^9$ K. The oxygen fusion reaction produces a compound ^{32}S nucleus that can decay following one of the following channels:

$$^{16}O + {}^{16}O \longrightarrow {}^{32}S \longrightarrow {}^{31}S + n$$
$$\longrightarrow {}^{31}P + p$$
$$\longrightarrow {}^{30}P + {}^2D$$
$$\longrightarrow {}^{28}Si + \alpha$$

At low temperatures the channel producing deuterium is inhibited, whereas at the high temperatures characteristic of oxygen burning, all the produced deuterium is promptly photo-disintegrated into a neutron and a proton.

Oxygen burning always occurs in a convective core, owing to the nuclear burning rate being able to overcome the neutrino losses. Its mass is $\sim 1M_\odot$, regardless of the total stellar mass. The O-burning lifetime decreases as the total mass increases, a consequence of the fact that more massive stars burn oxygen at higher core temperatures. In any case, the core O-burning lifetime is so short that at this stage – and from now on – the various burning shells present in the star remain practically frozen out until the final core collapse. The only exception to this general rule (see also previous discussion) is represented by the Ne-burning shell, which being located in a zone strongly affected by gravitational contraction and, in turn, by heating during the time between two successive core burning processes, can move outwards.

Core O-burning produces neutron-rich nuclei such as ^{30}S, ^{35}S and ^{37}Cl as a consequence of the large efficiency of weak interactions such as ^{30}P$(e^+, \nu)^{30}$S, ^{35}Cl$(e^-, \nu)^{35}$S, and ^{37}Ar$(e^-, \nu)^{37}$Cl. The production of neutron-rich nuclei affect the so-called level of *neutronization* of the stellar matter, that is an indicator of how neutron rich the constituents of the gas are, and is related to the difference between the number density of neutrons and the number density of protons, both bound in nuclei and free. The higher the number density of neutrons with respect to protons, the higher the level of neutronization.

At the exhaustion of oxygen in the core, the burning moves outwards and a shell is activated. As for the C-burning shell, this O-burning shell can experience one (or two) convective episode(s). Near the end of the core O-burning, another interesting feature is that as the central temperature is very high, $T \sim (2.5\text{--}2.8) \times 10^9$ K, a number of isolated *quasi-equilibrium clusters* of nuclei exist. These are groups of nuclei coupled by strong electromagnetic interactions which occur at rates nearly balanced by their inverse, and the time derivative of their abundances is approximately zero; for example the rate of the $^{29}\text{Si}(\text{p, }\gamma)^{30}\text{P}$ reaction is equal to the rate of the reversed reaction $^{30}\text{P}(\gamma,\text{ p})^{29}\text{Si}$. During the subsequent evolutionary phases, as the temperature increases, these initially separate groups merge into a single one and involve more chemical species.

7.5.4 The silicon-burning stage

When the central temperature attains the value $T \sim 2.3 \times 10^9$ K, silicon-burning can start. Silicon-burning does not occur predominantly as a nuclear fusion reaction, but it occurs in a unique fashion: a portion of ^{28}Si experiences a sequence of photo-disintegration reactions by the chain

$$^{28}\text{Si}(\gamma, \alpha)^{24}\text{Mg}(\gamma, \alpha)^{20}\text{Ne}(\gamma, \alpha)^{16}\text{O}(\gamma, \alpha)^{12}\text{C}(\gamma, 2\alpha)\alpha$$

The available alpha particles, protons and neutrons add onto the remaining silicon and heavier nuclei, gradually increasing the mean atomic weight of the core, until it is dominated by nuclei of the iron group. Si-burning is strongly affected by the level of neutronization in the stellar core, as this largely affects the nucleosynthesis of the various nuclear species. It is worth noting that, since the oxygen-burning ignition, the neutron density increases significantly as a consequence of the increasing efficiency of weak interactions.

Numerical simulations ([130]) disclose that the core Si-burning stage can be divided in two distinct episodes: a radiative one and a later convective one. During the first radiative episode, depending on the initial total mass, ^{28}Si is largely destroyed and ^{30}Si produced (the lower the mass, the more complete the ^{28}Si destruction is). At a certain moment, when the nuclear energy release overcomes the neutrino energy losses (which in this phase are no longer dominated by pair production but receive a large contribution from weak interactions – mainly electron captures on heavy elements) a convective core appears. The convective core mass at its maximum is of the order of $\sim 1 M_\odot$, being practically independent of the stellar mass. During this convective core episode, regardless of the stellar mass, the ^{28}Si abundance increases. This is because the convection penetrates into a region where the abundance of this element is quite high, due to the previous core O-burning, and it becomes the most abundant Si isotope. As a consequence of Si-burning, the core of the star is composed mainly of ^{56}Fe and ^{52}Cr, with similar abundances by mass.

At the exhaustion of Si in the core, the burning moves into a shell surrounding the Si-exhausted core. This Si-burning shell usually experiences recurrent convective episodes, the first ones always extending at most up to the border of the previous convective core. However, the final episode(s), whose exact behaviour strongly depend(s) on the stellar mass, can extend beyond the border of the former convective core. This occurrence is very important as the maximum extension of this convective region fixes the location of the boundary between the part of the star that has experienced a strong neutronization process (inside the boundary) and the one that has not been affected significantly (if any) by this process (outside the boundary). This boundary determines the mass of the so called *iron core*: this is a fundamental quantity for the subsequent explosive phase, determining the mass within which the shock wave (see Section 7.5.5) loses the largest part of its energy through the photo-disintegration of heavy elements. Thus, the lower the mass of this iron core is, the larger the possibility of obtaining an explosive outcome.

At the end of Si burning, all nuclear reactions between different elements become fully balanced by their inverses. The last reaction to achieve equilibrium is the 3α reaction, which finally occurs at a rate balancing that of carbon photo-disintegration. When this occurs, the stellar material has achieved the *Nuclear Statistical Equilibrium* (NSE).

7.5.5 The collapse of the core and the final explosion

Once NSE has been achieved in the stellar core, the star has really reached the very end of its evolution. Only the final explosion is still missing. The physical process which sets off the final collapse of the inner core can easily be understood by considering the Virial theorem. It has already been shown in Section 3.1.8 that a *stable* physical configuration is allowed until the specific heat ratio ($\gamma = c_P/c_V$) is larger than 4/3. In the core of a massive star after reaching NSE, both photo-disintegration processes and relativistic effects decrease the value of γ below the critical value 4/3, and as a consequence the core becomes dynamically unstable and begins to collapse.

At the beginning, the core collapse is not a dramatic event. It occurs on a Kelvin–Helmholtz timescale due to the huge neutrino flux which carries away the core binding energy. However, soon after, the occurrence of two instabilities causes the collapse to accelerate strongly: (1) due to the increase in density, the neutronization process becomes increasingly efficient, this process removing free electrons which are the main contributors to the pressure, and (2) photo-disintegration becomes more efficient, producing a large number of free α particles. The binding energy of the core resulting from this new composition, is lower than before, and so the star cannot gain enough energy from its contraction. As a consequence, the collapse accelerates. The collapse process is very short, lasting only a few milliseconds, the exact value depending on the initial core density.

When the core density becomes of the order of 10^{14}g cm^{-3}, which is the density of a neutron star, the material becomes incompressible and the collapse of the central regions stops. As the compression phase ends abruptly, the core rebounds and the (supersonic) impact with the external layers that are continuing to contract produces a shockwave. In a perfectly elastic collision, the energy of the infalling outer portion of the core would bounce back after reflection to its position before the collapse. If one compares this energy ($\sim 10^{52-53}$ erg) with the binding energy of the outer envelope ($\sim 10^{50-51}$ erg) it seems possible that the rebounding core could be responsible for the expulsion of the whole stellar envelope, the so-called *prompt explosion scenario*. There are, however, two reasons why this is impossible. As the shock wave moves through the infalling layers, it heats them up and induces photo-disintegration; owing to this process the shock wave loses about 10^{51} erg for each $0.1M_\odot$ crossed. In addition, the emission of neutrinos from behind the shockwave is a quite efficient shock-wave cooling process.

For a successful explosion, an additional energy source is required. As first proposed by [59], it is now commonly assumed that this essential energy source is provided by the neutrino energy deposition (the *neutrino-powered explosion*). Due to the high density in the collapsing core, the mean free path of neutrinos becomes comparable to the core radius. This means that the neutrinos produced via the neutronization processes cannot escape from the star without interacting. With increasing density, the material in the collapsing core becomes more and more opaque to neutrinos, which experience multiple scattering processes. In the inner core, the density is so large that the neutrino diffusion velocity becomes less than the velocity of the collapse. Under these conditions, the neutrinos cannot leave the star, they are trapped. The point at which the neutrino diffusion velocity equals the velocity of the collapsing core is defined as the *neutrino-trapping surface*: below this point neutrinos are forced to deposit their own energy in the material, whereas in the more external layers they can diffuse outwards until they reach the so called *neutrino photosphere*. Similar to the photons photosphere, this represents the surface at which neutrinos have a probability exactly equal to one of experiencing an interaction. This process is considered to be responsible for providing the outgoing shock wave with the energy necessary to produce an explosive event with the characteristics of a *core-collapse* – or alternatively Type II supernova.

The details of the shock propagation inside the structure are very complicated, due to the interplay between the neutrino energy flux and convection as well as the occurrence of mixing instabilities such as the Rayleigh–Taylor instability. Due to the uncertainties associated with the explosion mechanism and the related physical processes, it is a difficult task to predict firmly the mass of the remnant and the products of the explosive nucleosynthesis, given the mass of the progenitor. Roughly speaking, it is commonly estimated that stellar progenitors less massive than $\sim 25M_\odot$ produce as a remnant a neutron star whose mass should be about that of the iron core at the beginning of the core collapse. More massive stars should produce black holes.

The scenario outlined above holds for massive stars with mass less than about $100M_\odot$. More massive structures, and in general supernova progenitors which produce a He core with mass larger than about $40M_\odot$, can experience a different explosive process, one sustained by nuclear burning ([232]). After He-burning, due to the peculiar thermal conditions, the production of electron–positron pairs can occur. This process significantly reduces the value of the specific heat ratio below 4/3 and a runaway collapse develops. The advanced burning processes described previously are not able to halt the collapse which has become dynamic. The final fate of these *pair-instability supernovae* is to experience a prompt explosion (due to the huge kinematic energy of the rebounding core) and to produce a black hole. It is worth emphasizing that this kind of supernovae is receiving remarkable attention in the literature, as it is a common belief that super-massive stars could be the major component of the primeval stellar population, the Population III.

Concerning the chemical composition of the material burnt during the explosion of a Type II supernovae, one has to note that this burning is explosive, and the conditions for explosive nucleosynthesis are mainly determined by the maximum value of the temperature achieved during the passage of the shock wave and by how long the material remains at this temperature. The products of explosive burning of O, Si, Ne and C are quite similar to those produced during the corresponding hydrostatic burning stages, the only significant difference being that the isotopic ratio between different species is modified by the different burning timescales. It is also worth noticing that during the explosion, the presence of a huge flux of neutrinos can produce a transmutation of certain elements – obviously, of those elements present in large amount, being the cross section of these reactions very small. In particular, detailed numerical simulations have shown that a significant production of ^{7}Li, ^{11}B, ^{19}F and ^{26}Al is possible via neutrino processes. Core collapse supernovae, essentially the most massive ones, are also considered a possible site for the production of r-elements, i.e. those elements which are produced by r-processes which require a very large neutron flux.

A firm description of the chemical composition of the material ejected into the interstellar medium by the explosion as well as the exact abundances of the various elements is very difficult, due to our poor understanding of the details of the explosion. In particular, for each given supernova progenitor, we are not able to properly estimate the location inside the star of the point beyond which the whole envelope is ejected during the explosion; the so-called *mass-cut* parameter. Present evaluations of supernovae yields (see for instance [129]) are based on a free-parametrization of this quantity. Without entering into details, one can safely estimate that core-collapse supernovae are the main contributors to the production of α-elements (O, Ne, Mg, Si, S, Ca, Ti) and helium.

7.6 Type Ia supernovae

A supernova event is classified spectroscopically as a Type I supernova if it does not show any hydrogen feature in its spectrum. The Type I supernova class is further divided in three additional sub-classes: Type Ia, Ib and Ic, according to the features observed in their early spectra, i.e. the spectra obtained a few days after the explosion. In more detail, Type Ia supernovae are characterized by the presence of a clear Si II absorption line around 6150Å and their late spectra show many lines associated with Fe emission; Type Ib and Ic supernovae do not show this Si II absorption line and are distinguished according to the presence or not, respectively, of moderately strong He I lines around 5876Å. The favoured scenario for Type Ib and Ic supernovae is that both are a consequence of the explosion of massive stars that, owing to mass loss by strong stellar winds and/or mass exchange in a binary system, have lost the whole H-rich envelope.

To date, there is no doubt that Type Ia supernovae originate from CO WDs which have been able to accrete mass from a companion in a binary system, until the critical mass for triggering a thermonuclear runaway is achieved.

Even if Type Ia supernovae are, perhaps, the most important distance indicators at large distances (this issue is addressed in Section 7.6.3) one has to bear in mind that a number of outstanding issues related both to the progenitor evolution and the explosion mechanism(s) still remain to be solved. They include: (1) the double-degenerate versus the single-degenerate scenario, which means that it has to be fully understood if a Type Ia supernovae originates from the binary evolution and coalescence of two degenerate WDs or from the mass exchange between a WD and a non-degenerate star (an MS or RGB star); (2) the exact critical mass required to trigger the thermonuclear runaway, i.e. Chandrasekhar mass models versus sub-Chandrasekhar mass ones; (3) the explosive mechanism(s). In any case, it is largely agreed that a thermonuclear runaway is at the origin of the disruption process. In this section all of these topics will be briefly addressed. For more details, we refer the interested reader to several reviews addressing these important topics, e.g. [232] and [126].

7.6.1 The Type Ia supernova progenitors

As pointed out earlier, stars less massive than M^{up} develop a CO core, becoming WDs with luminosity decreasing with time as the core cools. In a close binary system, however, the WD evolution can be remarkably different. When the companion star expands, as will occur at the end of the core H-burning phase or during the RGB and the AGB, it can efficiently transfer material to the WD. Depending on its initial mass, on the accretion rate and the chemical composition of the accreted matter, the accreting WD will face a number of different evolutionary channels.

The accretion of H-rich matter onto WDs is the commonly adopted scenario for the production of classical and symbiotic novae. It was also suggested ([230]) as a promising evolutionary channel for Type Ia supernovae. The observed absence of H lines in Type Ia supernovae spectra, which does not necessarily imply the complete lack of H in the progenitor, has motivated the search for progenitor systems in which the H-rich material has been lost during the prior evolution. One scenario ([105]) that has emerged consists of a primordial system formed by two intermediate-mass stars that evolve through a series of common-envelope episodes (i.e. a process during which the envelopes of the two stars are physically in contact, and a sizeable amount of envelope mass can be lost from the system) into a final system of two very close CO WDs whose combined mass is larger than the Chandrasekhar mass limit. A subsequent merging of the two WDs will occur if there is sufficient angular momentum loss through the emission of gravitational waves. As the resulting object is more massive than the Chandrasekhar mass, the object cannot stay in hydrostatic equilibrium and will be disrupted. This is the so-called *double degenerate* (DD) scenario.

There are several difficulties in modelling this scenario, mainly related to the necessary fine tuning of the properties of the progenitor system (such as initial masses, initial separation and efficiency of the mass-loss process during the common-envelope episodes) and to the explosive mechanism (quiescent C-burning could result as a consequence of the merger, resulting in turn in a core-collapse supernovae with observational properties different to those of a Type Ia supernovae). However, the most widespread doubts about the DD scenario arise from the lack of empirical evidence for DD systems ([20]) with a combined mass greater than the Chandrasekhar mass. Until recently, this scenario has had an undisputable advantage: it naturally explained the empirical evidence that *all* Type Ia supernovae have almost the same observed spectra and light curves. However, in the last decade ([127]) it has became clear that Type Ia supernovae are, in fact, not as homogenous as generally believed. The morphology of the light curve and the peak brightness can show remarkable differences between various supernovae (this issue is discussed in Section 7.6.3). These latest observations have stimulated the investigation of alternative evolutionary scenarios.

In the so-called *single degenerate* (SD) scenario, accretion of H- or He-rich material at an appropriate rate from the companion star onto the WD would be responsible for increasing the mass of the WD above the Chandrasekhar limit, through either quiet or violent burning adding fresh CO-rich material to the WD. Accretion of H-rich matter onto massive ($\sim 1 M_\odot$) WDs has been extensively investigated. It has been found that for quite low accretion rates, $dM/dt \leq 10^{-9} M_\odot$ yr^{-1}, the accreting WD experiences strong He-pulses – stronger than those occurring during the AGB phase – which cause so much mass to be ejected during the thermonuclear runaway that the time needed to bring the WD to the Chandrasekhar mass will exceed the Hubble time. In this case, the accreting WD behaves as a strong *Nova* object. The physical reason for this behaviour is that, due the very low accretion rate, H-burning on top of the He-rich layer proceeds so slowly that the helium shell builds up as a strongly electron degenerate layer. At larger accretion rates, an explosion can occur. The resulting peculiar explosive nucleosynthesis and the large abundance of H which should be accumulated on the surface of the WD and, in turn, be observable in the early spectra of the supernova are, however, incompatible with Type Ia supernova observations. In spite of this, current uncertainties on the accretion and mass loss during the explosive nova-like outbursts and on the explosive mechanisms, do not yet allow definitive conclusions on this issue. It is important to notice that it is improbable to find WDs with initial mass (before accretion) larger than about $\sim 1.1 M_\odot$. This is because the progenitor would have been able to ignite C-burning before a definitive CO core WD configuration is attained, although a possible way out, accounting for core rotation, has been proposed by [66]).

In recent years, the search for possible progenitors of Type Ia supernovae has focused on CO WDs of low-mass accreting H-rich material at an appropriate rate. Numerical computations show that it is possible to obtain a thermonuclear explosion that delivers some 10^{51} erg, which is the typical energy released by a Type Ia supernova, without the requirement that the mass of the accreting WD exceeds the Chandrasekhar limit. This is the so-called *sub-Chandrasekhar* scenario. It was shown as early as 1977 ([141]) that this outcome is possible when the total mass of a very cold WD accreting He at an appropriate (see below) rate exceeds ~ 0.65–$0.8 M_\odot$. If He is accreted onto a CO WD at a rate $\sim 3 \times 10^{-8} M_\odot$ yr^{-1}, a violent He-ignition occurs after the accretion of $\sim 0.15 M_\odot$ of He-rich material, nearly independent of the initial mass of the WD. If the WD was also initially massive enough, this He-burning could result in a detonation, and an inward moving compression wave would then lead to the detonation of carbon in the core ([233]).

The development of a critical He layer above the CO core can be achieved as a consequence of the burning of accreted H-rich material, as well as of the direct accretion of helium from a He-rich star. The mass of the critical He layer, as well as the violence of the explosion differ in the two cases because of the injection of energy from the H-burning shell into the accreted layer, which modifies the thermal properties of the He-rich layer.

Figure 7.15 shows a diagram summarizing the different outcomes experienced by WD accreting H-rich matter, as a function of the initial mass and of the accretion rate. For H-accretion rates near to, or larger than, the so-called *Eddington* limit[1] the accreted matter achieves an expanded configuration, typical of an RGB star ([101]).

As the accretion rate is lowered, a range of accretion rates is encountered where H is burned at the base of the accreted layer at the same rate as it is accreted. Lowering the accretion rate still further, a regime is encountered where recurrent H-shell flashes take place. The accretion rate at the borderline between steady-state burning and flashing behaviour decreases as the initial mass of the WD is decreased. As the accretion rate is lowered below the borderline, the flashes become stronger and stronger, changing from mild, non-dynamical events to strong, nova-like outbursts. During these violent, nova-like outbursts, a sizeable amount of material is ejected from the system (through a combination of dynamical acceleration, wind mass loss and, perhaps, common envelope action). For about two-thirds of the observed events for which spectroscopic measurements are available, huge overabundances of elements heavier than helium are found in the ejecta. This is a clear evidence that the mass of the CO WD is decreasing with time.

---

[1] The Eddington luminosity limit represents the maximum surface luminosity of a star with the outer layers in radiative equilibrium: $L_{Edd} \propto (M/\kappa)$ where M is the total stellar mass and κ is the mean radiative opacity in the external layers. Larger luminosities imply a radiative pressure too strong to be balanced by the gravitational force.

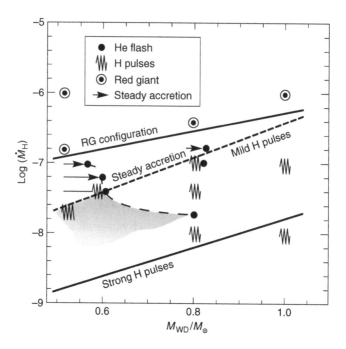

Figure 7.15 Accretion rate – initial WD mass diagram for H-accreting stars, from [41]. Various symbols mark the different possible outcomes of the accretion process, that depend on the initial WD mass and the accretion rate. The shaded area marks the region where, according to [151], the H-accreting WD would experience an explosive event resembling a sub-Chandrasekhar supernova. (courtesy of L. Piersanti)

Extensive evolutionary investigations of the possible outcomes experienced by low-mass H-accreting WDs were performed by [41]. The main result was that within the explored parameter space, H-accreting WDs were not able to reach the Chandrasekhar mass or to experience a sub-Chandrasekhar detonation in their He layers. However, it was conjectured that WDs of small mass, accreting hydrogen to rates which place them in the mild H-flash regime, but approaching the strong H-flash regime, might eventually experience a dynamical He-shell flash. This scenario was later validated ([151]) by following the long-term evolution of a WD of initial mass equal to $\sim 0.516 M_\odot$ accreting H at a rate of $\sim 2 \times 10^{-8} M_\odot$ yr^{-1}, whose final outcome will be an explosive event resembling a sub-Chandrasekhar supernova (note the shaded area in Figure 7.15).

To discriminate between the DD and SD scenarios for both Chandrasekhar and sub-Chandrasekhar events, a fundamental diagnostic is provided by photometry and spectroscopy: if any H or He were to be detected, the DD model could be ruled out. Even if hydrogen has never been observed in Type Ia supernovae spectra, the upper limit to the H abundance ($\sim 10^{-4} M_\odot$) is still too large to discard the SD scenario. Therefore, no firm conclusion has yet been achieved.

7.6.2 The explosion mechanisms

Once an explosive thermonuclear runaway is ignited, it induces the thermonuclear disruption of the WD. The properties of this explosion strongly depend on how the shock wave propagates inside the WD, bearing in mind that these characteristics are different for the case of a Chandrasekhar progenitor or sub-Chandrasekhar progenitor. Due to the difficulties in treating several physical processes involved in the explosion as well as to the numerical difficulties in performing a detailed modelling of the explosion, we are still faced with non-negligible uncertainties on what really occurs in an exploding WD.

The fundamental properties which have to be properly accounted for by an explosion model are: (1) the amount of energy which has to be delivered, which is of the order of $\sim 10^{51}$ erg; (2) the nucleosynthesis of a sizeable amount (in the range 0.4–$1.0 M_\odot$) of ^{56}Ni, necessary for powering the energy of the observed light-curves; (3) the production of a significant amount of intermediate-mass elements moving with an expansion velocity of the order of 10 000 km s^{-1} when the supernova is at its maximum brightness.

In the case of a Chandrasekhar WD progenitor, carbon burning is ignited in the central regions of the structure at a density of the order of 10^9 g cm^{-3}. Owing to the extremely strong level of electron degeneracy, the burning of carbon is explosive and it causes the incineration on the material into Fe-peak elements. Soon after, the explosive burning flame starts to propagate outwards; since its velocity is subsonic this process has the characteristics of a deflagration. During its crossing of the structure, the deflagration wave is affected by several kinds of instabilities, such as the Rayleigh–Taylor instability. The properties of the flame propagation are strongly influenced by these instabilities, and on how their presence is managed in the numerical simulations (which are quite complicated and not yet completely reliable). Behind the flame front, the stellar material undergoes explosive nuclear burning of Si, O, Ne and C. The exact composition of the nuclear processed material depends strongly on the maximum temperature achieved inside the burning front, which is strongly related to the density of the layers crossed by the explosive wave. For density values in the range $\sim 10^{10}$–10^6 g cm^{-3}, the chemical composition of the burnt material ranges from elements of the Fe-peak (mainly ^{56}Ni) to intermediate-mass elements such as S and Si, and to C and O in the more external layers.

During its propagation, the explosive flame encounters regions whose density is decreasing as a consequence of the expansion of the WD structure. So it is evident that the densities encountered by the shock wave and, in turn, the resulting explosive nucleosynthesis depend on the flame velocity, which is indeed quite uncertain. Since sizeable changes to the explosive nucleosynthesis, mostly in the ^{56}Ni mass, can be achieved by changing the velocity of the outgoing flame front, various mechanisms for changing the flame velocity have been investigated in order to reproduce the empirical evidence better.

It has been suggested that the characteristics of the propagating flame could change from deflagration to a detonation, i.e. achieve a supersonic velocity, due to a compression shock resulting from the crossing of the low-density outer layers. This process could be of some help in explaining the existence of extremely bright supernovae such as SN1991T, as they form a larger amount of ^{56}Ni.

In contrast, in the so-called *delayed detonation* scenario, the initial speed of the flame is so low that only a very little amount of ^{56}Ni ($\sim 0.1 M_\odot$) is produced, and the transition from deflagration to detonation occurs at densities of the order of $\sim 10^7$ g cm^{-3} when the star has already significantly expanded. This scenario allows a better reproduction of the typical velocities of intermediate-mass elements observed in the spectra of typical Type Ia supernovae.

In the *pulsating delayed-detonation* scenario, it is assumed that the transformation to a detonation does not occur during the propagation of the deflagration wave. If the initial flame speed is low enough, burning is quenched by the expansion of the outer layers before the binding energy of the structure becomes positive. As a consequence, the WD remains bound and experiences a strong pulsation. At maximum compression, the burning can be reignited and the speed of the new-born burning flame is now supersonic, so it becomes a detonation. As in the delayed-detonation, this scenario predicts that the detonation flame starts at low density ($\sim 10^7$ g cm^{-3} or lower) so allowing better reproduction of the velocities of intermediate-mass elements in the spectra.

In the case of a sub-Chandrasekhar progenitor for Type Ia supernovae, it is commonly assumed that the central explosive C-burning, which should have the characteristics of a detonation, is induced by the shock wave generated by the He-detonation on top of the degenerate CO core.

From these considerations, it is evident that a detailed prediction of the nucleosynthesis produced during the Type Ia supernovae explosion is strongly dependent on several issues which still have to be fully investigated. Nevertheless, the nucleosynthesis expected for a pure deflagration model has been calculated by many groups (see [215] and references therein). The dominant products are iron peak nuclei. Important minor synthesis of lighter elements such as silicon, sulphur, argon and calcium, also occurs.

7.6.3 The light curves of Type Ia supernovae and their use as distance indicators

Light curves, e.g. the change of brightness with time, are one of the main sources of information on supernovae. In recent years, a large effort has been devoted to collecting Type Ia supernova light curves in many photometric bands (we refer the reader to Chapter 8 for a detailed discussion on the photometric systems and magnitudes) from the near-UV to the near-IR. In order to show the dependence of the light curve shapes on the adopted photometric bands, typical light curves in different bands are shown in Figure 7.16. The most commonly used feature of Type Ia supernova light curves is the value of the maximum brightness in the optical *B* band.

The rising time of the light curve is very fast and, for this reason, very few observations have been collected so far. It is clear from the existing empirical data

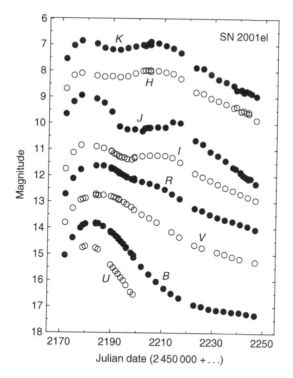

Figure 7.16 *UBVRIJHK* light curve of SN2001el (data from [119]). The different bands are offset for the sake of clarity. Note that maxima in red and near-infrared bands occur earlier than in the *UBV* bands (courtesy of K. Krisciunas)

that they rise to maximum *B* magnitude in about 18 days. The rise is very steep, with about ~ 0.5 mag day^{-1} increase until about ~ 10 days before maximum.

The maximum bolometric magnitude does not coincide with the maximum *B* magnitude, anticipating it by about 5 days, in coincidence with the maximum brightness in the near-IR bands. The colours evolve very quickly and non-monotonically around the maximum: they appear almost constant during the rising phase, but change from blue, $(B - V) \sim -0.1$ at 10 days before, to red, $(B - V) \sim 1.1$ about 30 days after the maximum *B* magnitude

After reaching the maximum *B* magnitude, the supernova starts to fade slowly and becomes increasing fainter in the near-UV and optical *B* bands. At longer wavelengths, the behaviour is different: in the *V* band the fading slows down after about 20 days, in the *R* band a plateau appears, and in the *IJHK* bands a secondary maximum occurs. The rise of this secondary maximum, from the dip to the peak, is quite large, being of the order of ~ 0.5 mag in the near-IR. The explanation of this empirical evidence is still uncertain.

After about 50 days, regardless of the considered photometric band, the supernova fades monotonically with time. The decline rate in the interval ~ 50–120 days, is

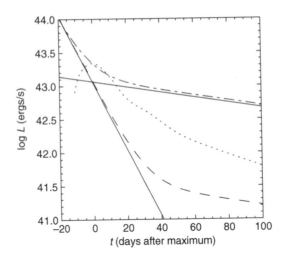

Figure 7.17 Bolometric light curve of a Type Ia supernova. The dotted line is the observed bolometric light curve. The solid lines represent the ^{56}Ni and ^{56}Co decay lines. The dashed line represents the expected light curve if all γ-rays from the radioactive decay escape the SN ejecta and only positrons are converted into optical emission. The dash-dotted line shows the light curve expected in the case where all photons resulting from the radioactive decay are fully trapped in the ejecta ([60]) (courtesy of B. Leibundgut)

almost the same for all Type Ia objects and is: $\sim$0.014 mag day^{-1} in the B band, $\sim$0.028 mag day^{-1} in the V band, and $\sim$0.04 mag day^{-1} in the I band.

Due to the significant dependence of the light curve characteristics on the adopted photometric band, which is an evidence of the interaction between the expanding ejecta and the outgoing radiative flux, it is more meaningful to analyse the behaviour with time of the bolometric light curve (see Figure 7.17). The peak of the bolometric light curve is slightly asymmetric, with the rising branch steeper than the fading branch. The rise to and fall from maximum is slowest for the brightest SNe (see the discussion below). In the later stages, the decline rate is similar for all supernovae (but some exceptions do exist) amounting to $\sim$0.026 mag day^{-1} in the time interval $\sim$50–80 days after maximum.

The most important parameter is the luminosity at maximum, as it is the essential ingredient for the use of these objects as distance indicators. Since their identification, Type Ia supernovae have been considered to be good standard candles ([128]) due to the constancy of their maximum B magnitude. However, with the increasing amount of observational data and with the more accurate distance evaluations of galaxy hosting supernovae of this type, it has became evident that some objects (up to $\sim$30 per cent of the sample) show sizeable deviations from the common behaviour, appearing as sub-luminous or super-luminous. The scatter of the maximum B magnitude is in the range 0.2–0.4 mag. Nevertheless, despite these differences, Type Ia supernovae have a few invariants in their appearance: the most important one is the linear correlation between the absolute magnitude at maximum and the decline rate Δm_{15} (the magnitude difference in B between the time of the maximum and 15 days later) discovered by [150]

$$M_B = -21.726 + 2.698\Delta m_{15}(B)$$

$$M_V = -20.883 + 1.949\Delta m_{15}(B)$$

$$M_I = -19.591 + 1.076\Delta m_{15}(B)$$

The application of these (or similar) correlations has allowed the continued use of Type Ia supernovae as distance indicators. Different implementations of this correlation between brightness at maximum and light curve shapes do not provide the same correction to the maximum luminosity (for a detailed discussion on this important issue we refer to reviews by [126] and [127]) with offsets of the order of 0.25 mag.

As envisaged by [58], the light curves of these supernovae are powered by the decay of radioactive ^{56}Ni and its radioactive daughter nucleus ^{56}Co: ^{56}Ni is synthesized during the explosive burning, and decays through electron capture with a half-life time of 6.1 days to ^{56}Co. ^{56}Co decays mostly (~ 80 per cent) via electron capture and also (~ 20 per cent) through β^+ decay to ^{56}Fe, with a half-life time of 77 days. The final shape of the light curve (see Figure 7.17) depends on the balance between the release of photons generated as γ-rays in the radioactive decays and the fraction of γ-rays that escape (a fraction that increases with time as the ejecta expand and become more and more optically thin). During the late phases, it also depends on the fraction of positrons (coming from the minor channel of the ^{56}Co decay) that are annihilated in the ejecta after losing their kinetic energy.

7.7 Neutron stars

Neutron stars are the remnant of the supernova explosion of massive stars, with initial mass between $\sim 11 M_\odot$ and $\sim 25 M_\odot$. Neutron star masses, determined by the evolutionary history of the supernova progenitor, are generally above $\sim 1.2 M_\odot$, up to $\sim 2.5 M_\odot$; the average mass of neutron stars detected in binary systems is of about $1.4 M_\odot$. A typical $1.4 M_\odot$ neutron star has a radius of 10–15 km, central density of the order of 10^{14}–10^{15} g cm^{-3} and temperatures below $\sim 5 \times 10^6$ K.

The main source of the pressure needed to counterbalance the gravitational force is the degenerate neutrons, since electron degeneracy cannot support objects more massive than M_{ch}. Being degenerate objects, neutron stars follow a mass – radius relationship like WDs; reasonable approximations are given by

$$R = 14.6 \left(\frac{\rho_c}{10^{15} \text{g cm}^{-3}} \right)^{-1/6} \text{km}$$

$$M = \left(\frac{15.2 \text{ km}}{R} \right)^3 M_\odot$$

where ρ_c is the central density. In the following we will give just a very brief summary of the main properties of neutron stars. A thorough discussion can be found in [199] and [238].

In order to understand qualitatively the internal structure of neutron stars, we can notice that at densities of the order of $10^7 \, \text{g cm}^{-3}$ and temperatures typical of the iron cores of massive supernova progenitors, the reaction

$$ e^- + p \to n $$

is energetically favourable, because the electrons have a total energy larger than the energy associated to the mass difference between neutrons and protons. Having isolated neutrons seems to lead to an unstable situation, since free neutrons decay into a proton–electron pair after about 12 minutes in a vacuum. However, at this stage electrons are degenerate and, due to their high Fermi energy, when released by the neutron decay they would not be able to find an empty and sufficiently low-energy state, to make the decay energetically favourable. This means that the neutron decay cannot happen and the level of neutronization of the stellar matter strongly increases.

If matter is squeezed even more, at densities above $\sim 10^{15} \, \text{g cm}^{-3}$ the Fermi energy of the degenerate neutrons and electrons is high enough that reactions between neutrons and electrons leading to the production of hyperons are energetically favourable. At even higher densities more massive particles can be produced (possibly leading to the formation of quark matter).

Neutron stars have an extremely thin atmosphere and a non-degenerate envelope, whose extension is of the order of 1m. The temperature drops by a factor of about 100 over the length of the envelope. Below this envelope typical densities are already in the WD regime, and the stellar matter is made of ionized nuclei arranged in a lattice configuration and degenerate electrons. One kilometer inside the star the density has already reached $\sim 10^{11} \, \text{g cm}^{-3}$, steadily increasing moving towards the centre, so that first one encounters layers where neutron production has been efficient, and deeper inside matter where hyperons have been produced. It is possible that in the cores even heavier particles are present, like quarks. Overall, one expects a composition made predominantly of free neutrons and a small percentage of protons and electrons, apart from the surface, dominated by iron nuclei. Electrons and neutrons are highly degenerate, hence the conductivity is high and the star is practically isothermal, apart from the non-degenerate atmospheric layers. The maximum mass for having a stable neutron star is of the order of $2.5 M_\odot$ (not the value predicted by Equation (7.6) for $\mu_e = 1$, i.e. $\sim 5.8 M_\odot$, because of the effect of nuclear forces and general relativity corrections) the uncertainty mainly due to uncertainties in the EOS at very high density.

It is not difficult to grasp the complexity of modelling the structure of neutron stars, in particular its EOS. An additional complication is that, since the gravitational energy of a neutron star is comparable to its rest-mass energy, general relativity corrections to the hydrostatic equilibrium equation are necessary. The equation of hydrostatic equilibrium in case of a neutron star becomes

$$\frac{dP}{dr} = -G\frac{(m_r + 4\pi r^3 P/c^2)(\rho + P/c^2)}{r^2[1 - 2Gm_r/(rc^2)]}$$

Comparing this equation with Equation (3.4) we can see that the mass m_r and the local density are effectively increased, whereas the local radius is decreased. Thus, for a given mass the general relativistic hydrostatic equilibrium condition requires a higher pressure gradient compared to the classical hydrostatic equilibrium.

The evolution of neutron stars is a cooling process, like WDs. At formation, neutron stars have temperatures of the order of 10^{11}–10^{12} K; due to neutrino emission, the temperature drops to $\sim 10^{10}$ K within minutes, and below 10^6 K in about 10^5 years. When $T > 10^9$ K neutrinos are produced mainly through the URCA reactions

$$n \rightarrow p + e^- + \bar{\nu}$$

$$e^- + p \rightarrow n + \nu$$

where neutrinos are electron neutrinos. When the temperature drops below $\sim 10^9$ K the main reactions producing (electron) neutrinos are the so-called modified URCA reactions

$$n + n \rightarrow n + p + e^- + \bar{\nu}$$

$$n + p + e^- \rightarrow n + n + \nu$$

The same reactions involving muons and muon neutrinos in place of electrons and electron neutrinos can be also efficient. A discussion of additional neutrino production processes that are possibly efficient in neutron stars can be found in [199]. One can write an equation for the evolution of the luminosity like for WDs case, i.e.

$$L_\gamma + L_\nu = -\frac{dE}{dt}$$

where we have added the neutrino luminosity to the photon one, since the large rate of neutrino emission in the initial cooling phase. After about 10^5–10^6 years it is the photon emission that is the main mechanism of cooling, and L_ν can be neglected. The energy E is largely the thermal energy of the degenerate (mainly non-relativistic) neutrons and (relativistic) electrons. This energy contribution stems from the fact that neutrons and electrons are not a zero temperature degenerate gas, therefore there are still some available energy states below the Fermi energy (see Figure 2.1); in the absence of a consistent contribution of thermal energy from non-degenerate ions (as in WDs) this is the main energy source for neutron stars. The cooling timescale is shorter compared to the WD cooling; typically, the luminosity of a $1.4M_\odot$ neutron star goes down to $\log(L/L_\odot) \sim -6$ in just $\sim 10^7$ years.

7.8 Black holes

When the remnant of the supernova event is above $\sim 2.5 M_\odot$ (initial progenitor mass above $\sim 25 M_\odot$) matter is too compact to establish a hydrostatic equilibrium configuration of finite density. General relativity applied to these compact objects predicts the existence of a so-called black hole, i.e. an object (or a place in space) into which anything can fall but out of which nothing can escape, because the density of matter is so high that space is curled around itself carrying matter, light and any other form of energy with it. This is equivalent to saying that the escape velocity v_{esc} out of the gravitational well generated by a mass M of radius R is larger than the speed of light, i.e.

$$v_{esc}^2 = 2\frac{GM}{R} > c^2$$

this relationship provides the typical radius R_s (Schwarzschild radius) for which the escape velocity equals the speed of light

$$R_s = 2\frac{GM}{c^2} = 2.95 \times 10^5 \frac{M}{M_\odot} \tag{7.11}$$

in cgs units.

If the mass of the remnant of a supernova event exceeds the neutron star limit, nothing can stop the continuous collapse of the core; the ever increasing density steadily increases the curvature of space until the escape velocity at the collapsing object exceeds the speed of light. This happens when the object radius decreases below R_s. The Schwarzschild radius is a surface in the geometry of space–time beyond which we can see no events; due to this property it is also called an event horizon. The remnant will continue collapsing within its Schwarzschild radius (and eventually will produce a point singularity of zero radius) but the information we may get can never come from within R_s.

The space–time metric in the empty space outside a spherical non-rotating object (we won't discuss here the more general case of rotating black holes) can be described by the Schwarzschild metric

$$ds^2 = \left(1 - \frac{R_s}{r}\right)c^2 dt^2 - \frac{dr^2}{\left(1 - \frac{R_s}{r}\right)} - r^2 d\theta^2 - r^2 \sin^2\theta d\phi^2 \tag{7.12}$$

where r, θ, ϕ are spherical space coordinates whose origin is at the centre of the object.

Consider the proper time $d\tau_1$ read by a standard clock stationary ($dr = d\theta = d\phi = 0$) at radial coordinate r_1; if the same standard clock were to be located at r_2, this time interval $d\tau_1$ would correspond to a time interval $d\tau_2$ given by:

$$\frac{d\tau_2}{d\tau_1} = \left(\frac{1 - (R_s/r_2)}{1 - (R_s/r_1)}\right)^{1/2} \tag{7.13}$$

We can notice from Equation (7.13) that $d\tau_2/d\tau_1 \to 0$ when $r_2 \to R_s$ and $r_1 \to \infty$, i.e. time slows down to a complete stop at the Schwarzschild radius with respect to a standard clock at infinite distance. This occurrence has very important observational consequences. Suppose that a light source at r_2 emits signals at regular intervals $d\tau_2$, so that its frequency is $\nu_2 = 1/d\tau_2$; a receiver at location r_1 will measure a frequency $\nu_1 = 1/d\tau_1$. If we again consider $r_1 \to \infty$ (i.e. an observer very far from the black hole) and $r_2 \to R_s$, we have that the resulting frequency redshift z is given by

$$z \equiv \frac{\nu_2 - \nu_1}{\nu_1} = \frac{\nu_2}{\nu_1} - 1 = \left(1 - \frac{R_s}{r_2}\right)^{-1/2} - 1 \to \infty$$

i.e. a distant observer cannot detect signals coming from R_s.

In addition to this redshift caused by the curvature of space, one has to add the redshift due to the acceleration of the infalling particles; if one applies these two combined effects to the light emitted by a collapsing stellar remnant, the timescale for the object to reach the typical limiting magnitude of the largest telescopes is very short, much smaller than 1 s. Therefore, although in principle the object never reaches R_s, in practical terms it becomes undetectable almost instantaneously after the collapse begins.

An external observer cannot receive any information from within the Schwarzschild radius; however, it is actually possible to describe mathematically the region with $r < R_s$ with other coordinate systems (see, for example, [11] and [115] for elementary discussions on this subject) and hence follow the approach to the black-hole singularity from the point of view of the infalling object that will smoothly cross the event horizon around the black hole. One can also demonstrate again in this different coordinate system that no information can reach the external world from $r \leq R_s$.

To summarize, from a viewpoint of a distant observer a black hole does not emit any signal and interacts with the external world only through its gravitational field. This is why black holes can be detected only indirectly, from the X-ray radiation of infalling accelerated ionized hot gas accreted from a binary companion or from the surrounding interstellar matter.

8 From Theory to Observations

8.1 Spectroscopic notation of the stellar chemical composition

Until now we have always specified the chemical composition in terms of X, Y and Z and a given heavy element abundance distribution. This is a customary and convenient choice from the theoretical point of view which is not, however, directly related to what is determined from spectroscopy. The helium abundance, for example, cannot be determined for all stars, since low-mass objects are generally too cold to show helium spectral lines, and the metal abundances are usually determined differentially with respect to the Sun. The traditional metal abundance indicator is the quantity $[\mathrm{Fe/H}] \equiv \log(N(\mathrm{Fe})/N(\mathrm{H}))_* - \log(N(\mathrm{Fe})/N(\mathrm{H}))_\odot$, i.e. the difference of the logarithm of the Fe/H number abundance ratios observed in the atmosphere of the target star and in the solar one. The choice of iron as the metal abundance indicator (in spite of the fact that in the Sun iron is not one of the most abundant metals – see Table 2.1) stems from the fact that iron lines are prominent and easy to measure. For the Sun $[\mathrm{Fe/H}] = 0$, whereas stars more metal poor than the Sun have $[\mathrm{Fe/H}] < 0$. If one assumes that the solar heavy element distribution is universal, the conversion from Z to $[\mathrm{Fe/H}]$ is given by

$$[\mathrm{Fe/H}] = \log\left(\frac{Z}{X}\right)_* - \log\left(\frac{Z}{X}\right)_\odot \tag{8.1}$$

because X_{Fe}/Z is the same in the Sun and in the target star. When accounting for the solar value of (Z/X) it becomes

$$[\mathrm{Fe/H}] = \log\left(\frac{Z}{X}\right)_* + 1.61 = \log\left(\frac{Z}{1-Y-Z}\right)_* + 1.61 \tag{8.2}$$

Evolution of Stars and Stellar Populations   Maurizio Salaris and Santi Cassisi
© 2005 John Wiley & Sons, Ltd

The previous equation provides the observational counterpart of the chemical abundances used as input for a given stellar model, in terms of the chosen Z and Y, assuming a solar metal distribution. As an example, $Z = 0.001$ and $Y = 0.25$ provide $[Fe/H] = -1.26$, whereas $Z = 0.04$ and $Y = 0.30$ provide $[Fe/H] = 0.39$. Generally, the helium and heavy element abundance variations are negligible with respect to the hydrogen abundance that largely dominates the stellar atmospheres apart from specific cases (like non-DA WDs). This means that X can be considered approximately a constant and Equation (8.2) can be simplified into

$$[Fe/H] \sim \log\left(\frac{Z_*}{Z_\odot}\right) \tag{8.3}$$

With this approximation the two previous numerical examples provide $[Fe/H] = -1.24$ and $[Fe/H] = 0.36$, respectively. Typical errors of the spectroscopic determinations of $[Fe/H]$ are of the order of at least 0.10 dex.

If one relaxes the assumption of a universal scaled solar heavy element distribution, the correspondence between $[Fe/H]$ and X, Y, Z obviously changes because the ratio between the iron abundance and Z is different from that in the Sun, and depends on the exact metal distribution in the observed star. In this case, one can still use Equation (8.1), but the left-hand side refers to the ratio of the 'total' abundance of metals to hydrogen

$$[M/H] = \log\left(\frac{Z}{X}\right)_* - \log\left(\frac{Z}{X}\right)_\odot$$

which is equal to Equation (8.1) for a scaled solar metal mixture.

A very important case is the α-enhanced metal distribution typical of old metal-poor ($[Fe/H]$ below about -0.6) objects in the halo of our galaxy (and presumably in the haloes of all spiral galaxies) and metal-poor objects in the Magellanic Clouds. Determinations of O, Ne, Mg, Si, S, Ca and Ti (the so-called α elements) abundances in the halo of our galaxy disclose that these elements (denoted collectively by α) display a distribution characterized by $[\alpha/Fe] \sim 0.3$–0.4, while for metal-poor (spectroscopic $[Fe/H]$ values lower than ~ -1.0) Magellanic Cloud stars $[\alpha/Fe] \sim 0.2$ (here we employed the spectroscopic notation described before, but considering the number ratio of α elements to Fe instead of Fe to H), i.e. they are enhanced with respect to Fe compared with the scaled solar distribution, by a factor approximately constant (i.e. a factor 2–3) for each of these elements.

The simplest explanation for this occurrence is related to the chemical composition of the ejecta of Type II and Type Ia supernovae. The former inject into the interstellar medium matter rich in α-elements, with a minor amount of Fe, while Type Ia supernovae eject mainly Fe, with a minor α-element component. Since Type II supernovae progenitors are massive, short-lived objects, the metal mixture of the most metal-poor stars born at the beginning of the formation of the Galaxy was produced essentially by Type II supernovae; at later times Type Ia supernovae started to explode and contribute to the chemical composition of the interstellar medium, so that younger stellar generations (like our Sun) are characterized by a metal mixture with a smaller α/Fe ratio with respect to the oldest stars.

For these α-enhanced mixtures the general relationship between [M/H] and [Fe/H] is well approximated by

$$[M/H] \sim [Fe/H] + \log(0.694 f_\alpha + 0.306)$$

where $f_\alpha = 10^{[\alpha/Fe]}$. For $[\alpha/Fe] \sim 0.3$ this equation gives $[M/H] \sim [Fe/H] + 0.2$. It is easy to understand why with this specific non-scaled solar metal mixture [M/H] > [Fe/H]. The ratio α/Fe is larger than the solar counterpart at a given Fe abundance, and therefore the same Fe abundance corresponds to a total metal abundance larger than the case of a scaled solar one.

8.2 From stellar models to observed spectra and magnitudes

Stellar evolution models provide the run of physical and chemical quantities from the centre up to the photosphere of a star of given initial mass, and initial chemical composition, and their evolution with time. Almost all the information we gather from observations of stars comes from the detection and analysis of the photons they emit; observations of the stellar radiation provide low- and high-resolution spectra and broadband photometry that have to be related to the properties predicted by stellar models.

A fundamental tool in stellar evolution is the HRD, i.e. the plot of a star bolometric luminosity versus its effective temperature, which we have widely used in the previous chapters. The observational counterpart of the HRD is the Colour Magnitude Diagram (CMD), i.e. the plot of a star magnitude in a given photometric band versus a colour index (that is the difference between the magnitudes in two different photometric bands). In this section we discuss the definition of magnitudes and colour indices, and how they are related to the bolometric luminosity and effective temperatures predicted by the stellar evolution models.

Given the monochromatic flux f_λ (energy per unit time, unit area and unit wavelength) received at the top of the Earth's atmosphere, what is generally observed at the telescope is the portion of the flux within a generic photometric band A that covers the wavelength range between λ_1 and λ_2, from which a quantity called apparent magnitude is defined as

$$m_A = -2.5 \log \left(\frac{\int_{\lambda_1}^{\lambda_2} f_\lambda S_\lambda d\lambda}{\int_{\lambda_1}^{\lambda_2} f_\lambda^0 S_\lambda d\lambda} \right) + m_A^0 \tag{8.4}$$

Here f_λ^0 denotes the spectrum of a reference star that produces a known apparent magnitude m_A^0 and S_λ is the response function of a given photometric filter (i.e. a measure of the efficiency of photon detection within the filter wavelength range) in the wavelength interval (λ_1, λ_2). It is clear from this definition of m_A that the apparent magnitudes are relative quantities defined with respect to some reference star. This reflects the historical development of stellar photometry.

Many photometric systems exist, covering several different wavelength ranges, from the ultraviolet to the infrared part of the spectrum.[1] Figure 8.1 shows the

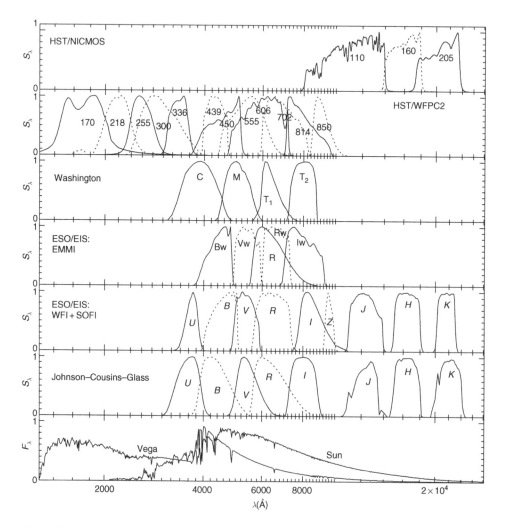

Figure 8.1 Response functions $S(\lambda)$ for filters of various photometric systems. The bottom panel shows the spectra (in arbitrary units) of Vega and the Sun (courtesy of L. Girardi)

---

[1] See http://ulisse.pd.astro.it/Astro/ADPS/Systems/index.html.

response functions $S(\lambda)$ of some widely used filters. A measure of the 'representative' wavelength of a given photometric band is the effective wavelength λ_{eff}, given by

$$\lambda_{\text{eff}} = \frac{\int_{\lambda_1}^{\lambda_2} \lambda f_\lambda S_\lambda \, d\lambda}{\int_{\lambda_1}^{\lambda_2} f_\lambda S_\lambda \, d\lambda}$$

It is interesting to notice that the effective wavelength depends not only on the response function S_λ, but also on f_λ, i.e. on the spectrum of the star under scrutiny (in the following, when we consider λ_{eff} for some photometric system, we refer to the case of a solar-like star).

The difference between the apparent magnitudes of a star in two different bands A and B, $m_A - m_B \equiv (A - B)$, is called colour index and it is completely defined when the flux and apparent magnitude of the chosen reference star in the A and B bands are fixed.

It is obvious that the apparent magnitude of a star depends on its distance, and therefore it is not a quantity directly associated with the intrinsic properties of the star itself. To overcome this problem one defines the absolute magnitude in a given band A, M_A, as the apparent magnitude the star would have at a distance of 10 pc, that is given by (if the radiation travels undisturbed from the source to the observer)

$$M_A = m_A - 5 \log(d) + 5 \tag{8.5}$$

where d is the distance in parsec. The difference $(m_A - M_A) \equiv (m - M)_A$ is called the distance modulus. A comparison of the absolute magnitudes provides a measure of the stellar intrinsic luminosities in a given wavelength range. The colour index $(A - B)$ is obviously unaffected by the distance. The distance obtained from Equation (8.5) is what is called luminosity distance d_L in a cosmological context (see Chapter 1 and Equation (1.9)). In the case of distances to objects co-moving with the Hubble flow one can combine Equation (1.10) with the definition of magnitude in order to express d_L in terms of a distance modulus.

The surface properties predicted by theoretical stellar evolution models are T_{eff} and bolometric luminosity L (plus chemical element abundances) which are, however, not directly determined by observations. In fact, as shown above, observations measure the flux of a star (energy per unit area and unit time) received at the top of the Earth's atmosphere, in a given wavelength range determined by the detector and filter employed. Here we describe the transformation from effective temperatures and bolometric luminosities to observed magnitudes and colour indices following the simple and general formalism presented in [84].

If light travels undisturbed from the star to Earth, the flux f_λ at a given wavelength λ received at Earth is related to the flux F_λ at the stellar surface by

$$f_\lambda = F_\lambda \left(\frac{R}{d}\right)^2 \tag{8.6}$$

When transforming the prediction of stellar models into observational magnitudes, we first need to determine from the surface luminosity L and T_{eff} the flux F_λ at the stellar surface. This is done by using results from model atmosphere computations; a model atmosphere is uniquely determined by its chemical composition, gravity and effective temperature, and provides the monochromatic fluxes F_λ (i.e. the stellar spectrum at a given resolution) at the stellar surface that can be transformed into f_λ for a given distance d according to Equation (8.6). The L, T_{eff} and surface gravity g of the stellar model can therefore be used to determine the appropriate model atmosphere, hence the expected F_λ for the star. Notice that if stars were to radiate as perfect black bodies, F_λ would depend only on T_{eff}.

Figure 8.2 displays the fluxes (in units of erg cm^{-2} s^{-1} hz^{-1} ster^{-1}) computed from theoretical model atmospheres with the same chemical composition (solar) the same gravity (the solar gravity, $\log(g) \sim 4.5$) and two different T_{eff} values. The hotter spectrum is shifted towards shorter wavelengths and has in general higher fluxes as expected from black-body radiation; however, the shape is not exactly like a black body, due to the absorption lines produced by the chemical elements in the stellar atmospheres. These absorption lines subtract energy from some wavelengths and

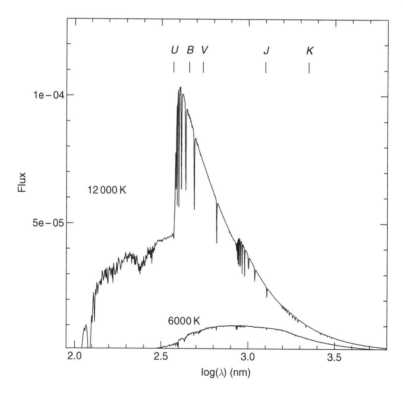

Figure 8.2 Comparison of the energy fluxes (units of erg cm^{-2} s^{-1} hz^{-1} ster^{-1}) emitted by two stars with the same solar chemical composition and solar gravity, and two different values of T_{eff}. The effective wavelength of some photometric filters is also marked

redistribute it to other parts of the spectrum. One can, in general, think of the stellar spectrum as a black-body spectrum emerging from the stellar photosphere, that is then modified during the crossing of the less dense and cooler atmospheric layers, that also induce a dependence of F_λ on both metallicity and surface gravity.

The absolute magnitude of a star, as discussed before, is equal to the apparent magnitude of the object when it is located at $d = 10$ pc, i.e. (see Equation (8.5))

$$M_A = -2.5 \log \left[\left(\frac{R}{10 \text{ pc}} \right)^2 \frac{\int_{\lambda_1}^{\lambda_2} F_\lambda S_\lambda d\lambda}{\int_{\lambda_1}^{\lambda_2} f_\lambda^0 S_\lambda d\lambda} \right] + m_A^0 \qquad (8.7)$$

By using Equations (8.7), assuming $S_\lambda = 1$ at all wavelengths, $\lambda_1 = 0$ and $\lambda_2 = \infty$, and adopting as reference star the Sun, we introduce the absolute bolometric magnitude of a star as

$$M_{\text{bol}} = M_{\text{bol},\odot} - 2.5 \log(L/L_\odot) = M_{\text{bol},\odot} - 2.5 \log(4\pi R^2 F_{\text{bol}}/L_\odot) \qquad (8.8)$$

where $M_{\text{bol},\odot}$ is the solar absolute bolometric magnitude (see below). The surface bolometric flux of a star is related to its effective temperature according to the relation

$$F_{\text{bol}} = \int_0^\infty F_\lambda d\lambda = \frac{ac}{4} T_{\text{eff}}^4$$

We define the bolometric correction BC_A to a given photometric band as

$$BC_A \equiv M_{\text{bol}} - M_A \qquad (8.9)$$

If we now substitute Equations (8.7) and (8.8) into Equation (8.9) we obtain

$$BC_A = M_{\text{bol},\odot} - 2.5 \log \left[4\pi (10 \text{ pc})^2 \frac{ac T_{\text{eff}}^4}{4L_\odot} \right] + 2.5 \log \left(\frac{\int_{\lambda_1}^{\lambda_2} F_\lambda S_\lambda d\lambda}{\int_{\lambda_1}^{\lambda_2} f_\lambda^0 S_\lambda d\lambda} \right) - m_A^0 \quad (8.10)$$

The bolometric correction to a given wavelength band A depends therefore on the stellar energy distribution F_λ (that depends on surface gravity, effective temperature and chemical composition) its effective temperature, the solar luminosity and solar bolometric magnitude, spectral distribution and apparent magnitude of the reference star that defines the apparent magnitude scale.

The value of $L_\odot$ is known and, once the bolometric magnitude of the Sun is fixed (e.g. $M_{\text{bol},\odot} = 4.74$ following [12]) what is left in order to convert T_{eff} and L into magnitudes is the choice of f_λ^0 and m_A^0, which we call (following [84]) 'zero points' of a given photometric system. Various choices are possible, here we discuss briefly the more widely used one.

The very popular Johnson–Cousins–Glass *UBVRIJHKLMN* – that will be called simply the Johnson system in what follows – and HST/WFPC2 VEGAmag systems (see Figure 8.1) for example, make use of the star Vega to fix the zero points.

The Johnson system assumes that the apparent magnitude of Vega in the V band is $m_V = 0$, and all its colour indices are equal to zero, whereas the *HST* system defines some colours with slightly different values. These choices determine the m_A^0 constant in Equation (8.10). We stress that this definition of Vega apparent magnitudes is purely operative; it sets the zero point of the various magnitude scales but it does not mean that the fluxes received from Vega are the same at all wavelengths, since they of course change with λ. As for Vega spectral energy distributions f_λ^0, there are observations that do not, however, cover the full wavelength range of the entire photometric system. What is often done is to take the spectral distribution F_λ^{Vega} from an appropriate model atmosphere for the star ($T_{\text{eff}} = 9550$ K, $\log(g) = 3.95$ in cgs units, [Fe/H] $= -0.5$). The F_λ^{Vega} has then to be multiplied by the factor $(R/d)^2$ in order to obtain the value at the Earth's atmosphere; this factor is easily derived by taking the ratio of the observed to the theoretical flux at a given wavelength – e.g. at 5556 Å, where there is a precise empirical determination – that provides $(R/d)^2 = 6.247 \times 10^{-17}$.

The procedure outlined above allows us to determine the bolometric correction BC_A to a given photometric band A; hence, if the stellar bolometric luminosity is known, one can determine the corresponding M_A from Equations (8.9) and (8.8). In practice, tables of bolometric corrections and colour indices (it is easy to realize from Equation (8.9) that a given colour index $(A - B)$ is simply equal to $BC_B - BC_A$) are available, for a grid of gravities and T_{eff} that cover all the major phases of stellar evolution, and for a number of chemical compositions (e.g. [122]). Interpolations among the grid points provide the sought BC_A for the surface gravity, T_{eff} and chemical composition at the surface of the theoretical stellar model. From the model luminosity one determines M_{bol} and then M_A is immediately computed using the BC_A obtained before.

Figure 8.3 displays, as an example, the evolution of a $1 M_\odot$ star with chemical composition typical of the halo of the Galaxy, from the ZAMS up to the He flash ignition. Figure 8.3(c) shows the evolutionary track in the HRD, whereas Figure (8.3)(a),(b) and (d) display various CMDs obtained with different filter combinations. As a general rule the colour index (displayed on the horizontal axis) tracks the model T_{eff}, whereas the magnitude in a given wavelength band (vertical axis) tracks the stellar luminosity; however, the shape of the evolution in the CMD is completely dependent on the filter selection. In these plots the $V - (B - V)$ combination follows closely the HRD, whereas the $U - (B - V)$ and especially the $V - (U - B)$ CMD produce a different shape. The reason for these differences is twofold. On the one hand, once the chemical composition of the stellar surface, surface gravity and T_{eff} are fixed, the bolometric correction depends on the filter considered, since the wavelength band coverage (and hence the portion of the stellar spectrum included in Equations (8.7) and (8.10)) changes with the filter selection. The other reason is that the sensitivity of a generic colour index to T_{eff} is a function of T_{eff} itself. This is qualitatively illustrated by Figure 8.2, that shows two spectra of different T_{eff} and the effective wavelength of some widely used filter. Consider a decrease of T_{eff} and, for example, the $(U - B)$ colour index. The spectrum is shifted to longer wavelengths, more flux is found in

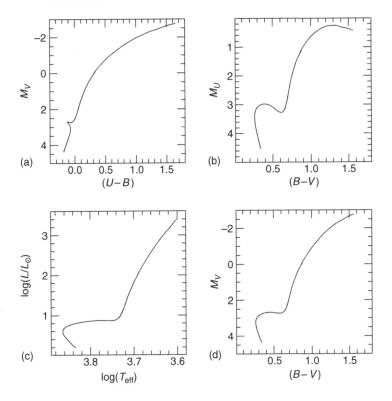

Figure 8.3 Evolution of a $1 M_\odot$ star with $Z = 0.001$ and $Y = 0.246$, from the ZAMS until the tip of the RGB, displayed in the HRD and various CMDs

the B band with respect to the U band, and the value of $(U - B)$ increases (remember that more flux means lower magnitude). If, however, the temperature becomes too low, the U and B filters will sample a wavelength region where the flux is negligible and then the contribution of the integral in Equation (8.10) to the final values of BC_U and BC_B tends to zero. This means that $(U - B)$ loses sensitivity to T_{eff}, the only dependence left being the term in T_{eff}^4 in Equation (8.10). The same can be said of, for example, $(J - K)$ when the temperature is too high.

In Figure 8.4 we show analogous CMDs for two DA WDs of masses equal to 0.55 and $1 M_\odot$. The less massive WD is always the cooler one in the bright part of the CMD, because its radius is larger. Three of these colour indices show a pronounced turn to the blue (lower values of the colour index) at low luminosities, in spite of the fact that the model T_{eff} is constantly decreasing and therefore one expects a steady increase of the colours with decreasing luminosity. Only the $(B - V)$ index closely mimics the behaviour of the track in the HRD. The reason for this turn to the blue is the blocking effect in the infrared of the H_2 molecule collision-induced absorption ([91], [193]) that redistributes the energy towards lower wavelengths (for the flux conservation). This is another important example of how the shape of evolutionary

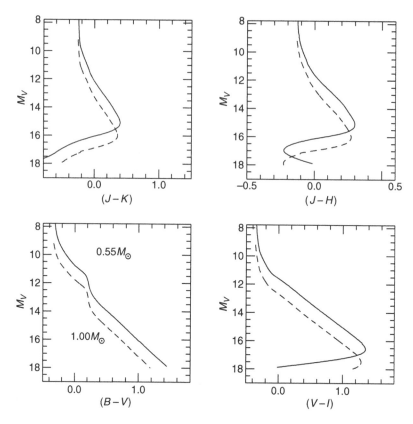

Figure 8.4 Evolution of two DA WDs with $0.55M_\odot$ and $1M_\odot$, respectively, displayed in various CMDs

tracks in the CMD can be substantially altered with respect to the HRD counterpart, when using specific filter combinations.

8.2.1 Theoretical versus empirical spectra

The transformation from theoretical luminosity and T_{eff} to observed magnitudes and colour indices has been described assuming that F_λ at the stellar surface is obtained from the appropriate theoretical model atmospheres. It is, however, well known that current theoretical model atmospheres suffer from at least two main shortcomings as follows.

- Although broadband colours (like the Johnson discussed before) of stars with solar chemical compositions appear to be reasonably reproduced, many spectral lines predicted by the models are not observed in the Sun, and also the relative strength of many lines is not well reproduced. This affects narrowband filters (like the so called Strömgren filters, that cover a spectral range a factor of ~10 smaller than the *UBV* system) in which individual metallic lines can significantly affect the bolometric corrections.

- In convective model atmospheres the energy transport is usually treated with the mixing length theory; this approximation introduces an uncertainty in the predicted spectra hence bolometric corrections and colour indices. Recent two- and three-dimensional hydrodynamical simulations of stellar model atmospheres aim at addressing this shortcoming, but they have not yet produced libraries of stellar spectra that cover all the relevant evolutionary phases, mass ranges and chemical compositions.

These shortcomings cause an uncertainty of several hundredths of magnitude on the BC_A values. An alternative solution is to use empirical spectra of a sample of nearby stars with independently determined T_{eff}, gravities and chemical composition. A fundamental problem with this approach is that stars for which empirical T_{eff} values can be determined are local objects, that cover a narrow range of chemical compositions, masses and evolutionary phases (reflecting the local population of the Galactic disk) and would not allow the modelling of different stellar populations. The only direct method (i.e. without the use of some information from theoretical model atmospheres) for the empirical determination of T_{eff} is based on the relationship between bolometric luminosity L and effective temperature of a star of radius R (see the extensive discussion about the effective temperature scale in [15])

$$L = 4\pi R^2 \sigma T_{eff}^4$$

and the relationship between the bolometric flux F_{bol} measured at the Earth and the star distance d

$$F_{bol} = \frac{L}{4\pi d^2}$$

From these two relationships one obtains

$$T_{eff} = \left(\frac{F_{bol}}{\sigma}\right)^{1/4} \left(\frac{d}{R}\right)^{1/2} = \left(\frac{4}{\sigma}\right)^{1/4} F_{bol}^{1/4} \theta^{-1/2} \tag{8.11}$$

where θ is the angular diameter of the star. Interferometric observations can in principle determine θ for local stars, while spectrophotometric observations (ground- and space-based) covering the relevant portion of the wavelength spectrum can provide an empirical value for F_{bol}; once these two quantities are known, Equation (8.11) is employed to obtain T_{eff}. This method (observationally very expensive) can be applied only to a very small sample of local object. Moreover, the determination of θ and F_{bol} need (apart from the case of the Sun) both some input from model atmospheres to estimate, respectively, the limb-darkening correction, and account for interstellar extinction (see next section).

8.3 The effect of interstellar extinction

The previous discussion about the calibration of photometric systems has assumed that the starlight travels undisturbed from the source to Earth. This is essentially the case of the reference star Vega, and also the theoretical absolute magnitudes M_A must be computed with this assumption if they were to be a measure of the intrinsic luminosity of the model in the A photometric band.

However, interstellar space is not empty, instead it is permeated by the ISM. The ISM interacts with the stellar radiation and its effect on the observed magnitudes and colours of stars must be allowed for if the stellar intrinsic properties are to be recovered. We have already seen that the main components of the ISM are gas and dust. Interstellar gas tends to absorb and radiate at a different wavelength and direction, and dust to scatter the stellar radiation. These effective losses are known collectively as extinction. In general, the observed flux f_λ is related to the intrinsic one in the case of no interaction with the ISM, $f_{\lambda,0}$, by

$$f_\lambda = f_{\lambda,0}\, e^{-\tau_\lambda}$$

where τ_λ is the optical depth of the ISM at the observed wavelength. Extinction is not uniform across the whole spectrum, because τ_λ varies approximately as λ^{-1} in the visual part of the spectrum. The observed apparent magnitude in some photometric band is the sum of its intrinsic apparent magnitude and a factor A_A (in units of magnitude) called extinction, which is dependent on the wavelength of observation. Stars within a distance of about 70–100 pc from us (like Vega) are largely unaffected by extinction; however, the light from more distant stars and in external galaxies may cross a substantial amount of ISM which affects its spectral energy distribution.

It is common to denote by $m_{A,0}$ the apparent magnitude of a star corrected for the effect of extinction, and by m_A its observed value. The extinction A_λ at a given wavelength is defined by the following relation:

$$m_A = -2.5\log\left(\frac{\int_{\lambda_1}^{\lambda_2} f_\lambda 10^{-0.4A_\lambda} S_\lambda\, d\lambda}{\int_{\lambda_1}^{\lambda_2} f_\lambda^0 S_\lambda\, d\lambda}\right) + m_A^0 \tag{8.12}$$

If A_λ is constant in the wavelength interval (λ_1, λ_2) (the shorter the wavelength range the better this approximation is) then the relationship between intrinsic and observed apparent magnitude is simply

$$m_A = m_{A,0} + A_A \tag{8.13}$$

where with A_A we denote the constant value of A_λ within the wavelength range covered by the filter A. In the presence of extinction, the absolute magnitude of a star (Equation (8.5)) has to be rewritten as

$$M_A = m_A - A_A - 5\log(d) + 5 = m_{A,0} - 5\log(d) + 5 \tag{8.14}$$

The difference between M_A and the apparent magnitude m_A uncorrected for the extinction, $(m - M)_A$, is called the apparent distance modulus, while the real distance modulus $(m - M)_0$ is given by

$$(m - M)_0 = m_{A,0} - M_A = 5\log(d) - 5 = (m - M)_A - A_A \qquad (8.15)$$

The effect of extinction on a colour index $(A - B)$ is

$$(A - B) = m_A - m_B = m_{A,0} - m_{B,0} + A_A - A_B \equiv (A - B)_0 + E(A - B) \quad (8.16)$$

where $E(A - B) = A_A - A_B$ is called colour excess or reddening. What is usually determined empirically is the so-called extinction law, i.e. the values of the ratio A_λ/A_V of the extinction at wavelength λ to that in the Johnson V band. For example, in the Johnson system mentioned before the reddening law is well approximated by

$$A_U = 1.53A_V, A_B = 1.32A_V, A_R = 0.82A_V, A_I = 0.60A_V$$
$$A_J = 0.29A_V, A_H = 0.17A_V, A_K = 0.11A_V, A_L = 0.06A_V$$

The ratio $A_V/E(B - V)$ is usually denoted by R_V. Moving from the blue part of the spectrum (B) to the infrared (L) the extinction greatly decreases, because extinction affects the shorter wavelengths preferentially. The observed flux is overall shifted to the red part of the spectrum with respect to the intrinsic one. This means that observed colours of stars are usually 'redder' than their intrinsic ones due to the wavelength dependent effect of extinction.[2]

In practice one can determine empirically for example $E(B - V)$ or some other colour excess (we will discuss some methods to obtain this information in the next section) and then use $A_V = R_V E(B - V)$ and the previous relationships to compute the extinction in the desired wavelength band. For $E(B - V)$ values up to a few tenths of a magnitude the assumption of constant A_λ within a given photometric band in the Johnson system does not produce substantial errors in the computation of the extinctions. However, for larger reddenings one has to take into account the non-constancy of A_λ in the given filter wavelength range. According to [12] this adds a colour term to the equation for A_V, leaving unchanged the extinction ratios given above.

A generic extinction law valid for our galaxy that covers a very large wavelength range and assumes $A_V = 3.12E(B - V)$ is displayed in Figure 8.5 (obtained from the analytical formulae given in [38]. See also [163].). The effective wavelength of some filters is also displayed. The bump in the extinction law seen at $\lambda \sim 218$ nm is attributed to absorption by graphite particles.

---

[2] In the rest of this chapter, to simplify the notation and unless otherwise specified in the text, we will adhere to the following guidelines. We generally omit the subscript '0' when referring to the intrinsic colours and magnitudes of theoretical models, unless otherwise specified; observed colours and magnitudes without the subscript '0' are referred to as quantities not corrected for extinction.

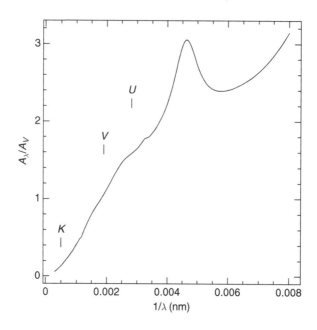

Figure 8.5 Extinction law valid for our galaxy. The effective wavelengths of some Johnson filters are also marked

In a few specific directions R_V differs from the standard value given above, and especially the extinction in the ultraviolet is affected substantially. It is not known in general how different the extinction laws produced by the ISM of other galaxies are; in the specific case of the Magellanic Clouds there appear to be variations with respect to the case of our galaxy, especially for the ultraviolet extinction ([85]).

We will discuss in the next chapter various methods to determine reddening and extinction of observed stellar populations. Here we mention briefly the classical method that employs UBV colour–colour diagrams. Figure 8.6 displays the theoretical $(U - B)_0 - (B - V)_0$ diagram of MS stars with solar chemical composition; the mass of the stars increases moving from the lower-right corner towards the upper-left one. The open circle displays the position of a hypothetical reddened MS star of solar chemical composition.

Since the chemical composition is the same, if this object is an MS star, its intrinsic colours have to lie on the standard sequence (in the position marked by the symbol ⋆) described by the unreddened stars. Its reddening can therefore be easily determined by shifting the observed $(U - B)$ and $(B - V)$ colours until the standard sequence is reached. The direction of this shift (the so-called reddening vector) must correspond to the ratio $E(U - B)/E(B - V) = 0.66$, that satisfies the reddening law given above. The amount of the shift, e.g. in $(B - V)$, provides $E(B - V)$.

The location of the MS in this colour–colour plane is strongly affected by the chemical composition of the parent stars. Metal-poor sequences are located above the solar standard sequence displayed in Figure 8.6, i.e. their $(U - B)$ colour at fixed

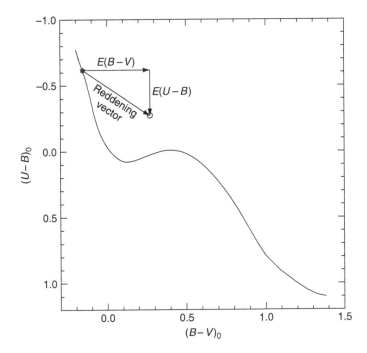

Figure 8.6 The theoretical $(U - B)_0 - (B - V)_0$ diagram of MS stars with solar chemical composition. The open circle marks the location of a hypothetical reddened MS star with solar chemical composition, the symbol $\star$ displays its unreddened location after the correction for interstellar extinction has been made, while the arrows show the direction of the reddening vector and the amounts of extinction in $(B - V)$ and $(U - B)$

$(B - V)$ is smaller. The colour differences due to the metallicity have therefore to be taken into account when this technique is applied to stars with a non-solar chemical composition.

8.4 *K*-correction for high-redshift objects

There is an important effect that has to be taken into account when observing objects (typically galaxies) whose cosmological redshift z is not negligible. If the object is moving rapidly away from us, the photons received in a generic photometric band A have been emitted at shorter wavelengths, and then redshifted by the expansion of the universe to the wavelength range covered by the filter A. To give a practical example, light emitted in the wavelength range of the filter U by an object located at redshift $z = 1.21$, is detected by us in the I filter. For the same object, light emitted in the V wavelength range is received in the J filter. The filter mismatch obviously worsens with increasing redshift; i.e. light emitted in the U wavelength range at redshift $z = 5.0$ is detected at Earth in the K filter.

As a consequence, the apparent magnitude observed within the filter A is not immediately related to the absolute magnitude M_A through luminosity distance d_L as shown by Equation (8.5), but a correction has to be added, the so-called *K-correction*. In the case of our fictitious object at $z = 1.21$, with a measured apparent magnitude m_V, the K-correction takes into account the difference between the fluxes in V (what we need to compare with the theoretical M_V) and J (what we observe now as m_V) emitted at the source; after its inclusion, the observed apparent magnitude in V will be the real counterpart of the flux emitted at the source in the V wavelength range.

It is also obvious from this example that, in order to determine appropriate K-corrections, one needs to know the shape of the object intrinsic spectrum at its redshift (i.e. at its age). Given that the stellar content of galaxies evolves with time, using local galaxies as templates to determine K-corrections for distant objects is not fully adequate. A theoretical modelling of integrated galaxy spectra and their evolution with time is therefore required to determine accurate K-corrections for various redshifts and galaxy types (see, for example, the discussion in [153]). We will discuss the modelling of integrated galaxy spectra in the final chapter of this book.

8.5 Some general comments about colour–magnitude diagrams (CMDs)

Figures 8.7 and 8.8 display the CMDs of two different stellar populations (respectively, the open cluster Praesepe using data by [111], and the globular cluster M3 using data by [24]) that belong to our galaxy, employing the widely-used BV filters. The spatial extension of these stellar systems (as for the case of galaxies) is much smaller than their distance from Earth, so that their stars can safely be assumed to be located all at the same distance from the observer.

The two most striking features are probably the fact that stars occupy a well-defined locus in the $m_V - (B - V)$ plane without being spread randomly all over the diagram, and that the morphologies of the two CMDs are very different. The first occurrence is related to the fact that, as shown in the previous chapters about stellar evolution, stars move in the HRD following well-defined paths. The difference in the CMD morphology (setting aside the fact that the faintest limit is dictated by the magnitude limit of the photometry, and the vertical and horizontal scales are affected by the distance modulus and, potentially, reddening and extinction) is caused by the different properties of two stellar systems, that stellar evolution has to be able to explain in terms of different physical and/or chemical parameters. Figures 8.9 and 8.10 show the Praesepe CMD in other photometric bands; the morphology of observed CMDs displays a strong dependence on the chosen photometric filters, exactly as expected from the theory. There is also a third dimension to the CMDs, i.e. the number of stars populating its various sections. This number is not constant, and this has also to be accounted for by some physical explanation rooted in stellar evolution theories.

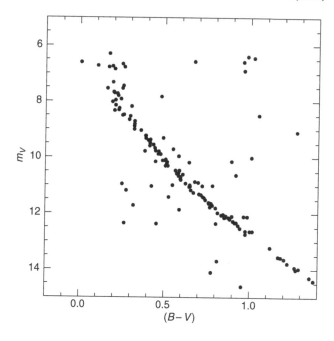

Figure 8.7   CMD of the open cluster Praesepe using the *BV* Johnson filters

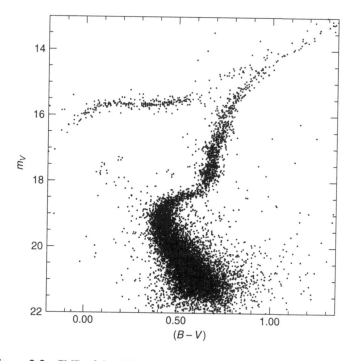

Figure 8.8   CMD of the globular cluster M3 using the Johnson *BV* filters

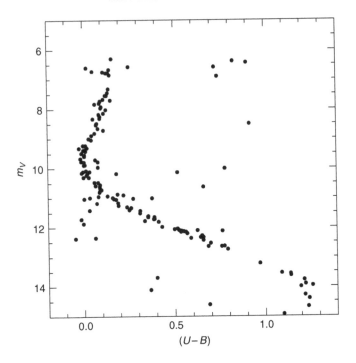

Figure 8.9 $m_V - (U - B)$ (Johnson) CMD of the open cluster Praesepe

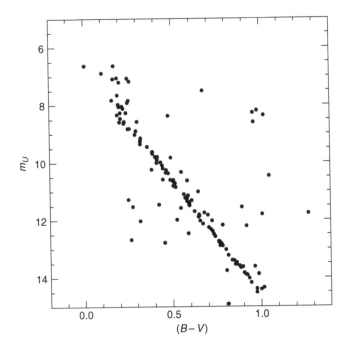

Figure 8.10 $m_U - (B - V)$ (Johnson) CMD of the open cluster Praesepe

In general, the interpretation of the observed CMDs of the stellar populations in our galaxy and in external galaxies in terms of the evolutionary properties of the parent stars is one of the main goals of stellar evolution. This, in turn, allows the determination of the star formation and chemical enrichment histories of the population under scrutiny, which provides a wealth of information about the formation and evolution of galaxies and of the universe in general. The next three chapters will present a number of techniques based on the comparison of observed CMDs and spectra with their theoretical counterparts, aiming at determining fundamental properties like distance, ages, chemical evolution and IMF of the observed populations.

9 Simple Stellar Populations

9.1 Theoretical isochrones

The most elementary population of stars is the so-called Simple Stellar Population (SSP) consisting of objects born at the same time in a burst of star formation activity of negligible duration, with the same initial chemical composition. Although this may seem just a theoretical toy model, there are very good observational counterparts of SSPs, namely globular and open clusters, elliptical galaxies and some dwarf galaxies.

The theoretical CMD for an SSP is called an isochrone, from the Greek word meaning 'same age'. The computation of an isochrone is conceptually very simple. Consider a set of evolutionary tracks of stars with the same initial chemical composition and various initial masses; different points along an individual track correspond to different values of the time t' and the same initial mass. An isochrone of age t is simply the line in the HRD that connects the points belonging to the various tracks (one point per track) where $t = t'$. This means that when we move along an isochrone, time is constant whereas the value of the initial mass of the star populating the isochrone at each point is changing. A generic point along an isochrone of age t is therefore determined by three quantities: bolometric luminosity, effective temperature and the value of the evolving mass. Once an isochrone of a given age and initial chemical composition is computed from stellar evolution tracks, it can be transferred to an observational CMD by applying to each point a set of appropriate bolometric corrections.

Figure 9.1 shows the HRD of two isochrones with solar metallicity and ages of 600 Myr and 10 Gyr, respectively. Selected tracks of the parent stellar models are also displayed. Different sections of a generic isochrone are named after the evolutionary phases experienced by the stellar masses evolving at that location. The MS of an isochrone is therefore the branch populated by objects that are still burning hydrogen in their cores. The bluest and brightest point along the isochrone MS is

---

Evolution of Stars and Stellar Populations   Maurizio Salaris and Santi Cassisi
© 2005 John Wiley & Sons, Ltd

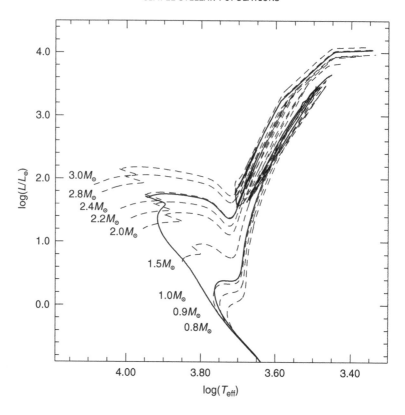

Figure 9.1 HRD of selected stellar evolutionary tracks (dashed lines) with the same initial solar chemical composition and the labelled masses (from [152]). The heavy solid lines display two isochrones for the same chemical composition and ages of 600 Myr (the brighter sequence) and 10 Gyr

called TO, because it is populated by objects in which the central hydrogen abundance is reduced to zero. The subsequent SGB, RGB, helium burning and AGB phases are also obviously represented and accordingly named.

It is clear from Figure 9.1 that the mass of the objects evolving at a given location along an isochrone (i.e. the mass of the individual evolutionary track intersecting the isochrone at that point) increases when moving towards more advanced phases. This increase is due to the fact that lower masses age more slowly, and at a given time they are less evolved than higher-mass objects. Another important property that is evident from Figure 9.1, is that the isochrone MS is populated by a large range of masses (the lower limit being the minimum possible stellar mass) whereas the mass evolving along the RGB and successive phases is approximately constant, given that the isochrone stays very close to the evolutionary track of the mass at the MS end-point (the TO).

A formal explanation of this occurrence is the following. Consider a curvilinear coordinate s (function of L and T_{eff}) whose value defines a position along a generic isochrone of age t, i.e. starting from zero at the bottom of the ZAMS and increasing when moving towards more advanced phases. The value of s at a given point is uniquely determined by the isochrone age t and the value of the evolving mass M at that point; this means that s is a function of only t and M, $s = s(t, M)$. We can invert this function and write t as a function of s and M. By definition of isochrone (constant age)

$$dt(M, s) = \left(\frac{dt}{dM}\right)_s dM + \left(\frac{dt}{ds}\right)_M ds = 0$$

from which we obtain

$$\left(\frac{dM}{ds}\right)_t = -\left(\frac{dM}{dt}\right)_s \left(\frac{dt}{ds}\right)_M \tag{9.1}$$

where the left-hand side represents the change of the mass for a change of position along the isochrone. If the right-hand side of this equation is close to zero somewhere, the mass evolving in that particular evolutionary phase is practically constant. Consider the derivative $(dt/ds)_M$; its inverse, $(ds/dt)_M$, represents the change of position of a star with initial mass M along the isochrone, when the age changes by an amount dt (for example, a change from the ZAMS to the TO, when age increases). This corresponds to the evolutionary speed of mass M along a given phase specified by the value of s; $(ds/dt)_M$ tends to large values for fast evolutionary speeds. In this case, a small variation of age moves the star a long way along the isochrone, for example down to the WD cooling phase, so that the variation of s is very large, i.e. $(ds/dt)_M \rightarrow \infty$ and its inverse $(dt/ds)_M \rightarrow 0$. As an example, consider a $1\,M_\odot$ star with solar chemical composition during its MS evolution: at any point along the MS an age change $dt = 1$ Gyr hardly changes the star luminosity and effective temperature, whereas the same age difference moves the model from the TO to approximately the ignition of central He. In addition, stellar evolution computations show that the value of $(dM/dt)_s$ is always finite. This means that for post-MS phases the range of evolving masses is small (see Figure 9.2) and a good approximation is to use a constant value of the post-MS evolving mass, equal to $M_{\rm TO}$ (the mass evolving at the isochrone TO).

In addition to the HRD and CMD location of an SSP, theoretical isochrones enable us to predict the relative number of stars along the different evolutionary stages. Given that each point along an isochrone is populated by stars with a certain value of the mass M, one needs to adopt an IMF that provides the number of stars dn born with mass between M and $M + dM$, hence the number of objects populating a generic interval between two consecutive points along the isochrone. When using an IMF of the form $dn = C\,M^{-x}dM$, the normalization constant C can be constrained by specifiying either the total mass ($M_{\rm tot}$) or the total number ($N_{\rm tot}$) of stars born

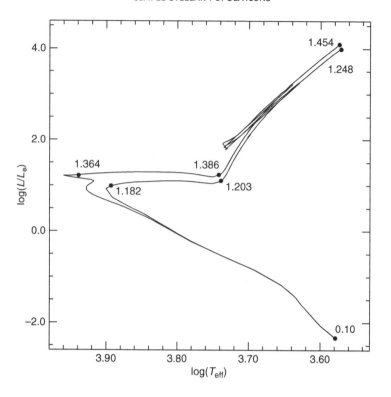

Figure 9.2 HRD of two isochrones (from the bottom of the MS until the end of the AGB phase) with ages equal to 2 and 3 Gyr ($Z = 0.001$). The value of the stellar mass (in solar mass units) evolving at representative points is also shown (isochrones from [84])

when the SSP formed. It is easy to show that when $x \neq 2$, the value of the constant C is given by

$$C = (2 - x)\frac{M_{tot}}{M_u^{2-x} - M_l^{2-x}} \tag{9.2}$$

where x is the exponent of the IMF, M_u and M_l are, respectively, the lower- and upper-mass limits of the stellar mass spectrum, e.g. ~ 0.1 and ~ 100–$200\, M_\odot$. If $x = 2$

$$C = \frac{M_{tot}}{\ln(M_u/M_l)} \tag{9.3}$$

This normalization guarantees that the total mass of stars formed stays constant, independent of the value of x, but the initial number of stars formed changes with changing value of the slope of the IMF. In the case where N_{tot} is given instead of M_{tot}, the previous relationships have to be rewritten as

$$C = (1 - x)\frac{N_{tot}}{M_u^{1-x} - M_l^{1-x}} \tag{9.4}$$

if $x \neq 1$ and

$$C = \frac{N_{\text{tot}}}{\ln(M_{\text{u}}/M_{\text{l}})} \tag{9.5}$$

if $x = 1$.

If we now use again the curvilinear coordinate s introduced before, the number of stars dN in a generic interval between s and $s + ds$ along an isochrone of age t is given by

$$dN = \frac{dn}{dM} \left(\frac{dM}{ds} \right)_t ds$$

which can be rewritten as

$$dN = -\frac{dn}{dM} \left(\frac{dM}{dt} \right)_s \left(\frac{dt}{ds} \right)_M ds \tag{9.6}$$

after using Equation (9.1). The term $(dt/ds)ds$ corresponds to the stellar lifetime in the interval ds. For a generic post-MS phase the first two terms in the right-hand side of Equation (9.6) are constant, because the evolving mass is to a good approximation equal to the mass at the isochrone TO. This means that the ratio between the number of stars in two different post-MS points along the isochrone will be simply equal to $t_{\text{PMS1}}/t_{\text{PMS2}}$ where t_{PMS1} and t_{PMS2} are the evolutionary timescales (of the TO mass) at these two locations.

For the purpose of comparing theory with observations one usually computes star number counts as a function of the magnitude in a given wavelength band along an isochrone, called differential luminosity function or simply luminosity function. From the previous discussion it is easy to see that the number of stars dN between magnitudes M_A and $M_A + dM_A$ along a given isochrone is given by

$$dN = \frac{dn}{dM} \frac{dM}{dM_A} dM_A = CM^{-x} \frac{dM}{dM_A} dM_A$$

where the derivative dM/dM_A is evaluated along the isochrone.

If mass-loss processes are included in the individual stellar tracks, the situation is slightly more complicated, because along each track the total mass is changing with time. The procedure to compute the isochrones is the same, i.e. one connects the points of equal age along tracks with various initial masses. However, the value of the mass evolving at a given point along the isochrone is now smaller than the initial mass of the parent track. It is important to remark that, when computing the number of stars populating a given point along an isochrone, one has to use in the IMF the value of the initial mass of the stellar track originating that point, not the actual mass. Also the entire discussion about the number of stars populating the isochrone makes use of the initial values of the mass, since the initial mass determines the evolutionary timescales and the IMF.

In the case of isochrones including the WD cooling, one has to consider necessarily the mass loss along the AGB phase, at least in the form of the initial–final mass relationship, even in case of isochrones where the mass loss is not included in the previous evolutionary phases. For the WD sequence the assumption of constant initial mass (corresponding to the assumption of constant progenitor mass equal approximately to the value of the TO mass) breaks down at low luminosities, due to the long WD cooling times and the finite age of the SSP, which ensures that at the bottom of the WD sequence of a given age one finds objects produced by the more massive low- and intermediate-mass stars evolved past the AGB phase in the earlier stages of the SSP evolution.

When determining the bolometric corrections to a generic photometric system it is always the value of the evolving mass that must be considered, since it is this quantity that, together with T_{eff} and L, determines the value of the actual surface gravity – needed to determine the appropriate bolometric corrections – not the initial mass.

In the following sections we will show in detail how quantitative and qualitative general properties of theoretical isochrones can be used to obtain relevant information about both old and young SSPs.

9.2 Old simple stellar populations (SSPs)

When studying the universe at any redshift, it is the properties of the oldest stellar populations that are the most relevant for unveiling the first stages of cosmic evolution, because they are the objects formed closer in time to the Big Bang. As a first step we will study the properties of old SSPs; 'old' here denotes ages larger than ~ 4 Gyr, corresponding to SSPs populated by low-mass stars far from the mass range of the RGB phase transition.[1] The CMD of these old SSPs is always characterized by prominent and well-populated SGB, RGB and HB phases. The age range of these old populations encompasses the age of the universe from redshift $z = 0$ up to $z \sim 1.5$–2.0, according to the parameters of the cosmological model shown in Table 1.1.

9.2.1 Properties of isochrones for old ages

Figure 9.3 shows a 10 Gyr isochrone for a metal-poor chemical composition (i.e. chemical composition with metallicity lower than the solar value) typical of the globular clusters in the Galaxy, from the MS until the end of the AGB phase. The isochrone has been computed including a Reimers mass-loss law ($dM/dt \propto \eta(LR)/M$) along the RGB with the free parameter η fixed at 0.2. Due to the mass loss, the mass of the objects evolving along the RGB is reduced by $\sim 0.1 M_{\odot}$ when

---

[1] The following alternative definitions can often be found in the literature. SSP ages older than ~ 10 Gyr are defined as 'old' ages, whereas the interval between ~ 1 Gyr and ~ 10 Gyr is defined as 'intermediate' age. Ages lower than ~ 1 Gyr are denoted as 'young' ages.

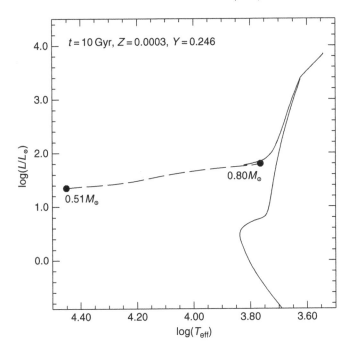

Figure 9.3   HRD of a 10 Gyr metal-poor isochrone. The ZAHB (dashed line) and the location of stars with selected masses along the ZAHB are also displayed

the tip of the RGB is reached. Increasing η leads to a higher mass loss, and while the RGB phase is practically unaffected, the portion of the isochrone corresponding to the beginning of the He-burning phase would move along the dashed line, towards lower masses, i.e. higher values of T_{eff} (see Section 6.3).

Figure 9.4 shows two pairs of isochrones (from the ZAMS until the ZAHB) in the HRD and the $M_V - (B - V)$ CMD, computed for $Z = 0.0001$ and $Z = 0.02$ respectively, and ages of 10 and 12.5 Gyr. For $(B - V)$ above ~ 0.0 the ZAHB is approximately horizontal, whereas at lower values of $(B - V)$ it becomes almost vertical, due to the steep increase of the bolometric correction to the V band with increasing T_{eff}. Along the ZAHB, in the colour range between $(B - V) \sim 0.2$ and ~ 0.5 lies the RR Lyrae instability strip, where stars pulsate radially (see Section 6.6.1). The brightness values obtained from stellar evolution models in hydrostatic equilibrium correspond to a good approximation to the average magnitudes over a pulsation cycle (this is the quantity we refer to, when discussing the magnitude of RR Lyrae or Cepheid stars). This average magnitude is obtained by determining the average value of the flux received from the star over a pulsation period, and transforming this flux into the corresponding magnitude.

A number of important properties of the isochrones displayed in the HRD and CMD of Figure 9.4 have to be noted. First, the lower MS – starting from ~ 2 mag below the TO – and RGB are unaffected by age, but are sensitive to the metallicity.

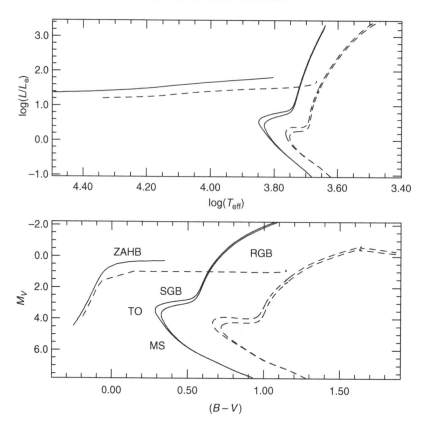

Figure 9.4 HRD and CMD of two pairs of isochrones from the ZAMS to the ZAHB, with ages $t = 10$ and 12.5 Gyr, $Z = 0.0001$ (solid line) and 0.02 (dashed line). The various evolutionary stages along the most metal-poor isochrone are marked

Second, the brightness of the ZAHB is unaffected by age but depends on Z. Third, the brightness and colour of the TO are affected by both age and metallicity. This qualitative behaviour of the isochrones with varying age and metallicity is the same in both the HRD and CMD (and also does not generally depend on the choice of the photometric bands used in the CMD). This means that it is related mainly to the properties of the stellar models (L and T_{eff}) although the variations of the bolometric corrections with metallicity also plays a part in determining the relative CMD location of the isochrones.

The independence of age of the lower MS is easily explained by the fact that objects populating this CMD region have very long evolutionary times and are still on the ZAMS. The location of the TO at a given chemical composition is determined by the value of the stellar mass evolving at the stage of central hydrogen exhaustion. Increasing the SSP age means that lower masses are in this evolutionary stage, hence the lower TO brightness. The metallicity dependence at a fixed age is caused by the lower luminosity of MS stars with higher metallicity, which more than compensates for the fact that higher metallicity SSPs of a given age display larger masses at the TO because of their longer evolutionary timescales.

As for the RGB, we have already discussed how its location is weakly dependent on the stellar mass, and this is true especially for the mass range corresponding to ages above a few Gyr. On the other hand, as remarked in previous chapters, the chemical composition strongly affects the temperatures of RGB stars, hence the dependence of the colours of the isochrone RGB on Z. The ZAHB brightness is mostly determined by the value of the He core mass at the He flash, which decreases with increasing metallicity, and therefore more metal-rich ZAHBs are fainter. Age does not appreciably affect the He core mass when the age of the evolving RGB star is above ~4 Gyr (corresponding to masses lower than 1.3–1.2 $M_\odot$, the precise value depending on the initial chemical composition) and therefore the ZAHB brightness is independent of age for old SSPs.

It is also important to discuss briefly the effect of different metal mixtures on theoretical isochrones. Figure 9.5 displays two pairs of 10 Gyr isochrones for two different values of the metallicity Z, one pair computed with a scaled solar mixture, the other with a $[\alpha/\text{Fe}] = 0.4$ metal mixture, typical of the halo of our galaxy. At low metallicities the scaled solar and α-enhanced isochrones are identical (as expected on the basis of the short discussion about single evolutionary tracks) whereas sizable

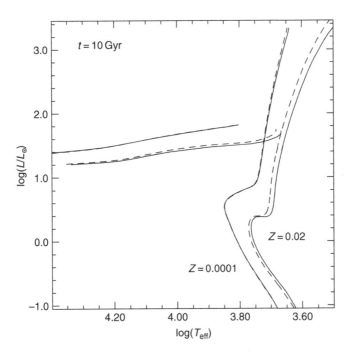

Figure 9.5 CMDs of two pairs of 10 Gyr isochrones from the ZAMS to the ZAHB, computed with metallicities $Z = 0.0001$ and 0.02, both scaled solar (solid line) and α-enhanced ($[\alpha/\text{Fe}] = 0.4$ – dashed line)

differences in both luminosities and effective temperatures are present when $Z = 0.02$ (see [175], [184]).

When the isochrones are transformed to an observational CMD by using bolometric corrections for the appropriate metal distribution, there is still good agreement at low metallicity between scaled solar and α-enhanced isochrones with the same Z, but this deteriorates when Z increases (large differences start to appear when $Z > 0.001$, [42]) scaled solar isochrones being generally redder and fainter at a given Z. This means that the use of $[\alpha/\mathrm{Fe}] = 0.4$ bolometric corrections further amplifies the differences found in the HRD. We also recall that a given value of Z corresponds to different values of [Fe/H] for scaled solar and α-enhanced mixtures.

Our ability to predict the observed CMD of SSPs of varying ages and initial chemical compositions, opens the door to the possibility of using quantitative and qualitative properties of theoretical isochrones as tools to determine a number of fundamental parameters of SSPs. In the next sections we will focus our attention on techniques aimed at determining ages, initial chemical abundances and distances to young and old SSPs; these are important parameters needed to study the processes of cosmic evolution that have lead to the formation of the structures we see in the universe today, and test the consistency of the cosmological model with the constraints coming from stellar evolution theory.

9.2.2 Age estimates

A direct lower limit to the age of the universe and information about the first stages of galaxy formation may be obtained by determining the age of the oldest objects in the Galaxy, that is, the metal-poor ([Fe/H]<0) stars located in the halo. Galactic globular clusters are particularly useful for this purpose, since they are SSPs located at distances much larger than their spatial extent, so that their stars are, to a very good approximation, all at the same distance from us. This enables us to apply simple methods based on the stellar evolution theory. The fact that globular clusters are SSPs is established by the striking correspondence between isochrones for old SSPs and globular cluster CMDs (see Figure 9.6; the ZAHB in the cluster CMD is the lower envelope of the observed HB star distribution) and spectroscopic measurements that provide remarkably uniform [Fe/H] values for stars belonging to the same cluster.

It is, however, also evident that observed CMDs at a given magnitude display a non-negligible colour range, at odds with theoretical isochrones. This broadening of the observed colour sequences is due to photometric errors and blending effects (due to unresolved neighbouring stars in the cluster) plus the presence of unresolved binaries. In the latter two cases the fluxes we receive from the unresolved components add up, producing brighter magnitudes at a given colour when compared with the isochrone prediction, hence widening the expected MS colour distribution at a given

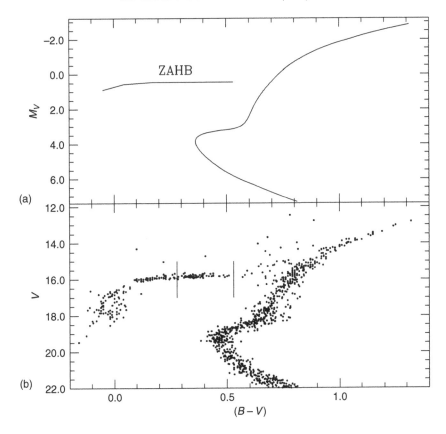

Figure 9.6 Comparison of the CMDs of (a) a 12 Gyr old metal-poor isochrone and (b) the globular cluster M15 (data from [67]). The position of the RR Lyrae instability strip is marked on the cluster CMD

magnitude (these unresolved objects are located to the right-hand side of the single star MS).

It is customary to determine from the empirical CMD a 'fiducial line' corresponding to the observed CMD in the case of negligible photometric errors, blending and unresolved stars. What is done in practice is to divide the observed CMD in magnitude bins (whose size is dictated by the need to have a large enough sample of stars in each bin to allow a statistical analysis) and determine the colour distribution of the objects in each bin. This colour distribution usually shows a very clearly-defined peak and the corresponding value (the mode of the distribution) is usually taken as representative of the fiducial point at that bin. In the case of the SGB, which is almost horizontal in the CMD, colour bins are considered and the mode of the magnitude distribution in each bin is assigned to the SGB fiducial line. In this way the observed CMD is reduced to a line, that can be more easily compared to theoretical isochrones (see Figure 9.7).

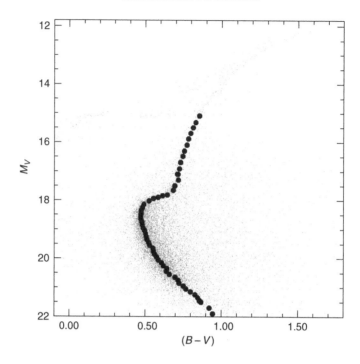

Figure 9.7 Fiducial line for the MS, SGB and RGB of the globular cluster M5 (filled circles) superimposed on the observed CMD (data from [190])

A natural approach to the age determination would be to fit the whole theoretical isochrone to the observed CMD of a generic SSP. If the stellar models were to be largely free of uncertainties, a global fit of isochrones to observations (for example through a χ^2 minimization procedure) would provide a series of parameters, i.e. age, distance and initial chemical composition. However, uncertainties hard to quantify especially on T_{eff} and colours of theoretical models still exist, related mainly to the treatment of superadiabatic convection (stars populating old SSPs are low-mass stars with convective envelopes) molecular opacities at low temperatures and theoretical spectra for cold stars. A global isochrone fit could therefore lead to inaccurate solutions for the parameters of the best-fitting isochrone if some of the theoretical predictions are largely in error. It is therefore customary to focus on some age-sensitive features of the models that are less affected by current shortcomings in the model computation. We will see in the next chapter that, for the specific case of composite stellar populations, one cannot avoid the use of the whole isochrones to gain information about the observed CMD.

Figure 9.4 has already shown the effect of varying age on isochrones. The TO region is clearly the point to be used for age determinations; it becomes redder and dimmer for increasing age. One could, in principle, compare either the observed TO brightness or the TO colour with the appropriate theoretical isochrones after correcting the observed quantities for the effect of distance and extinction. Differential quantities like the magnitude (usually V) difference between ZAHB and TO or the colour (usually either $(B - V)$ or $(V - I)$) difference between TO and the base of the RGB have been devised in order to overcome this problem. Since distance and extinction affect TO, RGB and ZAHB in the same way, magnitude or colour differences are insensitive to these parameters.

Recent results about absolute and relative age determinations of large samples of Galactic globular clusters ([170], [183], [226]) based on the methods described below provide a broadly consistent picture in which the metal-poorer clusters ([Fe/H] below ~ -1.6) are coeval (have the same age) within ~ 1 Gyr, their age being $\sim 12.5 \pm 1.5$ Gyr. More metal-rich clusters display an age spread up to ~ 4–5 Gyr. Among these non-coeval clusters an age–metallicity relationship is present, with the mean age decreasing by ~ 0.2 Gyr for a 0.1 dex increase in [Fe/H]. The oldest clusters in the Magellanic Clouds appear to be coeval with the clusters in our galaxy, and in general the age of the oldest star clusters for which we can detect the TO region is in good agreement with the age of the universe predicted by the currently favoured cosmological model (see Chapter 1).

The vertical method

The so-called vertical method (see, for example [53], [205]) for the age determination of old SSPs is based on the comparison between observed and theoretical values of the quantity $\Delta V = V_{TO} - V_{ZAHB}$ (see Figure 9.8) the magnitude difference between the TO point and the ZAHB at the instability strip region, around $\log(T_{eff}) = 3.85((B - V) \sim 0.3)$. The precise reference point along the ZAHB is not crucial, as long as the horizontal part of the ZAHB is considered.

Given that the ZAHB brightness is largely unaffected by age, a change of age at a given [Fe/H] changes the value of ΔV through the change of the TO brightness; for increasing ages ΔV increases, because the TO gets dimmer. This method works well mainly in the V band (or photometric bands at a similar wavelength) where the ZAHB is mostly horizontal. Figure 9.9 shows how in different wavelength bands the ZAHB is no longer horizontal, due to the behaviour of the corresponding bolometric correction with T_{eff}; in these cases the observed value of ΔV is affected by the choice of the reference colour along the ZAHB and, as a consequence, by the SSP reddening.

We have displayed in Figure 9.10 the run of ΔV with [Fe/H] for different ages of the SSP. At a given [Fe/H] a 0.1 mag variation of ΔV (that is, a 0.1 mag variation of V_{TO}) corresponds to ~ 1 Gyr age change in this age range typical of the oldest objects in our galaxy. Once an observed value of ΔV is fixed, an 0.4 dex uncertainty in the cluster [Fe/H] causes an age uncertainty by only ~ 1 Gyr. This makes the vertical method relatively insensitive to uncertainties in the cluster [Fe/H]. This occurrence stems from the fact that at a given age both the ZAHB and the TO brightness scale with [Fe/H] in approximately the same way. Typical errors of the observed ΔV values

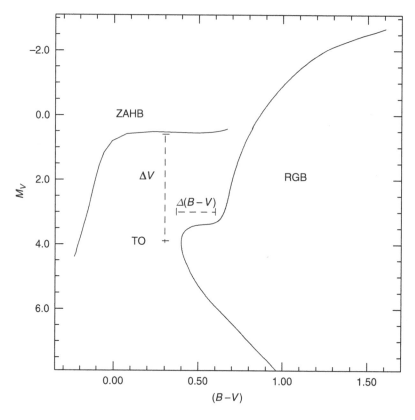

Figure 9.8 Graphical representation of the ΔV (vertical) and $\Delta(B-V)$ (horizontal) age indicators for old SSPs

for the best-observed clusters are of the order of 0.10 mag, mainly due to the fact that the TO region in the CMD is approximately vertical (large V range at almost constant colour) making somewhat difficult the precise detection of the TO point. Some authors (i.e. [25], [53]) try to circumvent this problem by using the magnitude of a point close to the TO but 0.05 mag redder, either on the MS or on the SGB. The brightness of this reference point along the MS or the SGB is still sensitive to the SSP age (see the isochrones in Figure 9.4) and in principle is better defined than the TO one. However, some information on the MS or SGB theoretical colours are being introduced in the calibration of its brightness with age, and they depend on more uncertain quantities, like the superadiabatic convection treatment and the colour transformations.

Uncertainties in the isochrone V_{TO} predictions are expected to be of the order of a few hundredths of a magnitude, due to both the bolometric luminosity and bolometric correction predictions.

Sometimes the observed HB magnitude considered does not correspond to the ZAHB, instead it is the mean level $< V_{HB} >$ of objects in the RR Lyrae instability

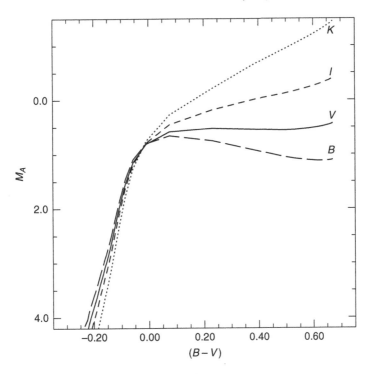

Figure 9.9   CMDs in various photometric filters of a ZAHB computed for a metallicity $Z = 0.001$

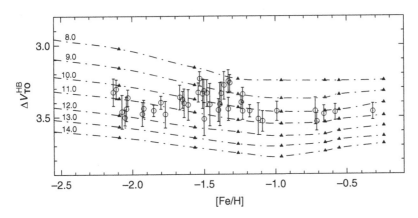

Figure 9.10   Theoretical values of ΔV as a function of [Fe/H] and age, compared with the observational counterpart data for a large sample of Galactic globular clusters

strip or at its red side. In this case one can use empirical relationships between the observed mean HB level and the ZAHB one, as in the following one (from the data published in [188]):

$$V_{ZAHB} = <V_{HB}> + 0.05[\text{Fe/H}] + 0.20 \tag{9.7}$$

This empirical relationship clearly shows that the HB becomes wider (in magnitude) when the metallicity increases.

Practical difficulties are encountered when trying to apply the ΔV technique to clusters with an HB populated only in the blue part (see the discussion about the second parameter problem later on) or when the observed CMD includes only a few HB stars. In the former case the ZAHB is almost vertical, thus making its use to define the ΔV parameter difficult, and again its value would depend on the choice of the reference colour at which the ZAHB level is determined. In the latter case, when only a few stars populate the observed HB, it is impossible to define the lower boundary of the star distribution that corresponds to the ZAHB, and also the determination of the mean magnitude of HB stars may be subject to biases due to the small star sample. In both these situations one can determine the age difference with respect to clusters with well-behaved HB, using the horizontal method discussed in the next section. Alternatively, in the case of a poorly populated (and non-blue) HB another possibility is to determine the relationship between the ZAHB (or mean HB) level and the observed mean value by performing Monte-Carlo (MC) simulations of the observed HB population ([143]). By employing an MC algorithm, it is possible to distribute randomly along the HB portion of the theoretical isochrone, a number of objects equal to the observed sample, and also include their photometric errors. In brief, the location of one star in the CMD of the MC simulation is obtained by drawing a mass value (within the range of the initial masses of the HB stars) according to a prescribed IMF, and interpolating between the colours and magnitudes of neighbouring points along the isochrone (corresponding to initial masses bracketing the value drawn with the MC algorithm). One can then determine magnitude and colour values that include photometric errors by drawing a number from a Gaussian distribution with mean value equal to the magnitude (or colour) determined from the isochrone, and σ equal to the prescribed photometric error.

After the mean magnitude of this synthetic sample is determined, simulations are repeated many times, and the final distributions of mean values (say 100) is analysed. The difference between the mean (or the mode) of this distribution and the theoretical ZAHB level provides the best estimate of the correction to be applied to the observed HB mean magnitude, to determine the observational value of ΔV.

The horizontal method

The horizontal method for age determination (see, for example [191], [227]) involves the comparison of the observed and theoretical values of the quantity $\Delta(B - V) = (B - V)_{RGB} - (B - V)_{TO}$ (or equivalently in $(V - I)$ colours), i.e. the colour difference between the TO and the base of the RGB (see Figure 9.8). There are various possible definitions of $(B - V)_{RGB}$; here we denote with $(B - V)_{RGB}$ the colour of the RGB 2.5 mag above the TO magnitude. This horizontal parameter is sensitive to age through the variation of $(B - V)_{TO}$ with t, given that the RGB colour is unaffected by age for old populations. This means that an increase of the SSP age decreases the value of

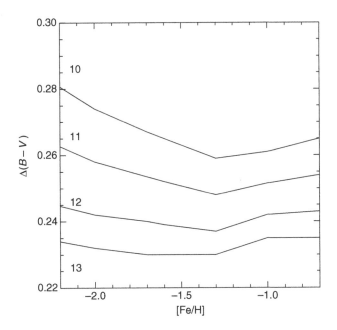

Figure 9.11 Theoretical values of $\Delta(B - V)$ as a function of [Fe/H] and age (from [181])

$\Delta(B - V)$ (or $\Delta(V - I)$) because the TO gets redder. Figure 9.11 displays the run of $\Delta(B - V)$ with [Fe/H] for some representative ages.

As in the case of the vertical method, $\Delta(B - V)$ at fixed age is weakly sensitive to the metallicity. This may appear surprising, given the strong dependence of the RGB colour on Z. However, the change of the RGB location with Z is compensated by the corresponding change of the TO colour; in fact the dependence of the TO colour on the metal content (at fixed age) has the same sign as for the RGB, e.g. they both become redder when the metallicity increases. The derivative $\Delta(B - V)/\Delta t$ is ~ 0.010–0.015 mag Gyr^{-1} around $t = 12$ Gyr (the precise value depends on the value of the absolute age); this means that one needs an extremely high accuracy in both the observational determination and theoretical prediction of this quantity to keep the error on the estimated age low. Theoretical uncertainties (due to colour transformations and treatment of superadiabatic convection) are surely higher than 0.01–0.02 mag, and therefore the horizontal method is hardly used for absolute age determinations. However, as discussed, for example, in [181] and [227], one can employ this horizontal method to determine age differences (for example in the case of clusters with a very blue vertical HB or with a few HB stars in the CMD, as discussed before) with respect to template clusters whose absolute age is well established using the ΔV technique. The reason is that the value of $\Delta(B - V)/\Delta t$ around a given age t appears to be weakly affected by the use of different sets of colour transformations, different values of the mixing length and also by a change of the initial He content of the stellar models.

Ages from Strömgren photometry

The Strömgren system has been designed to isolate parts of the stellar spectra from which to build colour indices sensitive to specific properties of the stars; it consists of four filters *uvby*, each covering a wavelength range of typically ≈ 200Å, plus a narrower pair β_n and β_w centred at 4860 Å, with a bandwidth of 30 Å and 150 Å, that measure the strength of the Balmer H_β line and its adjacent continuum. From these filters the colour indices $(b - y)$, $c_1 \equiv (u - v) - (v - b)$, $m_1 \equiv (v - b) - (b - y)$, $\beta \equiv \beta_w - \beta_n$ are usually computed.

The reddening effect on the $(b - y)$ colour due to extinction, $E(b - y)$, is related to $E(B - V)$ according to $E(B - V) = 1.4\ E(b - y)$; the dereddened values of c_1 and m_1 are related to the reddened ones by $(c_1)_0 = c_1 - 0.20\ E(b - y)$ and $(m_1)_0 = m_1 + 0.32 E(b - y)$. The β index is insensitive to reddening due to the very narrow spectral range covered.

Figure 9.12 displays two pairs of isochrones (MS and part of the RGB) for ages of 12 and 14 Gyr and two different values of [Fe/H], in the $c_1 - (b - y)$ plane. The morphology of the isochrones resembles the CMDs determined from the *BVI* filters; the MS is on the lower part of the diagram, the TO is again the age-sensitive region (lower c_1 for increasing age) whereas the RGB is on the right side of the diagram, at high values of $(b - y)$. The TO position can be used for age determination –

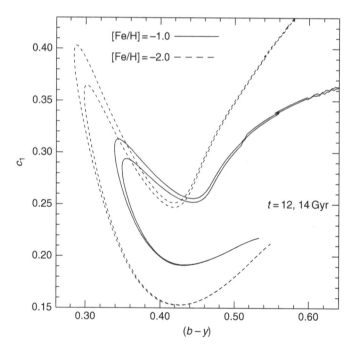

Figure 9.12 Isochrones in the $c_1 - (b - y)$ CMD for the labelled values of age and [Fe/H] (from [181]). The older isochrone at each metallicity has a lower value of c_1 at the TO

by comparing observed TO colours with predictions from theoretical isochrones – without the need to know the SSP distance (only reddening corrections are needed) since c_1 and $(b - y)$ are both colour indices. Transformations to Strömgren colours have still large uncertainties, larger than the case of wider filters like the Johnson's ones, and therefore ages derived from this kind of diagram have to be treated with caution.

Ages from WD isochrones

The age-dating methods discussed above are based on isochrones that cover the evolutionary phases from the MS until the HB. Isochrones can, however, be computed until the final WD stages if the thermal pulsing phases and associated mass-loss processes are accounted for. Figure 9.13 shows isochrones extended to the WD evolutionary stages (hereinafter WD isochrones; we consider here only DA objects). The brighter part of each WD isochrone overlaps with the cooling track of the single WD mass corresponding to the WDs produced by stars evolving at the end of the AGB phase. Due to the behaviour of the mass loss and evolution along the AGB phase, the mass of the WD produced by stars evolving along the AGB in old populations

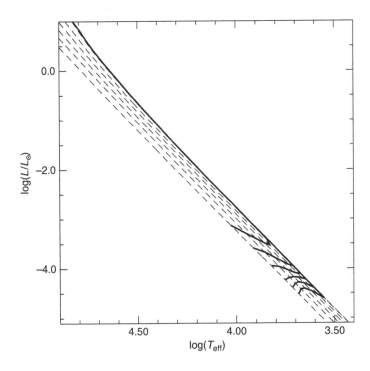

Figure 9.13 HRD of a set of WD cooling tracks with masses between 0.54 and 1.0 $M_\odot$ (dashed lines). The heavy solid lines display isochrones with ages of 2, 4, 6, 8, 10 and 12 Gyr (from [177])

is practically constant; this explains the fact that the bright part of all isochrones in the figure is exactly the same. At the bottom end of the isochrone one must finally recover the objects produced by all stars already evolved past the AGB phase, which entered the cooling sequence earlier. This explains the left turn of the WD isochrone at the bottom of the sequence, since higher-mass WDs (produced by larger progenitor masses) evolve at smaller radii.

Figure 9.14 displays the same WD isochrones in various CMDs. The shape of the isochrones mirrors the shape in the $\log(L/L_\odot) - \log(T_{eff})$ plane; for $(J - K)$ colours and high ages the situation is more complicated, and the overall isochrone shape is due to the combined effect of both a blue turn of individual cooling tracks when a certain low temperature is reached, due to the effect of the H_2 collision-induced absorption on the spectrum, and the increase of the WD mass at the bottom of the isochrone.

An important property is that the brightness of the bottom end of the WD isochrone decreases with increasing age, because of a more advanced cooling stage of the WDs. This property suggests the use of the bottom luminosity of the WD sequence of an SSP as an age indicator, if its distance is known. Matching the observed end of the sequence to theoretical WD isochrones provides an estimate of the SSP age.

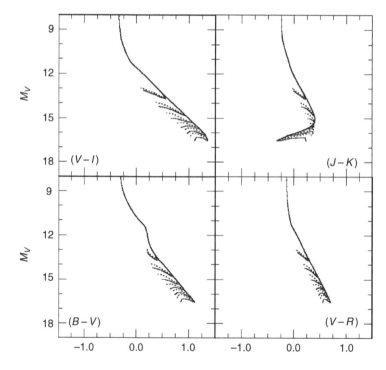

Figure 9.14   Various CMDs of the WD isochrones shown in Figure 9.13. The dashed lines display the effect of including chemical separation upon crystallization

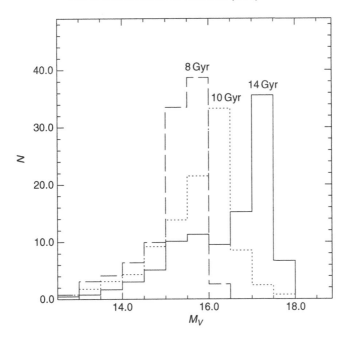

Figure 9.15 Luminosity functions for WD isochrones with ages of 8, 10 and 14 Gyr. The WD progenitors have solar metallicity, and the adopted IMF is the Salpeter one

This procedure is, however, performed better considering the LF of the WD population. From the theoretical point of view the LF is easily determined by computing the number of progenitor stars – hence the number of corresponding WDs – populating a given point along the WD isochrone, using an IMF. Figure 9.15 displays theoretical LFs (for an arbitrary total number of WDs) using a Salpeter IMF. The peak of the LF and the subsequent cut-off correspond approximately to the bottom end of the isochrone, where WDs of different masses pile up due to their finite cooling time. Increasing the age of the SSP moves the peak and cut-off towards fainter magnitudes, mirroring the behaviour of the underlying isochrones. A match of the position of the observed LF cut-off with theoretical LFs (once the SSP distance modulus is known) provides an estimate of the SSP age.

At present, this technique only provides an approximate age estimate for old SSPs, due to both theoretical and observational problems. On the observational side it is difficult to detect the bottom end of the WD sequence, due to its very low luminosity when ages are above a few Gyr; a clear-cut detection has only been possible up to now for a few open clusters. On the theoretical side there are still sizable uncertainties in the input physics (EOS of the CO core and envelope, opacities of the hydrogen and helium external layers, boundary conditions) for cool WDs. Also our very approximate knowledge of mass-loss mechanisms for RGB and AGB stars introduces a large uncertainty in the predictions of the relationship between the final WD mass and the initial mass of its MS progenitor, and the trend with metallicity.

The initial–final mass relation and the adopted IMF probably only have a small impact on the derived WD ages ([162]).

HB colour and second parameter problem

There is potentially another relative age indicator for old stellar populations, that makes use of HB stars. As mentioned before, as a first approximation one expects that the mass loss along the RGB can be described by the Reimers formula with a fixed mean value of η plus a given spread $\Delta\eta$ around this mean value. With this approximation one can determine the ZAHB location and HB evolution of the SSP stars after the He flash; once η and $\Delta\eta$ are fixed, for increasing metallicity and fixed age stars are located at lower T_{eff}, hence redder colour. The reason is that, although for a fixed η more metal rich RGB stars tend to lose relatively more mass (higher luminosities at the tip of the RGB and lower T_{eff} along the RGB) this effect is compensated for and reversed by the larger evolving mass in more metal-rich RGB isochrones of a given age (because of large TO masses at a given age, due to the longer MS lifetime of stars with a given mass when metallicity increases). The net result is that isochrones with larger metallicity have larger evolving masses (at a fixed isochrone age) on the HB phase, hence redder HBs. In addition, due to the higher envelope opacity, even if the HB mass were to be the same, more metal-rich stars are redder along the HB phase. Therefore, the metallicity is the main ('the first') parameter controlling the HB stellar distribution.

On the other hand, if metallicity is kept fixed and the age changed, the HB becomes redder (bluer) for an age decrease (increase) due to the larger (smaller) mass evolving along the HB (the mass loss along the the RGB is practically independent of age at fixed metallicity, for old ages).

This means that the colour distribution of HB stars in principle depends on age, once the metallicity is known. If at the same metallicity one cluster has a redder HB, one expects it to be younger. The HB colour is very often quantified by the parameter $HB_{type} = (N_B - N_R)/(N_B + N_V + N_R)$ where N_B, N_R and N_V denote the number of stars at the blue side, red side and within the RR Lyrae instability strip, respectively. The value of HB_{type} degenerates for clusters populated only at the red or blue side of the instability strip; in these cases different colour distributions provide the same value of this parameter. Figure 9.16 displays HB_{type} and [Fe/H] data for a large sample of globular clusters, together with theoretical predictions for various age differences with respect to a reference age of 13 Gyears.

The problem with the colour of HB stars is that some pairs of globular clusters with the same metallicity (inferred from spectroscopy) – like M3–M13 or NGC288-NGC362 – seem to have a negligible age difference (as estimated from the vertical or horizontal method) but not the same value of HB_{type}. This is the origin of the so-called 'second' parameter phenomenon, i.e. the fact that at a given metallicity clusters of apparently the same age have different HB colours. What is the second parameter that, beside age, can change the HB colour at a given [Fe/H]? A different

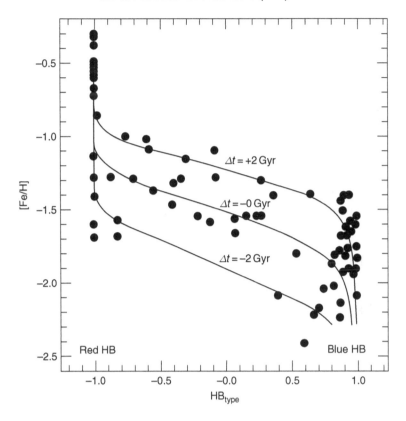

Figure 9.16 [Fe/H] vs HB_{type} for a large sample of Galactic globular clusters. Theoretical relationships corresponding to various age differences with respect to a reference age of 13 Gyr are also displayed. Notice the large range of HB_{type} at constant metallicity when [Fe/H] is between ~ -1 and ~ -2

mass-loss law, maybe caused by fast stellar rotation or dynamical interactions within the cluster? A different initial He abundance (the He abundance changes the T_{eff} location of a given stellar mass along the HB)? These questions are still unanswered, although probably the initial He abundance is not the main culprit.

9.2.3 Metallicity and reddening estimates

We have already discussed how the colours and slope of the RGB of an old SSP are affected by the cluster metallicity and are insensitive to age. In principle the comparison of the dereddened colours of RGB stars in an SSP with theoretical isochrones can provide an estimate of the SSP metallicity. Due to the mentioned uncertainties in the theoretical prediction of the colours of cold stars, it is customary to use empirical relationships – based on clusters with good photometry and spectroscopic determinations of [Fe/H] – linking [Fe/H] to appropriately chosen parameters that measure location and shape of the RGB. Recent calibrations (see [194]) based on $V - (V - I)$ photometries (similar relationships in infrared colours are given by [73]) of a sample of globular clusters provide a number of useful relationships between the parameters $S, \Delta V_{1.4}, (V - I)_{-3.5}, (V - I)_{-3.0}$ and [Fe/H]. S is the slope in the $V - (V - I)$ plane of the line connecting the RGB point at the level of the HB and a point 2.5 magnitudes brighter; it is reddening and distance independent. $\Delta V_{1.4}$ is the magnitude difference between the HB and the RGB at fixed (dereddened) colour $(V - I) = 1.4$; it needs an estimate of the reddening. $(V - I)_{-3.5}$ and $(V - I)_{-3.0}$ are the RGB colours at $M_I = -3.0$ and -3.5, respectively. These two parameters require the knowledge of reddening and distance; due to the approximately vertical shape of the RGB, a rough distance estimate is sufficient. These four parameters are related to [Fe/H] through the following numerical relationships (the σ error on the [Fe/H] estimate is also given):

$$[Fe/H] = -0.24 \, S + 0.28, \quad \sigma = 0.12$$

$$[Fe/H] = -0.85 \, \Delta V_{1.4} + 0.77, \quad \sigma = 0.16$$

$$[Fe/H] = -2.12(V - I)_{-3.5}^2 + 8.81(V - I)_{-3.5} - 9.75, \quad \sigma = 0.15$$

$$[Fe/H] = -3.34(V - I)_{-3.0}^2 + 12.37(V - I)_{-3.0} - 11.91, \quad \sigma = 0.15 \qquad (9.8)$$

valid for [Fe/H] between ~ -2.2 and ~ -0.7.

The Strömgren photometry is also very useful to determine separately the metallicity of stars, and their reddening. Reddening of low-mass MS stars can be determined by applying the following relationship derived empirically by [198] using a sample of about 250 local MS stars, close enough to be negligibly affected by interstellar extinction. After introducing the quantity $\Delta \beta \equiv 2.720 - \beta$, it is possible to determine $E(b - y)$ for a given star – hence $E(B - V)$ and all other extinctions – by comparing its observed $(b - y)$ with the intrinsic value $(b - y)_0$ given by the following empirical relationship ([198]).

$$(b - y)_0 = 0.579 + 1.541(m_1)_0 - 1.066(c_1)_0 - 2.965(\Delta \beta)$$
$$+ 9.64(\Delta \beta)^2 - 4.383(m_1)_0(\Delta \beta) - 3.821(m_1)_0(c_1)_0$$
$$+ 6.695(c_1)_0(\Delta \beta) + 7.763(m_1)_0(c_1)_0^2 \qquad (9.9)$$

This relationship covers the range of low-mass MS stars with $(b - y)_0$ between 0.254 and 0.550, $(m_1)_0$ ranging from 0.033 up to 0.470, $(c_1)_0$ between 0.116 and 0.540, β between 2.550 and 2.681. It is obvious that this equation has to be iterated in order to obtain a self-consistent value of $(b - y)_0$ since one cannot know from the start the dereddened counterpart of the observed c_1 and m_1. One usually starts with the observed c_1 and m_1. One usually starts with the observed c_1 and m_1, determines a first estimate of $(b - y)_0$, hence $E(b - y) = (b - y) - (b - y)_0$. With this first estimate of $E(b - y)$ one corrects the observed values of m_1 and c_1 and the procedure is repeated until $(b-y)_0$ changes by less than a specified (small) amount (i.e. 0.001 mag) between two consecutive iterations.

Starting from the dereddened Strömgren colours of MS stars, it is also possible to estimate their [Fe/H] making use of the following relationships derived by [94] calibrated on a sample of object with spectroscopic estimate of [Fe/H] ranging between -2.0 and 0.5 (in the following we drop the subscript 0 for conciseness, and all Strömgren colour indices are assumed to be extinction corrected):

$$[\text{Fe/H}] = -2.0 - 43.90m_1 + 353.4(b-y)m_1$$
$$+ 18.0(b-y)m_1^2 - 612.6(b-y)^2 m_1$$
$$+ [6.0 - 48.0m_1 - 7.85(b-y)][log(m_1 - c_3)] \qquad (9.10)$$

where $c_3 = 0.627 - 7.04(b-y) + 11.25(b-y)^2$. It is valid when $(b-y)$ is between 0.22 and 0.37.

$$[\text{Fe/H}] = -1.64 + 11.09m_1 - 29.29m_1^2 - 57.40(b-y)m_1 + 116.96m_1^2(b-y)$$
$$+ (128.0m_1 - 22.231c_1 - 206.48m_1^2)c_1 \qquad (9.11)$$

valid when $(b-y)$ is between 0.37 and 0.47.

$$[\text{Fe/H}] = -1.64 + 16.75m_1 - 12.61m_1^2 - 52.17(b-y)m_1 + 66.026m_1^2(b-y)$$
$$+ (47.98m_1 - 3.99c_1 - 65.06m_1^2)c_1 \qquad (9.12)$$

applicable when $(b-y)$ is between 0.47 and 0.59.

In the case of RGB stars with $(b-y)$ between 0.5 and 1.1 and [Fe/H] ranging from -2.0 to 0.0, the following relationship calibrated by [98] can be used

$$[\text{Fe/H}] = \frac{m_1 + a_1(b-y) + a_2}{a_3(b-y) + a_4} \qquad (9.13)$$

with $a_1 = -1.277$, $a_2 = 0.331$, $a_3 = 0.324$, $a_4 = -0.032$.

By employing these techniques based on SSP photometry one can derive an estimate for the initial [Fe/H] abundance of the parent stars. This value of [Fe/H] is also a measure of the total initial metallicity Z, once the metal distribution is given. It is very important to notice that these empirical relationships are based on local stars with their own trends of metal abundance ratios with [Fe/H]. In the case of the calibrating field halo population with [Fe/H] below ~ -1.0 and globular clusters, the parent stars are characterized by $[\alpha/\text{Fe}] > 0$. The formulae given above are appropriate only for objects with the same metal abundance ratios, since colours of scaled solar and α-enhanced SSPs are not the same for a fixed [Fe/H] value. In the case of a hypothetical scaled solar metal-poor star, the previous relationships would therefore not provide the correct value of [Fe/H].

9.2.4 Determination of the initial helium abundance

The initial He abundance is another important quantity that affects the evolution of stars, and in particular the properties of SSPs. In the previous sections we have discussed mainly the effect of a change of the initial metallicity, but a variation of the initial He mass fraction also affects the CMD of an SSP. In Figure 9.17 we display two isochrones with the same initial metallicity and the same age, but initial values of Y that differ by 0.02. The mass evolving at the TO and in post-MS phases is lower (by about $0.03M_\odot$) for the helium-rich isochrone; the helium increase also slightly shifts to the blue the overall location of the isochrone, from the MS up to the tip of the RGB. Even more importantly, an increase of Y increases the ZAHB brightness and decreases slightly the TO luminosity at a fixed age. This means that the age estimated from the observed ΔV value for an SSP decreases. A variation $\Delta Y = 0.02$ causes a decrease of the estimated age by ~ 1 Gyr for ages typical of globular clusters. It is therefore important to determine the initial helium abundance of an SSP, to employ isochrones with the correct chemical composition. An additional reason to estimate the initial helium content in old SSPs accurately is the fact that these estimates provide a strong constraint on the amount of helium synthesized during the primordial nucleosynthesis, and consequently on the cosmological baryon density.

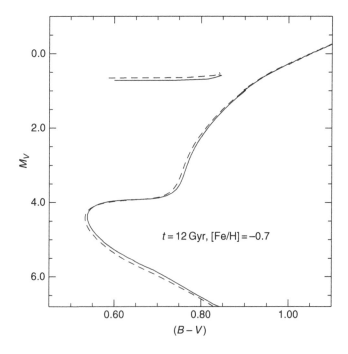

Figure 9.17   Two isochrones with the same age and metallicity, Y equal to 0.25 (solid line) and 0.27 (dashed line)

When dealing with old SSPs, it is generally not possible to measure spectroscopically the stellar surface helium abundance because stars are in general too cold to show He lines in their spectra. Estimates of Y in old SSPs make use of results from stellar evolution, taking advantage of the fact that the evolution of stars is affected by the initial Y value.

The parameter R

The so-called R parameter ([102]) is defined as the number ratio of HB stars to RGB stars brighter than the HB level (here we consider the ZAHB level) – $R = N_{HB}/N_{RGB}$ – and can be employed to determine Y. As discussed before, post-MS phases are populated by stars with approximately the same initial mass, and the star counts along a generic post-MS stage are proportional to the evolutionary timescale along that phase. Since the value of the masses evolving along the RGB and HB and their lifetime are weakly affected by realistic variations of Y, the basic idea behind the use of R as helium indicator is that a higher initial Y (at fixed metallicity) makes the HB brighter and, in turn, produces a lower value of N_{RGB} (a smaller fraction of the RGB is contained between the HB level and the tip of the RGB) with the consequent increase of R. The value of the derivative dR/dY is ~ 10.

As shown in Figure 9.18, at fixed age and Y the theoretical value of R is very slowly decreasing up to [Fe/H] ~ -1.15. Between [Fe/H] ~ -1.15 and [Fe/H] ~ -0.85, R increases steeply; this increase happens when the RGB bump, previously located at luminosities larger than the ZAHB, moves below the ZAHB level due to the higher metallicity, causing an abrupt decrease in the number of RGB stars brighter than the ZAHB (see Section 5.9.2). At higher [Fe/H] values R is again only very mildly decreasing with increasing [Fe/H]. It is also interesting to notice how the dependence of R on age is restricted to the interval ranging from [Fe/H] ~ -1.15 to [Fe/H] ~ -0.85, which is exactly the metallicity range where the RGB bump crosses the ZAHB level. This is easily explained by the fact that the RGB bump luminosity depends on the stellar age; higher ages shift the RGB bump location towards lower luminosities. Typical errors affecting the empirical determination of R are of the order of 0.1–0.2, due mainly to determination of the ZAHB level and to Poisson statistics for the number counts; these observational errors cause an uncertainty of the order of $\sigma(Y) \sim 0.01$–0.02 in the estimate of Y for a single cluster.

The theoretical values of R displayed in Figure 9.18 do not depend on the morphology of the HB as long as it is populated in the RR Lyrae instability strip or redward, because for this mass range the evolutionary timescales along the HB phase are practically constant. In the case of SSPs with an HB populated at the blue side of the RR Lyrae region, evolutionary timescales along the central He-burning phase increase with decreasing mass, and corrections to the theoretical values of R displayed in Figure 9.18 have to be applied. At the bluest end of a typical HB, corrections up to about 20 per cent have to be taken into account.

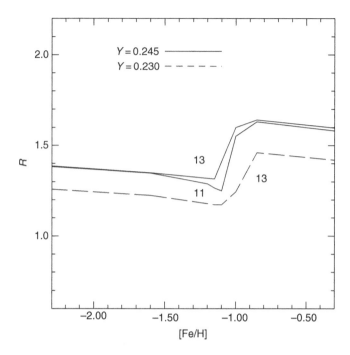

Figure 9.18 R-parameter values as a function of [Fe/H] for the labelled values of the helium mass fraction Y and age (in Gyr)

Recent analyses ([44], [178]) of the R parameter values for a large sample of Galactic globular clusters provide individual values of Y consistent (within the error bars) with a constant helium abundance whose weighted mean is $Y = 0.250 \pm 0.005$, in good agreement with the abundance predicted by the currently accepted value of the cosmological baryon density (see Chapter 1).

The parameter Δ

Figure 9.17 suggests the possibility of employing another parameter – independent of distance and reddening estimates – as a helium abundance indicator for old SSPs, i.e. the magnitude difference between the ZAHB and the MS at a given colour. This quantity, usually called the Δ parameter, is sensitive to Y because at a given colour the MS becomes fainter and the ZAHB brighter when Y increases. The exact definition of Δ varies among authors (see, for example, [189]) but it is important to consider a colour that corresponds to the lower MS, unaffected by age.

The value of $d\Delta/dY$ is ~6 mag, but its dependence on metallicity is quite high, $d\Delta/d[Fe/H] \sim -0.5$ mag dex^{-1}. A typical [Fe/H] uncertainty of 0.1–0.2 dex is therefore equivalent to a 0.01–0.02 change in the SSP helium mass fraction. This strong dependence on metallicity renders the Δ parameter badly suited for absolute Y determinations, although it is sometimes used for relative Y scaling.

The parameter A

The pulsational properties of RR Lyrae stars provide a third helium indicator for old SSPs populated in the instability strip region, the so-called mass-to-luminosity ratio A (see Section 6.6.1). The mass-to-luminosity ratio for stars inside the instability strip can be written as

$$A = \log(L/L_\odot) - 0.707 \log(M/M_\odot) \tag{9.14}$$

where L and M are the luminosity and mass of the single RR Lyrae star. When the chemical composition is fixed, A depends only on the mode of pulsation, i.e. fundamental or first overtone. If the mode of pulsation is known, A is affected by Y because increasing He increases the luminosity of the HB (hence the luminosity of RR Lyrae stars) and also the value of the mean mass populating the instability strip. The two effects tend to compensate for each other and A has a small sensitivity to Y, i.e. $dA/dY \sim 1.4$; however, in the case of large numbers of RR Lyrae stars, the mean value of A can be determined with a high accuracy, of the order of ~ 0.01, that allows estimates of Y with an accuracy of ~ 0.01. The dependence of A on [Fe/H] is practically negligible (an increase of metal abundance decreases the luminosity but also the mean mass at the instability strip, and the two effects compensate for each other) as shown in [35].

Determinations of A from RR Lyrae observations make use of the relationship between the stellar effective temperature and the pulsation period P

$$\log(P) = 11.627 + 0.823A - 3.506 \log(T_{eff}) \tag{9.15}$$

Equation (9.15) shows that in order to determine A from observations, one has to measure the pulsation periods (only marginally affected by systematic uncertainties) and estimate the effective temperature. It is mainly the uncertainties in the T_{eff} scale that hamper the use of A as a helium abundance indicator.

9.2.5 Determination of the initial lithium abundance

Empirical estimates of the amount of lithium produced during the primordial nucleosynthesis are another powerful constraint of the baryon mass density Ω_b. In the following discussion we denote the abundance of Li by [Li], defined as [Li] $= 12 + \log[N(Li)/N(H)]$, and consider only the main isotope ^7Li, since the cosmological production of ^6Li is negligible in comparison with ^7Li.[2]

---

[2] Spectroscopic observations of lithium abundances in stars determine the sum $^7Li + ^6Li$.

Spectroscopic measurements of Li abundance in the atmospheres of nearby old metal-poor MS field stars provide a qualitative picture that is basically consistent among the different authors, but which can differ in the details (see [171], [182], [217]). The main result (see Figure 9.19 with the data by [217] as an example) is that metal-poor MS stars with T_{eff} larger than approximately 5800 K show a remarkably constant [Li] value (Spite-plateau) while there is a larger depletion at lower temperatures, increasing for decreasing temperature. Moreover, in the plateau region, a handful of stars show a much lower [Li] than the plateau counterpart. The exact T_{eff} location, the extent of the plateau, as well as the existence of some weak trend of [Li] with T_{eff} and [Fe/H] are still debated. Also the absolute average value of [Li] for plateau stars shows differences between different authors, ranging between [Li] ~ 2.1 and [Li] ~ 2.4.

The simplest empirical explanation for the existence of this plateau is that it mirrors the primordial Li abundance (due to the low metallicity of the observed stars) and therefore the measured plateau [Li] should put strong constraints on the value of Ω_b. An accurate interpretation of these observed abundances must, however, take into

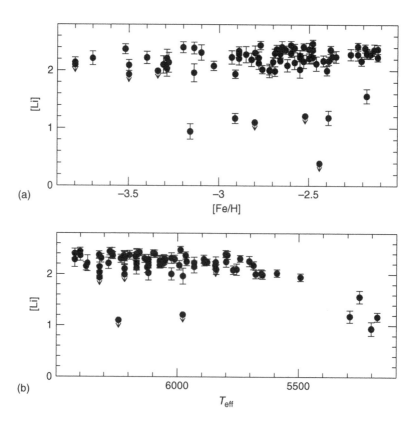

Figure 9.19 Trend of [Li] abundance values with, (a) [Fe/H] for a sample of metal-poor MS stars and (b) T_{eff}. The flat part of the [Li] distribution in (b) is called the 'Spite-plateau'

account the constraints posed by stellar evolution models. Lithium is a very fragile element, easily destroyed in stellar interiors when $T \geq 2.5 \times 10^6$ K. Such temperatures are already attained in the stellar cores during the contraction to the MS phase, and whenever surface convective regions extend down to these Li burning regions, the surface value of [Li] is rapidly decreased. For decreasing stellar mass at a fixed chemical composition, or for increasing metallicity at a given mass, PMS stellar models show that convection extends from the surface down to Li burning regions; although the bottom of this convective region is rapidly retreating towards the surface, there is time to burn a substantial amount of Li. If the stellar mass is low enough, the bottom of the convective envelope continues to overlap with the lithium burning regions during the MS phase as well, and the surface lithium depletion is much larger.

Table 9.1 shows the surface lithium depletion for selected masses and three metal-poor chemical compositions typical of Spite-plateau stars, as obtained from stellar models. Objects below $\sim 0.65 M_{\odot}$ continue to deplete the surface lithium during the MS, whereas masses larger than $\sim 0.70 M_{\odot}$ do not deplete their surface lithium appreciably even along the PMS. Figure 9.20 displays the run of [Li] as a function of T_{eff} along a metal poor isochrone with an age of 14 Gyr (a so-called lithium isochrone); it is obtained simply by displaying the Li abundance at the surface of the stars populating the various points of the isochrone MS, as a function of their T_{eff}. The shape of the Li isochrone closely mirrors the observations (the inclusion of isochrones with different metallicity spreads the theoretical points mainly in the horizontal direction) with an almost flat part and a sudden decrease of [Li] at the lowest temperatures. This sharp decrease can easily be explained in terms of lower-mass (cooler) MS stars that substantially deplete their surface Li during both the PMS and MS phases. The flatter part (that is the theoretical counterpart of the Spite-plateau) corresponds to those MS stars massive enough not to deplete Li substantially during the PMS phase, and therefore reflects the initial [Li] abundance in these objects. The uncertainties in the empirical [Li] values and the possibility of additional depletion due to non-canonical element transport mechanisms (see, for example, [182]) make it difficult to use the Spite-plateau as a constraint on Ω_b at present.

9.2.6 Distance determination techniques

Direct geometrical distance determinations are based on the concept of parallax. The idea is to employ measurements of a star angular position with respect to a stationary background of much more distant objects made 6 months apart. The parallax is defined as one-half of the change in angular position obtained from these observations, and the distance d to the star is easily obtained from trigonometry

$$d = \frac{1\text{ AU}}{\tan(p)} \sim \frac{1}{p}\text{AU}$$

i.e. the ratio between the Sun–Earth distance (1 Astronomical Unit, 1AU, is equal to 1.4960×10^{13} cm) and the parallax angle p (in radians). If the parallax is measured

SIMPLE STELLAR POPULATIONS

Table 9.1 Surface lithium depletion $\Delta[\text{Li}]$ (in dex) due to nuclear burning along the PMS and MS (from [182]) for various selected masses and initial metallicities (Z). The four lines for each case correspond to the depletion after 10 and 100 Myr (PMS) 1 and 10 Gyr (MS)

$\Delta[\text{Li}]$		
$Z = 2.5 \times 10^{-5}$	$Z = 2 \times 10^{-4}$	$Z = 6 \times 10^{-4}$
$0.6 M_\odot$		
−0.75	−0.63	−0.42
−1.03	−0.90	−0.79
−1.36	−0.96	−0.89
−1.83	−1.83	−1.76
$0.65 M_\odot$		
−0.19	−0.20	−0.19
−0.19	−0.21	−0.23
−0.19	−0.22	−0.25
−0.19	−0.22	−0.26
$0.7 M_\odot$		
−0.03	−0.06	−0.07
−0.04	−0.06	−0.07
−0.04	−0.06	−0.08
−0.04	−0.06	−0.08
$0.75 M_\odot$		
−0.01	−0.02	−0.03
−0.01	−0.02	−0.03
−0.01	−0.02	−0.03
−0.01	−0.02	−0.03
$0.8 M_\odot$		
−0.00	−0.01	−0.01
−0.00	−0.01	−0.01
−0.00	−0.01	−0.01
−0.00	−0.01	−0.01

in arcseconds, the unit distance is the parsec; the distance corresponding to a parallax of 1 arcsecond is 1 parsec, equal to 2.063×10^5 AU, or 3.26 light years. Parallax distances, even after the most recent results from the Hipparcos satellite ([148]) are limited to relatively nearby objects in our galaxy, including a handful of open

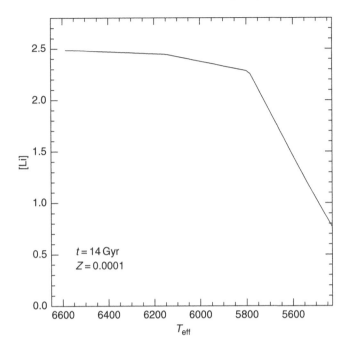

Figure 9.20 Trend of [Li] with T_{eff} along the MS of a theoretical isochrone with $Z = 0.0001$ and $t = 14\,\text{Gyr}$, for an initial lithium abundance [Li]=2.5

clusters, like Hyades and Pleiades. Among the methods presented in this section only the MS- and WD-fitting techniques can be calibrated empirically using direct parallax distances. The situation will, without doubt, improve in the near future with the launch of the GAIA satellite that will provide distances to about 1 billion stars in our galaxy, including a large sample of open clusters and even some globular clusters.

The techniques discussed in this section are based on properties of theoretical isochrones and make good use of the concept of a stellar standard candle. A perfect stellar standard candle is a class of objects for which the known absolute magnitude does not change with changing properties (e.g. age, metallicity) of the parent stellar population. When this class of objects is detected in another stellar system, the difference between their observed apparent magnitude and the absolute one provides the distance modulus of the system. Perfect standard candles may not really exist (although Cepheid stars are often considered to be a perfect candle) but what stellar evolution can do is to single out various classes of stellar objects whose brightness is predicted to depend only on the initial chemical composition – a quantity that can be estimated more easily than age, i.e. even when the TO region cannot be detected – and provide a calibration for their absolute magnitude. The classes of objects discussed below allow the determination of distances within our galaxy, between Local Group galaxies and out to the Virgo Cluster.

Main sequence fitting

Figure 9.4 shows how the brightness of the lower MS (i.e. when M_V is larger than $\sim$5.0–5.5) of old SSPs is unaffected by the age of the stellar population, even for old SSPs, because stars in this magnitude range are essentially still on their ZAMS location. It is only the initial chemical composition that determines the location of the lower MS, that becomes redder for increasing metallicity and/or decreasing helium mass fraction. Once the Y, Z values are fixed, the lower MS can be used as a template and compared to the observed MS in an SSP with the same initial chemical composition; the difference between the absolute magnitudes of the template MS and the apparent magnitudes of the observed one immediately provides the population distance modulus, hence the distance in parsec. This is the so-called MS-fitting method.

In order to determine SSP distances using the MS-fitting method, one needs the observed CMD of the SSP MS in some filter combination (e.g. V–$(B - V)$) and a template one for the same chemical composition of the SSP, which may be either theoretical or empirical. In order to derive accurate distances, anything which may systematically affect the intrinsic and observed colours and magnitudes of either the cluster or the template MS must be accounted for. The lower MS has a slope of about 5.0–5.5 in the much used V–$(B - V)$ or V–$(V - I)$ planes, and therefore even a small error on the colour quickly becomes a larger error on the magnitude, hence the derived distance modulus. One must therefore know in advance the SSP initial chemical composition, in order to select the appropriate template MS, plus the extinction along the line of sight, to correct the observed magnitudes and colours to the intrinsic values (see Figure 9.21).

The choice of the template MS is obviously extremely important, and the preferred method is to build an empirical template MS instead of using theoretical isochrones, due to the existing uncertainties in the stellar T_{eff} scale and colour transformations. In fact, an uncertainty of only 0.02 mag in colours (not due to the uncertainty in the reddening) translates into an uncertainty of $\sim$0.10 mag in the derived distance modulus (and a consequent uncertainty of about $\sim$1 Gyr in the ages obtained from the absolute magnitude of the observed TO). The uncertainty related to the error on the reddening estimate ($\Delta E(B - V)$) is smaller, of the order of $\Delta(m - M)_0 \sim 2.0 \Delta E(B - V)$; this value is derived taking into account the slope of the MS and the fact that a given value of $E(B - V)$ increases the observed m_V by a factor $\sim$3.2$E(B - V)$.

An empirical MS is built by considering local field stars of known [Fe/H] (e.g. determined from spectroscopy) with distances determined geometrically from parallax measurements, and known or negligible reddening (see, for example, the detailed discussions in [143], [145]).[3] The main problem with this approach is that only a few of these stars with parallax distances have the exact metallicity of the SSP under scrutiny; therefore it is difficult to determine the appropriate template MS. To

---

[3] Template empirical MS loci are often also used to compare theoretical MS isochrones and models with observations.

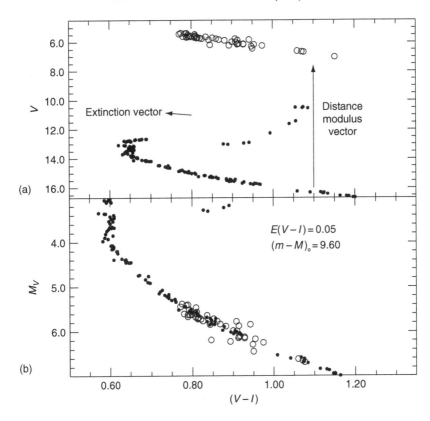

Figure 9.21 Example of MS-fitting distance determination applied to the old open cluster M 67 (see [143]). (a) This displays the absolute magnitude and intrinsic colours of the template MS (large open circles) for the cluster metallicity ([Fe/H]$= +0.02$) and the observed magnitudes and colours of the cluster (small filled circles). The effect of extinction on the cluster CMD ($E(V - I) = 0.05$) and the effect of the distance modulus are also shown. (b) This shows the best fit of the (reddening corrected) cluster MS to the template one, and the derived distance modulus

overcome this problem one has to shift the position of many template field stars of various [Fe/H] values to the location they would have at the metallicity of the SSP. The procedure of shifting these template stars should in principle preserve mass, in the sense that they have to be shifted both in colour and magnitude because stars of a given mass change both their ZAMS T_{eff} and luminosity when the metallicity is changed (e.g. cooler T_{eff} and fainter luminosity for increasing Z). However, for not too large metallicity ranges – i.e. ranges of the order of ~ 1 dex – the observed CMDs of globular and open clusters show that the shape of the lower MS is approximately constant; if this is true, one needs only to apply colour shifts – that take into account the change of T_{eff} – because, even without the appropriate magnitude correction, the shifted location will lie on the right MS. Making use of this property of constant MS slope, derivatives Δ (*colour*)/Δ [Fe/H] for the lower MS can be derived empirically (see [145], [144]) for metallicities around solar. At lower metallicities one has to determine those derivatives by using differentially theoretical isochrones, e.g. computing the colour difference between the MS of isochrones of different metallicities (hence hoping that systematic errors in isochrone colours are minimized) because of the very small number of field MS stars with known parallax (e.g. [87]). In the range $-0.45 \leq [Fe/H] \leq 0.35$ and $5.5 \leq M_V \leq 7.0$ one finds empirically

$$\Delta(B - V) = 0.154\Delta[Fe/H]$$

$$\Delta(V - I) = 0.103\Delta[Fe/H]$$

$$\Delta(V - K) = 0.190\Delta[Fe/H]$$

For $-2.0 \leq [Fe/H] \leq -0.7$ and $5.5 \leq M_V \leq 7.0$ theoretical isochrones (from [181]) provide

$$\frac{\Delta(B - V)}{\Delta[Fe/H]} = 0.062[Fe/H] + 0.207$$

A fundamental assumption behind the use of these colour shifts is that the initial helium content at a given metallicity and the metal distribution of template field stars is the same as that of the observed SSP.

Typical errors on the best MS-fitting distances to date are of the order of ~ 0.07–0.08 mag (neglecting possible systematic errors in reddening and metallicity of both the SSP under scrutiny and template MS stars).

White dwarf fitting

The WD-fitting ([158], [174]) is analogous to the MS-fitting, but in this case the bright part of the WD cooling sequence between $M_V \sim 10$ and ~ 12 is used ($10\,000 \leq T_{eff}(K) \leq 20\,000$) instead of the MS. The WD sequence in this magnitude range is independent of age for old stellar populations, as shown in Figure 9.14. In brief, a template local WD sequence made of objects with precise parallax-based distances is determined, and compared to the WD sequence in the SSP under scrutiny. The difference between the apparent magnitude of the SSP and the absolute magnitude of the template sequence provides the SSP distance modulus, and hence its distance. As in the case of the MS-fitting, the SSP WD sequence has to be corrected for the effect of extinction and reddening, before the determination of the distance modulus (the effect of reddening uncertainties on the derived distances is approximately the same as for the MS-fitting). The advantage of this technique over the MS-fitting is that it is in principle independent of the knowledge of the SSP initial metallicity, since all WDs are virtually metal-free at their surfaces, hence no colour corrections have to be applied.

There are, however, a number of points to be noticed. One has first to realize that the location of the WD sequence in this magnitude range can be affected by different initial–final mass relationships for the local WDs and the objects in the SSP under scrutiny. WDs in this bright part of the cooling sequence of old SSPs have been produced by stars just evolved out of the AGB phase, hence of mass $\sim 0.55 M_\odot$. The variation of the WD brightness M_V with mass is $\Delta M_V / \Delta(M/M_\odot) \sim 2.3$ mag at fixed colour (because of the smaller radius of more massive WDs) for masses around 0.5–0.6 $M_\odot$, typical of the lower end of the WD mass spectrum; therefore, a difference of $\sim 0.05 M_\odot$ between the template and SSP WD mass causes a bias of ~ 0.10 mag in the distance modulus.

A second potential problem relates to the existence of WDs with different envelope compositions. DA and non-DA objects do not share the same position in the CMD, and their relative location depends on the colour filters used (see Figure 9.22). This is due to the very different behaviour of the bolometric corrections between H and He atmospheres; in the HRD the non-DA object displayed in Figure 9.22 is slightly underluminous ($\Delta(L/L_\odot) \sim 0.1$) at a given T_{eff}, with respect to the DA one.

Local field WDs in this magnitude range display a number ratio of DA versus non-DA objects of the order of 4:1 and the template WD sequence used in globular cluster distance determinations ([158]) is usually a DA sequence. In a distant SSP

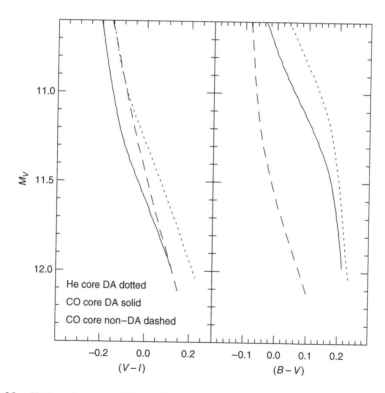

He core DA dotted

CO core DA solid

CO core non–DA dashed

Figure 9.22 CMDs of various WD cooling sequences, in the magnitude range used for the WD-fitting method. The mass of the DA and non-DA CO WDs is equal to $0.55 M_\odot$, while the mass of the He WD is $0.45 M_\odot$

like a globular cluster in our galaxy we are not yet able to determine spectroscopicaly the composition of the WD atmospheres, and typical 1σ photometric errors of ~0.05 mag suffice to overlap DA and non-DA sequences in the VI and BV plane. What we observe might therefore be a mixture of both WD types. This causes (in the case where the DA to non-DA ratio is the same as in the local neighbourhood) a very small systematic error by -0.03 mag in the WD-fitting distance modulus obtained from the VI plane, whereas the systematic error is of $+0.20$ mag in the BV plane. A larger fraction of non-DA objects would increase this bias.

A variation of the thickness of the surface hydrogen layer in DA objects can alter their radius, hence their CMD location. If we denote by $q(\mathrm{H})$ the ratio between the mass of the H envelope and the total WD mass, the magnitude M_V of the WD sequence scales with $q(\mathrm{H})$ as $\Delta M_V / \Delta \log(q(\mathrm{H})) \sim -0.035$, for $\log(q(\mathrm{H}))$ ranging between -4 and -7.

Contamination from WDs with He core can also introduce a systematic error, due to their brighter luminosity at fixed colours (because of larger radii). The amount of this bias is difficult to quantify, given that we are not able to predict the mass distribution of He core WDs in a generic SSP. Another potential source of uncertainty is the thickness of the H layer in the DA WDs, which is another quantity we cannot predict with confidence; different H masses in the template and SSP WDs can bias the distance modulus at the level of ~0.05 mag because the H envelope mass affects (albeit slightly) the WD mass radius relationship.

Tip of the RGB

The bolometric luminosity of the tip of the RGB (TRGB) is determined by the mass of the He core at the He flash, once the initial chemical composition is fixed. Since low-mass stars ignite He all with similar core masses (slightly increasing for decreasing initial mass) $M_{\mathrm{bol}}^{\mathrm{TRGB}}$ changes by a few hundredths of magnitudes in the age range between ~4 and 12–14 Gyr (see Figure 9.23). When approaching the RGB phase transition $M_{\mathrm{bol}}^{\mathrm{TRGB}}$ increases sharply (L decreases) due to the lifting of the electron degeneracy in the He core, that causes He ignition to occur at significantly smaller core masses. On the other hand, once age is fixed, $M_{\mathrm{bol}}^{\mathrm{TRGB}}$ decreases for increasing metallicity, in spite of the decrease of the He core mass at the TRGB (see Section 5.10.3). The net effect is that, for ages larger than ~4 Gyr, $M_{\mathrm{bol}}^{\mathrm{TRGB}} \propto -0.19$ [Fe/H] for metallicities well below solar, with an almost negligible dependence on the age.

Considering now the TRGB magnitude in various photometric systems for ages above ~4 Gyr, one notices that in the I-band the dependence of the TRGB brightness on [Fe/H] is minimized. In fact $dM_{\mathrm{bol}}^{\mathrm{TRGB}} / d[\mathrm{Fe/H}] \propto -0.19$, and from empirical determinations of bolometric corrections for bright RGB stars in globular clusters one obtains that $dBC_I / d(V-I) = -0.243$ ([64]). Empirically, the colour of bright RGB stars in globular clusters is related to their [Fe/H] according to $d(V-I)/d[\mathrm{Fe/H}] = 1.162[\mathrm{Fe/H}] + 2.472$ ([9]). Combining these relations one finds $dM_I^{\mathrm{TRGB}} / d[\mathrm{Fe/H}]$ $\propto -0.15$ about [Fe/H] $= -2.0$, and $dM_I^{\mathrm{TRGB}} / d[\mathrm{Fe/H}] \propto +0.20$ about [Fe/H] $= -0.7$,

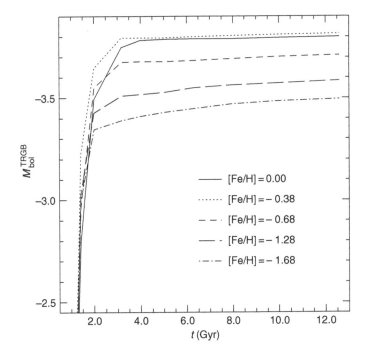

Figure 9.23 Bolometric magnitude of the TRGB for various ages and metallicities (data from [82])

with an almost negligible dependence on [Fe/H] at intermediate metallicities. For [Fe/H] > -0.7 the dependence of M_I^{TRGB} on [Fe/H] becomes very large so it cannot be used safely as a standard candle. Figure 9.24 shows the most recent empiric and semi-empiric calibrations of the I-band brightness of the TRGB as a function of the metallicity as well as a fully theoretical calibration of this standard candle. For a detailed discussion about the reason(s) of the significant disagreement between empiric calibrations and the theoretical one we refer the reader to the references quoted in the figure caption.

This kind of relationship between M_I^{TRGB} and [Fe/H] is matched by theoretical models coupled to theoretical bolometric corrections ([173]). At both shorter and longer wavelengths the TRGB brightness displays a stronger dependence on [Fe/H] and also on age. For example, in the K-band $dM_K^{TRGB}/d[Fe/H] = -0.60$ and a change of age from 12 to 6 Gyr at fixed [Fe/H] causes an ~ 0.10 mag increase of M_K^{TRGB}.

The weak and well-established dependence of M_I^{TRGB} on [Fe/H], plus the practically negligible influence of age (for ages larger than ~ 4 Gyr) has prompted the use of the TRGB as a distance indicator for old and metal poor SSPs.

The first step towards obtaining distances is the detection of the TRGB. From what we know about stellar evolution, one expects TRGB stars basically to overlap with AGB objects, that populate the CMD from below the TRGB level up to magnitudes brighter than the TRGB. However, AGB evolutionary times are shorter than RGB ones, and the LF of bright red stars in an old SSP is expected to show a discontinuity

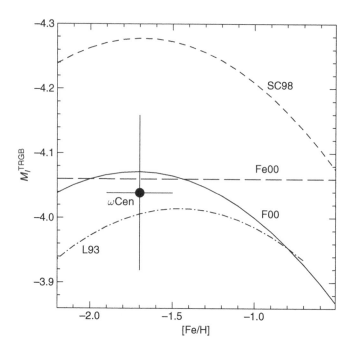

Figure 9.24 Various empirical and semi-empirical calibrations of the *I*-band brightness of the TRGB as a function of the metallicity. The different labels correspond to the works by [73] (F00), [72] (Fe00), [125] (L93). The circle with error bars denotes the empirical determination of the absolute magnitude of the TRGB in the Galactic globular cluster ω Cen obtained by [9]. The theoretical calibration by [173] (SC98) is also displayed

in correspondence of the TRGB location, as shown in Figure 9.25. The position of this discontinuity marks the level at which the contribution of the longer-lived (in comparison with AGB objects) RGB stars vanishes, and corresponds to the TRGB magnitude. A very popular technique to determine the location of the discontinuity is to employ an edge-detection algorithm, i.e. a kernel $[-1, -2, 0, +2, +1]$. Convolution of this kernel with the observed LF gives a spike in its output, at the magnitude where a discontinuity is present (see Figure 9.25). For a given magnitude M_I^i the result of the convolution is simply the weighted sum $\text{output}^i = -1N(i-2) - 2N(i-1) + 2N(i+1) + 1N(i+2)$, where $N(i)$ denotes the star counts at magnitude M_I^i.

Due to the fast evolution of stars along both the RGB and AGB phases, one needs large samples of stars to have the TRGB region populated with a sizable number of objects. Too small a star sample would leave the TRGB level almost or completely devoid of stars, and the LF would show a smooth decrease towards increasing brightness, without displaying any kind of discontinuity. An additional complication is that, when the number of objects in the magnitude bins is small, Poisson statistics causes large oscillations in the star counts between consecutive points of the LF. In this case the output of the edge-detection algorithm becomes very noisy, showing many large spikes not associated with the location of the TRGB. These constraints on the number of detected objects rules out the use of the TRGB to determine distances to individual globular clusters because the number of objects is too low. Only the most massive globular cluster of our galaxy, ω Centauri, has an unambiguous detection of the TRGB ([9]).

On the other hand, the TRGB can usefully be applied to the field halo population of external galaxies (generally supposed to be an old and metal-poor SSP) and with present observational capabilities it has been possible to detect the TRGB in objects belonging to the Virgo galaxy cluster.

Once the I-band apparent magnitude of the TRGB is determined from the discontinuity in the observed LF, and an extinction correction (to be independently determined) is applied to the observed magnitudes and colours, the distance modulus is determined from

$$(m - M)_0 = I_{0,\mathrm{TRGB}} - M_I^{\mathrm{TRGB}}$$

where $I_{0,\mathrm{TRGB}}$ is the extinction corrected TRGB magnitude, and M_I^{TRGB} is obtained using an M_I^{TRGB} – [Fe/H] calibration. In order to use this calibration one needs a metallicity estimate for the TRGB stars. Given that in general we cannot determine spectroscopically the chemical composition of RGB stars in external galaxies, one applies the following procedure. Assuming an arbitrary metallicity, we estimate a preliminary distance and determine the dereddened $(V - I)_0$ colour at the observed $M_I = -3.5$, denoted as $(V - I)_{0,-3.5}$. An empirical relationship between $(V - I)_{0,-3.5}$ and [Fe/H] (based on Galactic globular clusters) has been provided by [125]:

$$[\mathrm{Fe/H}] = -12.64 + 12.6(V - I)_{0,-3.5} - 3.3(V - I)_{0,-3.5}^2$$

This relationship, together with the first approximation of $(V - I)_{0,-3.5}$ gives an initial estimate of [Fe/H] that is then used in conjunction with the M_I^{TRGB} – [Fe/H] calibration to obtain a second approximation to the real distance modulus. This procedure is iterated until the difference between the distances moduli of two successive iterations is smaller than a prescribed amount e.g. 0.01–0.02 mag. Because of the weak dependence of M_I^{TRGB} on [Fe/H] and the almost vertical shape of the RGB, convergency is usually achieved within about three iterations. The case of TRGB distances when the observed population has an age and metallicity spread will be dealt with in the next chapter about composite SSPs.

Theoretical M_I^{TRGB} – [Fe/H] calibrations (see Figure 9.24) suffer from a zero point uncertainty of the order of $\sim$0.2 mag, related the prediction of the value of the He core mass at the TRGB and, to a lesser extent, to the BC_I scale. At [Fe/H] $= -1.3$ current results provide $M_I^{\mathrm{TRGB}} \sim -4.1 \pm 0.1$. The following relationship ([173]) is approximately at the brighter end of the existing calibrations

$$M_I^{\mathrm{TRGB}} = -3.953 + 0.437[\mathrm{Fe/H}] + 0.147[\mathrm{Fe/H}]^2$$

valid for [Fe/H] in the range between -2.35 and -0.3.

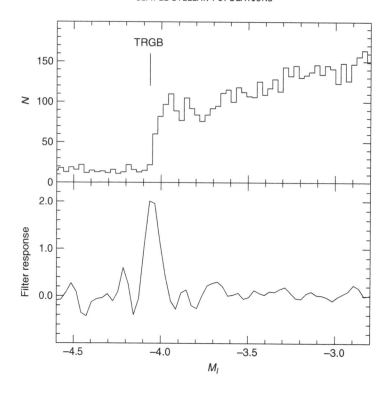

Figure 9.25 (a) Example of LF of a globular cluster-like stellar population around the TRGB plus (b) the response of the edge-detection algorithm applied to the same LF. The location of the TRGB is marked by the strong spike in the filter response, located at $M_I = -4.05$

Horizontal branch fitting

Horizontal branch fitting is historically one of the traditional methods to estimate the distance to old SSPs. The technique is conceptually very simple. The observed ZAHB, or the level of the ZAHB at a given colour (typically at the RR Lyrae instability strip) is compared to a theoretical counterpart; the magnitude difference between the observed and theoretical values provides the SSP distance modulus. Of course, the effect of extinction has also to be accounted for.

As in case of TRGB stars, the ZAHB brightness is determined by the value of the He core mass at the He flash; theory predicts a dependence only on the initial metallicity – and to a minor extent helium content – not on the age, for ages between ~4 Gyr and the age of the universe. To date, theoretical calibrations of the ZAHB luminosity suffer from the same zero point uncertainty as TRGB models, relating to the prediction of the value of the He core mass at the TRGB. The following theoretical relationship ([181])

$$M_V(\text{ZAHB}) = 0.17\,[\text{Fe/H}] + 0.78 \qquad (9.16)$$

has been shown to provide distances that agree with present MS-fitting distances ([39]) for a sample of globular clusters spanning the relevant [Fe/H] range of the globular cluster system of our galaxy.

9.2.7 Luminosity functions and estimates of the IMF

Luminosity functions of SSPs are a traditional tool to assess the level of agreement between theoretical stellar evolution models and real stars (see discussions in [43], [56], [159]). The reason is that star counts as a function of the stellar magnitudes contain information about the timescale of stellar evolution – that are not provided by the comparison of isochrones with observed CMDs – and are largely free of the uncertainties related to the treatment of the surface superadiabatic convection. In particular, the shape of the LF of post-MS evolutionary phases of an SSP is, as we have discussed before, fully determined by the evolutionary times of the single mass evolving along those phases, independently of the choice of the IMF.

Tests of post-MS phases involving the use of LFs are usually restricted to globular clusters, because they contain sizable samples of stars (typically 10^5–10^6 objects) and, being old, post-MS phases are relatively long and well populated. Figures 9.26

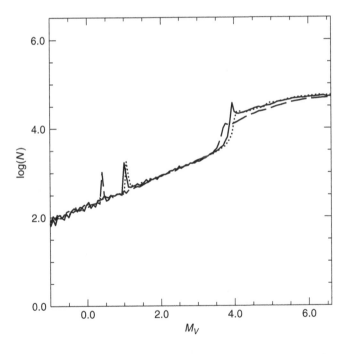

Figure 9.26 Luminosity functions covering MS, SGB and RGB for three isochrones (computed using a Salpeter IMF, $dn/dm \propto m^{-2.35}$) with, respectively, $t = 14$ Gyr, $Z = 0.008$, $Y = 0.254$ (dotted line); $t = 12$ Gyr, $Z = 0.008$, $Y = 0.254$ (solid line); $t = 14$ Gyr, $Z = 0.002$, $Y = 0.254$ (dashed line). All LFs are normalized to the same number of RGB stars at $M_V = 2.0$

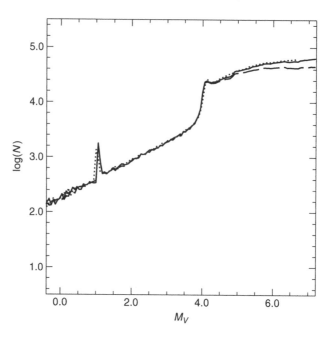

Figure 9.27   As in Figure 9.26 but for isochrones computed for, respectively, $t = 14\,\mathrm{Gyr}$, $Z = 0.008$, $Y = 0.254$ (solid line); $t = 14\,\mathrm{Gyr}$, $Z = 0.008$, $Y = 0.254$ and an IMF $dn/dm \propto m^{-1.35}$ (dashed line), $t = 14\,\mathrm{Gyr}$, $Z = 0.008$, $Y = 0.273$ (dotted line)

and 9.27 show the effect of changing Z, age, Y and IMF on the shape of theoretical LFs for old metal-poor SSPs, like Galactic globular clusters. We display the LF for MS, SGB and RGB, with the total number of stars normalized in order to have the same star counts at $M_V = 2$, a point along the RGB that is not affected by the LF age.

The MS phase is on the right-hand side of the figure; moving towards brighter magnitudes one encounters the TO region at around $M_V \sim 4$, before a steep drop in star counts representing the SGB phase. The shape of the TO region (and of course its brightness) are affected by the age; younger ages cause a more peaked shape before the drop corresponding to the SGB. The RGB phase is reached when the number count decrease follows a more gentle slope; the local maximum along the RGB is the so-called RGB bump (see Section 5.10.2). Comparison of the predicted bump brightness with SSP observations provides a powerful test for the extension of the surface convection ([43]).

The slopes of the various LFs are essentially independent of age and of the initial chemical composition. This reflects the universality of the relationship between electron degenerate He core mass and star luminosity along the RGB.

The observational counterpart of Figures 9.26 and 9.27 are the observed LF of Galactic globular clusters, like the one displayed in Figure 9.28. One cannot avoid noticing the extremely good agreement between the observed LF shape, and theoretical predictions for old SSPs. Also the RGB bump is found, as predicted by

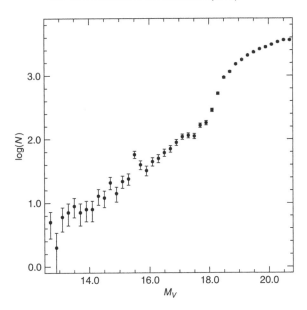

Figure 9.28   Observed LF for the MS, SGB and RGB of the globular cluster M3 ([168])

stellar evolution theory. Once the cluster distance modulus is fixed and a theoretical LF for the appropriate metallicity is computed, the agreement with theory is very good for an age of the order of ~12 Gyr ([168]).

It is very important to consider in Figure 9.27 the effect of the choice of the IMF. As expected, post-MS phases are insensitive to the IMF, but not the star counts along the MS, due to the large range of masses populating the MS. As we know, the value of the stellar mass evolving along the MS decreases towards fainter magnitudes; this explains the fact that a decrease of the exponent of a Salpeter-like IMF produces a flatter LF along the MS, because the number of stars increases more slowly when moving to fainter magnitudes. After the observed and theoretical LFs are normalized to the same number of stars along the RGB (the shape of the LF along the RGB phase is unaffected by age and choice of the IMF) the IMF of an SSP (old or young) can be determined by comparing the shape of the observed LF along the MS with the theoretical counterparts, computed for various types of IMF. This conceptually very simple technique rests on the accuracy of the theoretical mass–luminosity relationship for core H-burning stars.

In practical applications one has to be careful when interpreting the results of this method for the case of old and spatially compact SSPs. The basic problem is that, e.g. in globular clusters, dynamical interactions among the stars, over the course of time, tend to alter the initial spatial distribution of masses, slowly moving more-massive objects toward the cluster centre (mass segregation), and diffusing outwards less-massive stars. Dynamical interaction with the galactic gravitational field can also contribute to alter the star distribution within a stellar system. This means that the

LF of a particular cluster region has probably been affected by dynamical effects and instead of the IMF one determines the so-called Present Day Mass Function (PDMF), i.e. dn/dm at the present time. The relationship between PDMF and IMF has to be carefully deduced from the dynamical modeling of the observed population.

9.3 Young simple stellar populations

9.3.1 Age estimates

We define as 'young' SSPs those populations younger than ~4 Gyears. Figure 9.29 displays three young isochrones with solar metallicity in the $M_V - (B - V)$ CMD. A typical observational counterpart of young SSPs are Galactic open clusters like Praesepe, whose CMD was shown in Figure 8.7. The morphology of the TO is different from the case of globular cluster-like isochrones, because in this latter case MS stars burn hydrogen mainly through the p–p chain, whereas at young ages it is the CNO cycle that is relevant to the energy production, hence the appearance of the overall contraction (the hook-like feature at the isochrones' TOs). Also the TO luminosity is higher, because more massive stars are still evolving along the MS, and for very young ages the vertical TO region covers a large luminosity (hence

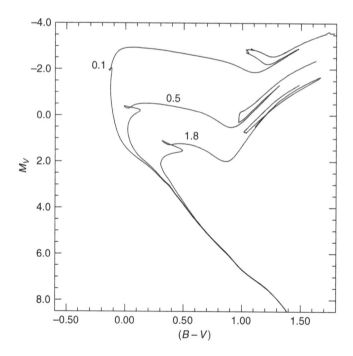

Figure 9.29   Three isochrones for a solar chemical composition and ages of 100 Myr, 500 Myr and 1.8 Gyr, respectively

magnitude) range. As for older SSPs the TO is the age indicator, although the location of the He burning phase is also now age dependent, due to the fact that for ages below 2–3 Gyr the He core along the RGB is progressively less degenerate until eventually the degeneracy is lifted. In these conditions the He core mass is no longer constant and the luminosity at the beginning of the core He-burning phase (for a fixed initial chemical composition) does depend on the age of the isochrone. The effect of Z and Y is the same as for older SSPs.

For ages between ~ 0.5 and ~ 4 Gyr the He-burning phase is usually a red clump of stars (not only for metal rich SSPs) close to the position of the (depopulated) RGB sequence, because the RGB evolution is very fast and stars lose a small amount of matter. When ages are below ~ 0.5 Gyr the helium burning phase moves progressively to the blue side of the CMD, because stars describe increasingly larger loops to the blue, and the longer evolutionary times are when He-burning objects are close to the blue end of the loops.

Precise age determinations for young SSPs are difficult, also due to the fact that the SGB and RGB phase is almost completely depopulated because of the much faster evolutionary timescale (the He core mass at the end of the central H-burning reaches the Schönberg–Chandrasekhar limit). In these conditions the horizontal method described before is of no use.

If the distance to the observed SSP can be determined, a fit of theoretical isochrones to the TO brightness provides an estimate of the age. Otherwise the vertical method can be used, considering the luminosity of the He-burning red clump stars, whenever a sizable sample is present; the dependence of ΔV on age and metallicity is different from the case of 'old' SSPs, because of the dependence on age of the He-burning phase (see Figure 9.30). A poorly populated cluster may suffer from the lack of a sizable number of stars in the He-burning phase (evolutionary timescales are shorter than for old SSPs) that hampers the use of the vertical method for age dating.

Another way to determine ages of young SSPs is, like for the old ones, to use the bottom end of the WD cooling sequence as an age indicator. This method can be applied to objects older than ≈ 50 Myr, the minimum age for the production of WDs by the more massive intermediate mass stars. In the case of SSP ages of the order of 50–200 Myr the use of this method is hampered by the possible paucity of WDs (because the IMF largely favours less massive stars) and also by the uncertain duration of the first cooling phases, that can be affected by the details of the transition from the AGB to the WD phase and is strongly dependent on the efficiency of neutrino energy losses (irrelevant at older cooling ages).

Young stellar populations also provide a mean to calibrate the efficiency of the overshoot from the convective core along the MS phase. As shown in Figure 9.31, the shape of the TO region in young SSPs can constrain the efficiency of the overshoot, because it is affected by the extension of the fully mixed central regions (also notice that ages obtained from isochrones including overshoot are systematically higher). Another possibility to constrain core overshoot is to compute the number ratio of MS stars to He-burning stars, N_{MS}/N_{He}. Due to the larger convective cores, models including core overshoot have larger He cores during the He-burning phase, hence

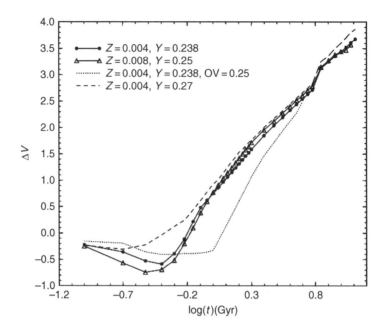

Figure 9.30 Values of the parameter ΔV as a function of age for young isochrones with various initial chemical compositions. Results for a set of isochrones including overshoot from the convective cores (extension of $0.25H_p$) are also shown (data from [47])

larger luminosities that cause shorter timescales in spite of the larger mass of the convective core; on the other hand the MS lifetime is longer because of the larger reservoir of hydrogen in the more extended central convective regions, and the ratio N_{MS}/N_{He} is higher than in the case of models without overshoot.

Age estimates from the lithium depletion boundary

The lithium depletion boundary (LDB) technique is another independent method to determine the age of young SSPs with ages between ~ 50 and ~ 200 Myr, not based on MS or WD stellar models. As discussed in the sections about PMS evolution and the Spite plateau, proton reactions destroy lithium around a temperature of $\sim 2.5 \times 10^6$ K. The low mass fully convective objects (M below $\sim 0.3 - 0.4 M_\odot$, VLM) reach the ZAMS having already completely destroyed their initial Li content due to the very fast and efficient mixing associated with convection, and therefore spectroscopic observations of their photosphere do not detect any lithium. However, if the cluster is young enough (age below ~ 200 Myr) these VLM objects are still evolving along the PMS phase, and the lithium might not have completely disappeared from their photosphere. The rate at which the core temperature increases to the Li burning temperature is a strong function of stellar mass; higher-mass stars reach this

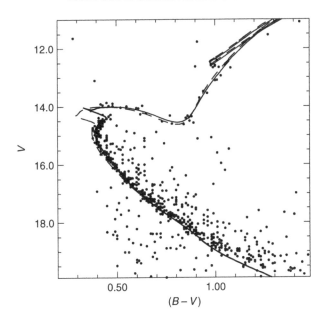

Figure 9.31 Comparison between two theoretical isochrones with [Fe/H] = −0.44 and the CMD of the open cluster NGC2420. The solid line represents a 3.2 Gyr isochrone computed including an overshoot of $0.2H_p$ from the boundary of the convective cores along the MS; the dashed line is a 2 Gyr old isochrone that does not include overshoot. A distance modulus $(m − M)_0 = 11.90$ and a reddening $E(B − V) = 0.06$ have been applied to the isochrones

temperature at an earlier age and higher luminosity than lower-mass stars (see, for example, [13]). Therefore, for a fixed value of the SSP age, one expects to find VLM PMS stars with surface fully depleted of Li only down to a certain luminosity (e.g. mass). Below this level – called the lithium depletion boundary – stars will again display some photospheric Li, because they are not hot enough to have burned this element in their interiors. Increasing the SSP age shifts the LDB to lower luminosities, because lower masses had time to reach the Li burning temperature. This property is not valid for ages lower than ∼50 Myr, because below this age limit VLM stars have not yet started to burn Li.

Figure 9.32 shows the CMD of PMS stars in the young open cluster α Persei (age around 100 Myr); the location of the LDB can be inferred by the brightest magnitude at which lithium is detected in the PMS objects. Once the LDB is detected and corrected for the cluster distance modulus, its location can be compared with theoretical results. In practice, one can compute theoretical isochrones of various ages for PMS stars (see Figure 9.33) and determine the expected level of the LDB from the Li abundances predicted by the models of the evolving masses; the theoretical LDB for various ages can then be compared with the observed one to find the best match. Analyses of the uncertainties involved in this age-dating method provide values of the order of 10–20 per cent, due mainly to the difficulty of detecting the LDB and theoretical bolometric corrections (see, for example, [28]).

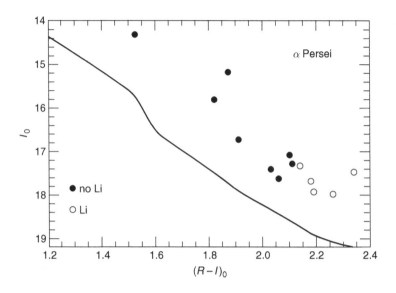

Figure 9.32 CMD of low-mass PMS stars in the young open cluster α Persei. Stars with and without lithium detection are plotted with different symbols (data from [204]). The age derived from the LDB is ~ 90 Myr when a distance modulus $(m - M)_0 = 6.23$ and $A_I = 0.17$ are assumed. The solid line shows the expected location of the ZAMS

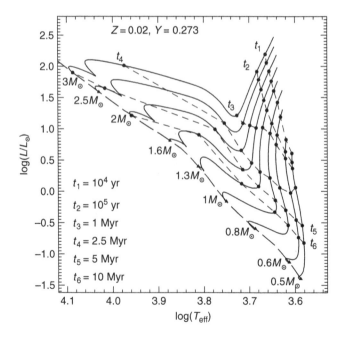

Figure 9.33 PMS tracks of stars of various total masses and solar metallicity. Thin dashed lines display PMS isochrones for selected ages. The heavy dashed line connects the ZAMS location of the different models

Comparison of LDB with TO ages is an additional method to constrain the extension of the overshoot from the convective cores of MS stars. At the present time the LDB ages of some clusters (e.g. Pleiades, IC 2391) are higher than the TO ages determined from isochrones without overshoot, suggesting a non-negligible value for this parameter.

9.3.2 Metallicity and reddening estimates

Metallicity and reddening of young SSPs can be estimated using the methods described before that make use of faint MS stars, since these objects are also present in young populations. Strömgren $uvby$ photometry provides an additional technique to determine the reddening of a young SSP. If the age of the SSP is of $\approx 10^7 - 10^8$ years, the CMD is populated below the TO by MS stars of masses between ~ 3 and $20 M_\odot$ that are still close to their ZAMS location (B-stars of luminosity class IV–V in the spectroscopists' terminology). Local objects in this age range describe in the $c_1 - (b - y)$ diagram a standard sequence (see Table 9.2) which, according to the theoretical isochrones, is largely independent of the star metallicity for [Fe/H] > -1 ([89]).

In Figure 9.34 we show the standard sequence obtained from local stars with negligible reddening and approximately solar metallicity, plus the direction of the reddening vector. This vector is nearly horizontal, given that $(c_1)_0 = c_1 - 0.20E$

Table 9.2 Standard sequence for B-stars in the Strömgren filters (from [147])

$(b - y)$	c_1	$(b - y)$	c_1
−0.134	−0.250	−0.050	0.578
−0.126	−0.128	−0.046	0.619
−0.120	−0.075	−0.044	0.656
−0.118	−0.025	−0.042	0.693
−0.114	0.022	−0.041	0.724
−0.109	0.065	−0.040	0.755
−0.105	0.108	−0.039	0.785
−0.100	0.150	−0.038	0.811
−0.096	0.192	−0.037	0.833
−0.091	0.235	−0.035	0.856
−0.086	0.278	−0.034	0.878
−0.080	0.321	−0.032	0.900
−0.075	0.362	−0.029	0.925
−0.070	0.404	−0.026	0.950
−0.065	0.448	−0.023	0.975
−0.061	0.491	−0.020	1.000
−0.055	0.535		

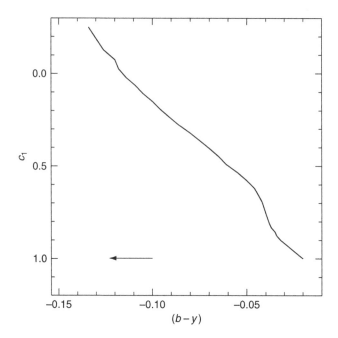

Figure 9.34 Standard sequence for unreddened local B-stars. The arrow shows the direction of the reddening vector

$(b - y)$, as obtained from the extinction curve. Once one or more B-stars are observed in the Strömgren filters, comparison of their position in the $c_1 - (b - y)$ CMD with the local sequence provides the reddening $E(b - y)$. The value of $E(b - y)$ can be obtained by shifting the colours of the observed stars along the reddening vector, until the standard sequence is reached. The amount of the shift, e.g. in $(b - y)$, provides $E(b - y)$. To the first order, $E(b - y)$ is simply the difference between the observed $(b - y)$ and the standard sequence value corresponding to the observed c_1, because the reddening vector is almost parallel to the horizontal axis.

Once $E(b - y)$ is known, extinctions and reddenings in all other photometric bands and colour indices can easily be derived according to the extinction curve; for example, $E(B - V) = 1.4E(b - y)$.

9.3.3 Distance determination techniques

Distances to young SSPs can be determined using the MS-fitting and WD-fitting techniques described before. The TRGB and He-burning phases are of less use because they are both affected by the SSP age (the TRGB does not exist for ages below ~ 0.5–1 Gyr).

A very important technique that can be applied to young SSP makes use of the Cepheid period–luminosity $(P\text{–}L)$ relationships, whenever a number of Cepheid variable stars is detected in the SSP under scrutiny (see [89] for an application to an LMC star cluster). The typical age range of SSPs harbouring Cepheid stars is between ≈10 and ≈100 Myr.

As already discussed, Cepheid stars show both empirically and theoretically well-defined relationships between their average intrinsic luminosity along a pulsation period (hence average absolute magnitudes in various wavelength bands) and the period itself. The main idea of this method is to measure the periods (from their observed light curves) of the Cepheid stars in the observed SSP and fit these values to a standard P–L relationship of the kind

$$M_A = a \log(P) + b$$

for the photometric band A employed. The magnitude shift to be applied to the SSP objects in order to fit the standard P–L relationship provides the SSP distance modulus. Distances out to ~ 25 Mpc ([75]) have been determined in this way using photometric data taken with the Hubble Space Telescope. The use of P–L relationships in two (or more) different photometric bands allows one to determine simultaneously the distance modulus and extinction. Consider, as an example, the much used V and I filters; the apparent magnitudes m_V and m_I of the observed Cepheids are related to their absolute magnitudes M_V and M_I as follows:

$$m_V = M_V + (m - M)_0 + A_V$$
$$m_I = M_I + (m - M)_0 + A_I$$

Computation of the apparent distance moduli in V and I provides

$$(m - M)_V = (m - M)_0 + A_V$$
$$(m - M)_I = (m - M)_0 + A_I = (m - M)_0 + 0.60 A_V$$

where we made use of the relationships between A_I and A_V coming from the extinction curve. Therefore, the difference between the two apparent distance moduli provides an estimate of A_V, that can be used to determine $(m - M)_0$ from $(m - M)_V$. If additional photometric bands (for example B, K or other filters) are available for both the SSP Cepheid observations and the P–L standard relationship, this procedure can be performed making use of additional P–L relationships and extinction coefficients.

Standard Cepheid P–L relationships

The key ingredient to determine distances using Cepheid stars is the P–L relationship that links measured periods to absolute mean magnitudes. One outstanding problem to date, is the following: what provides the most reliable calibration of the P–L relationships?

Historically it has been customary to employ empirical calibrations which, however, are not usually based on Galactic Cepheids. In fact, only a few Cepheids in the Galaxy have parallax error below 30 per cent, i.e. a photometric error below ~ 0.6 mag ($\sigma(M_X) = 2.17\sigma(p)/p$) and moreover Galactic Cepheids – which are necessarily located in the disk – have usually high and uncertain reddening. Independent distances to young clusters harbouring Cepheids are also difficult to estimate, because of the high reddening and the fact that these clusters are sparsely populated and highly contaminated by field disk stars.

The template P–L relationship traditionally used has been determined on the Cepheids in the LMC, which to a good approximation are all at the same distance, and reddenings are thought to be relatively low and easier to estimate. Assuming individual extinctions from reddening maps of the LMC, the following relationships are obtained

$$< V_0 > = -2.760 \log(P) + 17.042, \quad \sigma = 0.159 \tag{9.17}$$

$$< I_0 > = -2.962 \log(P) + 16.558, \quad \sigma = 0.109 \tag{9.18}$$

(the period is in days) where the dispersions around these mean relationships are given ([222]). These equations refer to Cepheid stars pulsating in the fundamental mode. For first overtone Cepheids the observed periods (P_1) can be transformed into a corresponding fundamental one (P_0) according to ([71]):

$$P_1/P_0 = 0.716 - 0.027 \log(P_1) \tag{9.19}$$

In this way one can use the calibrations given by Equations (9.17) and (9.18) for first overtone pulsators as well.

The non-negligible dispersion around the calibrations given by Equations (9.17) and (9.18) are due not only to photometric errors and some small depth effect of the LMC (i.e. the spatial dimensions of the galaxy are not completely negligible with respect to its distance) but also to an intrinsic property of Cepheid stars. The morphology of the Cepheid instability strip is shown in Figure 6.16, where lines of constant periods are also displayed. It is clear from this plot that at a fixed value of P there is a range of luminosities (hence magnitudes) allowed to the Cepheid variables, due to the intrinsic width of the strip and the slope of the constant period lines. This occurrence provides a natural explanation for the observed dispersion around the mean P–L relationship of the LMC Cepheids. As a general rule, the value of the P–L slope becomes more negative moving towards longer wavelength filters, and the dispersion decreases.

The dispersion around the P–L relationship can be reduced if one introduces a colour term which takes into account the width of the strip at constant period, effectively determining a period–luminosity–colour (P–L–C) relation. In the case of the LMC and for the VI filters one obtains ([222])

$$< I_0 > = -3.246 \log(P) + 1.409(V - I) + 15.884, \quad \sigma = 0.074 \tag{9.20}$$

with a reduced dispersion.

The use of the previous P–L empirical calibrations determines only the relative distance between the observed SSP and the LMC. The distance from Earth can be obtained only assuming a distance to the LMC, whose most accepted value to date is $(m - M)_0 = 18.50 \pm 0.1$, corresponding to $\sim 50\,\text{kpc}$.

On the theoretical side, we report here V and I P–L relationships for $\log(P) <$ 1.5 (the upper limit of the period range in the LMC empirical calibration) obtained from the sophisticated pulsational models published in [17], [37]

$$< M_{V,0} > = -2.94 \log(P) - 1.32; \quad \sigma = 0.17, \ [\text{Fe/H}] = -0.7$$

$$< M_{V,0} > = -2.75 \log(P) - 1.37, \quad \sigma = 0.18, \ [\text{Fe/H}] = -0.3$$

$$< M_{V,0} > = -2.20 \log(P) - 1.62, \quad \sigma = 0.14, \ [\text{Fe/H}] = 0.0$$

$$< M_{I,0} > = -3.11 \log(P) - 1.92, \quad \sigma = 0.12, \ [\text{Fe/H}] = -0.7$$

$$< M_{I,0} > = -2.98 \log(P) - 1.95, \quad \sigma = 0.13, \ [\text{Fe/H}] = -0.3$$

$$< M_{I,0} > = -2.58 \log(P) - 2.14, \quad \sigma = 0.10, \ [\text{Fe/H}] = 0.0$$

These relationships agree well with the LMC result for an LMC distance modulus of 18.50 mag, taking into account that the typical chemical composition of LMC Cepheids is $[\text{Fe/H}] \sim -0.4$.

We also note that the same theoretical models predict that the K-band P–L relation is almost completely unaffected by the metallicity and show a smaller dispersion due to the reduced width of the instability strip

$$< M_{K,0} > = -3.33 \log(P) - 2.61, \quad \sigma = 0.06, \ [\text{Fe/H}] = -0.7$$

$$< M_{K,0} > = -3.27 \log(P) - 2.61, \quad \sigma = 0.06, \ [\text{Fe/H}] = -0.3$$

$$< M_{K,0} > = -3.09 \log(P) - 2.67, \quad \sigma = 0.04, \ [\text{Fe/H}] = 0.0$$

A further advantage is that the K-band is also negligibly affected by interstellar extinction.

These theoretical models predict a metallicity dependence of the P–L relationships in V and I, in the sense that at a given period an increase of the metallicity decreases the Cepheid mean brightness. This means that one has to correct the distances obtained with LMC-based empirical relationships for this chemical composition effect, in case the target population displays a $[\text{Fe/H}] \neq -0.4$.

Empirical analyses of the metallicity corrections to Cepheids distances obtained assuming a universal LMC-based P–L ([116], [167], [172], [213]) have not yet achieved a firm results. To complicate the problem is the fact that not only a change in Z but also a change in Y appear to affect the theoretical P–L, and it is fair to say that to this day the question of chemical composition effects on Cepheid P–L relationships is far from being conclusively settled.

Ages from Cepheids

We briefly mention here another application of the Cepheid P–L relations. Together with the determination of distance and reddening, it is in principle also possible to estimate the age of a young SSP using the P–L relationships. The main idea is that – for a fixed chemical composition – at a given period there is a corresponding value of the stellar mass crossing the instability strip, with a certain age t. One can therefore determine a mean period–age (P–A) relationship using stellar evolution models ([19])

$$\log(t) = -0.79 \log(P) + 8.49, \quad \sigma = 0.09, \quad [\text{Fe}/\text{H}] = -0.7$$
$$\log(t) = -0.78 \log(P) + 8.41, \quad \sigma = 0.10, \quad [\text{Fe}/\text{H}] = -0.3$$
$$\log(t) = -0.67 \log(P) + 8.31, \quad \sigma = 0.08, \quad [\text{Fe}/\text{H}] = 0.0$$

where t is the Cepheid age in years.

10 Composite Stellar Populations

10.1 Definition and problems

A Composite Stellar Population (CSP) is a collection of stars formed at different times and with different initial chemical compositions. The observational counterparts of this theoretical concept are galaxies, that in many cases are made of multiple generations of stars, and often show clear signs of current star formation activity. Figure 10.1 displays the CMD of the solar neighbourhood, that appears clearly to be a CSP, due to the coexistence of a bright MS and well-populated SGB and RGB, that reveal the presence of both young (the bright MS objects) and old (the SGB and RGB objects) stars.

The fundamental information that characterizes a CSP is its Star Formation History (SFH), that is the evolution with time of the amount (i.e. total mass) of stars formed (Star Formation Rate – SFR) and their initial chemical composition (Age Metallicity Relation – AMR). We will formally denote the SFH as the function $\Upsilon(\Psi(t)\Phi(t))$ where $\Psi(t)$ is the SFR and $\Phi(t)$ is the AMR. The SFR and AMR are not independent, given that each generation of stars will eventually inject in the interstellar medium – through supernova explosions and mass-loss processes along the AGB and RGB phases of stars with the appropriate mass range – large quantities of gas chemically enriched by nuclear processes.

If the SFR is known, together with the IMF and the efficiency of accretion and gas loss from the galaxy, stellar evolution theory would then provide the necessary information to follow the chemical evolution of the gas and therefore the chemical composition of the various stellar generations, i.e. the function $\Phi(t)$. In the most generic case $2 + N$ quantities have to be calculated: the gas mass $g(t)$, the mass existing in form of stars $s(t)$, and the mass fraction $X_i(t)$ of the ith element ($i = 1, \ldots, N$). The evolution with time of the total mass of the system $M(t)$ (excluding

Evolution of Stars and Stellar Populations   Maurizio Salaris and Santi Cassisi
© 2005 John Wiley & Sons, Ltd

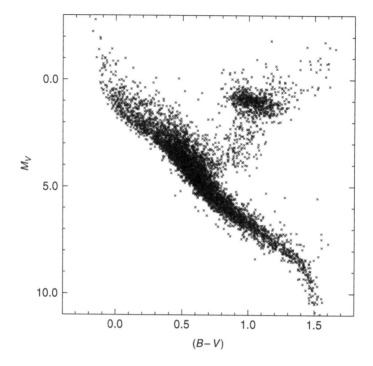

Figure 10.1 Extinction-corrected CMD of stars in the solar neighbourhood with precise parallax measurements from the Hipparcos satellite

the dark matter whose amount is not affected by the star formation activity) is described by the following equations

$$M(t) = g(t) + s(t) \tag{10.1}$$

$$\frac{dM(t)}{dt} = F(t) - E(t) \tag{10.2}$$

where $F(t)$ is the rate of accretion of gas from outside the system, and $E(t)$ the rate of ejection of gas from the system (neglecting possible loss or accretion of stars due to dynamical effects). The evolution of the gas mass $g(t)$ is described by

$$\frac{dg(t)}{dt} = F(t) - E(t) + e(t) - \Psi(t) \tag{10.3}$$

where $e(t)$ is the rate of ejection of gas from the stars due to mass loss during the stellar evolution and supernova events (an information coming from stellar evolution models), and $\Psi(t)$ is the SFR (in units of mass per year). The evolution of the mass in stars $s(t)$ can be written as

$$\frac{ds(t)}{dt} = \Psi(t) - e(t) \tag{10.4}$$

The evolution with time of the mass fraction $X_i(t)$ of a generic (non-radioactive) element i is described by

$$\frac{d(g(t)X_i(t))}{dt} = e_{X_i}(t) - X_i(t)\Psi(t) + X_i^F(t)F(t) - X_i(t)E(t) \qquad (10.5)$$

where $e_{X_i}(t)$ is the total mass of element i ejected from stars during their evolution, $X_i(t)\Psi(t)$ is the mass locked into stars due to star formation (that depends on the IMF, the stellar evolutionary timescales and the stellar evolution processes), $X_i^F(t)F(t)$ is the addition of mass due to inflowing material (the abundance of element i in this inflowing material, $X_i^F(t)$ is a priori different from the abundance within the system) and $X_iE(t)$ is the mass lost due to ejection from the system. These equations can be applied to the galaxy as a whole – and in this case all the quantities involved are mean values over the whole galaxy – or the system can be divided into various regions (like a stellar model is divided into many layers) and the equations applied to each individual region. The chemical abundances at a given time t are then used as input parameters to compute the evolution of stars formed out of the interstellar medium at t.

A simplified toy model for the chemical evolution of a CSP is the simple model with instantaneous recycling, also called the closed-box model (see, for example, [212]) that we will briefly discuss for heuristic purposes. Its simplicity makes it a useful standard for comparison with more sophisticated models, even if it is of limited applicability in galactic haloes and possibly elliptical galaxies and bulges.

This model is based on a series of simplifying assumptions. The first one is that $F(t)$ and $E(t)$ are equal to zero, i.e. no material leaves and enters the system ($M(t) = M$ is constant) and at time $t = 0$ all matter is in the form of a gas ($g(0) = M$). Second, the gas is always assumed to be homogeneous (i.e. well mixed at any time, like gas in a stellar convective zone that always has a homogeneous chemical stratification). Third, the delay between the formation of a generation of stars and the injection of freshly produced heavy elements into the interstellar medium by those stars is assumed to be negligible. This latter assumption is called the instantaneous recycling approximation and it is reasonable at least for Type II supernovae, whose progenitors have short lifetimes. We define the yield p_i of a generation of stars as the mass of a generic element i freshly produced, divided by the amount of stellar mass that remains locked in long-lived low-mass objects and compact remnants (i.e. white dwarfs, neutron stars). Typical values of p_i are of the order of 10^{-1}–10^{-2}. In the closed-box model the yield p_i of any element is assumed to be the same for any generation of stars (i.e. independent of their initial chemical composition).

Suppose that at a certain time t there is a mass of stars s, a gas mass g and a mass Z_i of a generic element i in the gas. The mass fraction X_i of element i in the interstellar gas is given by $X_i = Z_i/g$. Consider now the effect of forming a mass $\delta s'$ of stars. Because of the instantaneous recycling approximation, a certain amount of matter chemically enriched by this generation of objects is instantaneously injected into the interstellar medium. If the amount of mass remained locked into low-mass stars and remnants is denoted by δs, the total change δZ_i in the mass of element i arising from these new stars is

$$\delta Z_i = p_i \delta s - X_i \delta s \tag{10.6}$$

where the first term on the right-hand side gives the amount produced and given back to the interstellar medium by the new generation of stars, whilst the second term is the negative contribution due to the amount of gas locked into the long-lived object. A second equation needed to close the system is the fact that the total mass of the system has to be constant, hence

$$\delta s = -\delta g \tag{10.7}$$

The third equation is obtained by writing down the change of the abundance X_i in terms of the variation of the ratio Z_i/g, i.e.

$$\delta X_i = \delta \left(\frac{Z_i}{g}\right) = \frac{\delta Z_i}{g} - \frac{Z_i}{g^2} \delta g = \frac{1}{g}(\delta Z_i - X_i \delta g) \tag{10.8}$$

Combining Equations (10.6), (10.7) and (10.8) we finally obtain the very simple equation

$$\delta X_i = \frac{1}{g}(p_i \delta s - X_i \delta s - X_i \delta g) = -p_i \frac{\delta g}{g}$$

that connects the change of $X_i(t)$ to the gas content of the system. Integration of this equation, assuming a constant p_i (constant at least in the time interval of integration) provides the mass fraction of element i at a generic time t

$$X_i(t) - X_i(0) = p_i \ln \left(\frac{g(0)}{g(t)}\right) \tag{10.9}$$

Since $g(0)$ is the total mass of the system (no stars are formed at time $t=0$) according to the closed-box model the mass fraction of a generic element i in the gas and in stars formed at time t increases with decreasing gas fraction. This result is not surprising since, as time goes on, stars are formed from the interstellar gas and massive stars return chemical elements to the interstellar medium. The supply of interstellar gas is steadily consumed and the remaining gas is enriched in chemical elements. Equation (10.9) is obviously subject to the constraint that $X_i(t)$ can be at the most equal to one. This means that the yields cannot always stay constant, but have to become extremely small when the gas fraction is approaching zero.

One can also determine the X_i abundance distribution of the stars still surviving at time t. The mass of stars formed with abundance smaller than $X_i(t)$ is given by

$$s(X_i < X_i(t)) = g(0) - g(t) = g(0)(1 - e^{-X_i(t)/p_i})$$

where we have used Equation (10.9) to express $g(t)$ as a function of $g(0)$ (in this example we assumed $X_i(0) = 0$, as in the case of metals). Application of this closed-box model to the solar neighbourhood reveals a severe lack of long-lived metal-poor stars (the so-called G-dwarf problem) with respect to the theoretical predictions. More sophisticated chemical evolution models have therefore to be used for the local Galactic disk, relaxing at least some of the simplifying assumptions built into the closed-box model.

In principle, the full set of Equations (10.1)–(10.5), coupled to stellar evolution calculations, should be able to describe both the chemical and spectrophotometric evolution with time of a generic CSP. In practice, because of the lack of solid predictions concerning the efficiency of galactic star formation, inflow and outflow processes, and also uncertainties in the stellar evolution prescriptions for the yields of each stellar generation, the chemical evolution equations are actually only used to partially constrain the form of the functions $\Psi(t)$, $F(t)$ and $E(t)$ in real CSPs. This is done by comparing the predicted chemical abundance trends with observed ones, whenever spectroscopic abundance determinations for samples of stars belonging to the CSP are available.

A very powerful way to provide the empirical foundations upon which to build a comprehensive theory of galaxy formation, is to determine the SFR and AMR of a galaxy stellar population from their observed CMD, when available. Stellar evolution theory provides the appropriate tools to unveil the SFH of a generic CSP that can be resolved into its parent stars. The necessary ingredients are a CMD of the observed population, and a set of theoretical isochrones spanning a large range of ages and initial chemical compositions. The main idea[1] is to simulate theoretically the CMD of the observed CSP as a linear combination of CMDs of elementary populations with a homogeneous age and metallicity distribution within a small age and metallicity range, centred around discrete values of t and Z. The contribution of these elementary populations to the observed CMD is then determined by varying the coefficients of the linear combination until the best match of the synthetic CMD with the observed one is achieved, taking advantage of the fact that to a large extent stars of different ages and initial chemical compositions cover different regions in the CMD. Of course some degeneracies are in principle possible, but can be avoided by detecting stars in both the MS and more advanced evolutionary phases. As an example, consider an observed CMD that samples only the upper RGB of an unknown stellar population. If the chemical composition of the observed stars is the same but their age spans a range between for example, 13 and 8 Gyr, the observed RGB will be almost indistinguishable from an RGB produced by, for example, 13 Gyr old coeval stars with the same chemical composition. This is due to the very weak dependence of the RGB properties on age, at least in the age range of globular clusters. If, however, the MS phase is also sampled, an age range translates into multiple TOs, hence an SGB

---

[1] A different approach from the one described in this section is described in [97]; it is essentially based on searching the SFH that has the maximum likelihood to produce the observed CMD.

that covers a large magnitude range, whereas the coeval population will have a very narrow (in magnitude) SGB.

The approach briefly sketched above, which we call the 'inverse approach' to the determination of the SFH of a CSP, i.e. of a galaxy, will be the main subject of the rest of this chapter.

10.2 Determination of the star formation history (SFH)

Methods to determine the SFH of a generic CSP are based on the following general assumptions.

1. The stellar models accurately predict the observed properties of stars of different masses as a function of their age and metallicity.

2. The IMF used in the computations – either independent of age and metallicity or variable – is a realistic counterpart of the true IMF.

3. The observational errors can be accurately measured and modelled.

4. The theoretical stellar populations employed in the analysis represent all the populations present in the observed CSP.

We will give in the following more details about the method to determine the SFH of a CSP using photometric observations of its stellar content. Details and implementation of this technique vary from author to author (see, for example, [65], [76], [92], [219]) and a comparison of the results of different implementations applied to the same galaxy (the dwarf irregular IC 1613 in the Local Group) can be found in [200].

First, the observed CMD is divided onto a grid with N cells of a specified width (a variable cell size is often used, as discussed at the end of this section) in both the colour and magnitude directions (see Figure 10.2); the width of the cells – e.g. 0.1 mag – is dictated by the need to have a sizable sample of stars in each cell, and to resolve with more than a few cells the various evolutionary phases displayed by the observed CMD. The number of stars in a generic grid cell (i) of the observed CMD is denoted $N_0(i)$; the index i runs from one to the total number of grid cells N.

Second, a number of synthetic CMDs of elementary stellar populations is created. Each elementary population j is computed considering a collection of stars (for a specified value of their total mass) with a uniform distribution of ages and metallicities within intervals Δt and ΔZ, centred around n discrete values of age t and m metallicities. This corresponds to constant SFR and AMR within each Δt and ΔZ bin. The synthetic CMDs can easily be computed using MC techniques to draw randomly stellar mass values (according to the prescribed IMF) plus the corresponding

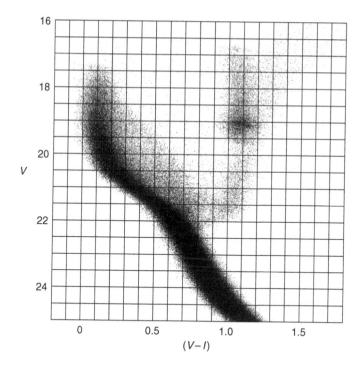

Figure 10.2   Observed CMD of a generic CSP, divided onto a grid with bin size equal to 0.1 mag in $(V - I)$ and 0.5 mag in V

age and metallicity (with a uniform distribution) and determining their CMD location by interpolating within a grid of isochrones. It is important to remark that these elementary stellar populations have to be determined using a very large number of synthetic stars (hence a large value of the total initial mass of the population) in order to avoid non-negligible statistical fluctuations of the number of objects populating the more advanced, short-lived evolutionary phases. As an example, using only 10 000 stars in a theoretical SSP with ages above $\sim 10^8$ yr would produce only a handful of objects along the AGB, and this number will be subject to large variations (in some cases will even be zero) if different MC realizations of the same population are computed.

The discrete values of t and Z need in principle to span the largest possible range; ideally t should range from a few Myr (to describe the youngest stellar generations) to the age of the universe (13–14 Gyr) and Z from ~ 0 to a few times solar. It is, however, possible to put some limits to the range of t and Z of the elementary populations, when an approximate distance and reddening are available; in this case the brightness of the youngest TO and the colour range of the MS and RGB displayed by the observed CMD can constrain the minimum age and the metallicity range spanned by the CSP under scrutiny. Even without any prior knowledge, the morphology of the CMD can tell us if the CSP is made exclusively of stars with ages of at least a few

Gyr (i.e. it displays a well-populated SGB and RGB) or young objects (i.e. there is no evident RGB).

For each of the n age values a set of CMDs for all m metallicities has to be produced. The total amount of elementary populations will therefore be equal to $n \times m$.

Given that one has to be able to match the distribution of stars in the observed CMD, it is necessary to include in the synthetic CMDs of the elementary populations the photometric errors and completeness fractions determined from the reduction of the photometric data.[2] This can again be done using MC techniques. The size of the photometric errors constrains the width Δt and ΔZ of the elementary populations. In fact, if the age and metallicity resolution is too fine compared with photometric errors, elementary populations with adjacent ages (and or metallicities) would be completely degenerate. This point is well illustrated by Figure 10.3, which shows an example of elementary populations before and after the inclusion of the appropriate photometric errors. Notice also how the morphology and distribution of the star counts along the various evolutionary phases is affected by the age range covered. The oldest population displays an extended HB and all evolutionary phases are well populated, while with decreasing age the SGB and RGB tend progressively to be depopulated until they virtually disappear (because of the Schönberg-Chandrasekhar limit). The location of He-burning stars is also greatly affected by age; it first becomes a red clump for ages down to 1–2 Gyr, then it is reduced to sparsely populated groups

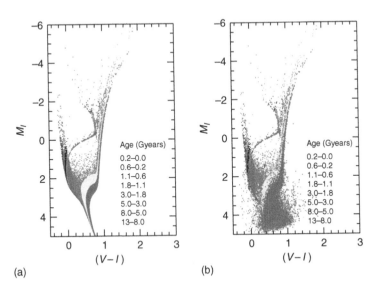

Figure 10.3 CMD of a set of elementary stellar populations covering the labelled age ranges, (a) without and (b) with the inclusion of photometric errors (courtesy of C. Gallart)

---

[2] For each given magnitude, the completeness fraction provides a measurement of the number fraction of stars recovered with the photometric data reduction with respect the actual number in the observed field.

of stars (due to very short timescales) that move towards increasing brightness and bluer colours.

The effect of unresolved binaries and blending must also be included in the synthetic populations, because, as an example, they broaden the observed MS. If the fraction of unresolved binaries is treated as a free parameter to be determined through the minimization procedure described below, additional CMDs with varying unresolved binary fractions for each elementary population j have to be produced. One possibility to include unresolved binaries is to select randomly some of the stars generated according to the prescribed IMF; the total number of the selected binary stars is fixed by the assumed value of the binary fraction. To these selected stars (whose mass is denoted by M_{pr}) it is assigned a companion (with the same age and metallicity of M_{pr}) whose mass M_{comp} is drawn according to

$$M_{comp} = [ran(1 - q_c) + q_c]M_{pr}$$

where q_c is the minimum value of the ratio M_{comp}/M_{pr}, whose typical value is usually assumed to be ~ 0.7 (see below). The variable ran is a random number with a flat distribution between zero and one, so that the previous relationship provides values of M_{comp} with a uniform distribution between M_{pr} and $0.7M_{pr}$. For unresolved binaries or blended stars the magnitude of the composite object is evaluated by simply adding the fluxes of the components; in a generic band A the magnitude of an unresolved binary is given by $M_A = -2.5\log(10^{-0.4M_A(1)} + 10^{-0.4M_A(2)})$, where $M_A(1)$ and $M_A(2)$ are the magnitudes of the two system components M_{pr} and M_{comp}. Unresolved binaries with $M_{comp} < 0.7M_{pr}$ would have magnitudes and colours almost indistinguishable from the values appropriate for M_{pr}.

It is also possible to account for foreground contamination in a consistent manner. The common procedure is to observe a second field (of the same size) near (in terms of position in the sky) the object field, but well beyond the limits of the object being studied; the resulting CMD is then divided into the same grid as the object CMD, and the number of objects $f(i)$ in a generic cell i is determined. The number counts $f(i)$ of this field CMD can then be added to the contribution of the synthetic partial populations, as described below.

The distance modulus and extinction of the observed CSP are often assumed a priori, but can in principle be determined through the minimization procedure. In this case they appear as constants added to the magnitudes and colours of the stars belonging to the elementary populations.

A synthetic CMD for the observed CSP is created as a linear combination of the elementary CMDs described above (plus the foreground population) and divided into the same grid as the observed CMD. The number of stars in grid cell i of the synthetic CMD is denoted by $N_s(i)$ and is given by

$$N_s(i) = \sum_j a_j N_e^j(i) + f(i)$$

where $N_e^i(i)$ is the contribution of elementary population j to the grid cell i; the index j runs from 1 to $n \times m$. The number of observed and synthetic stars present in each cell are compared using some merit function, e.g.

$$\chi^2 = \sum_i \frac{(N_o(i) - N_s(i))^2}{N_o(i)} \qquad (10.10)$$

with the index i running from 1 to N. This comparison is done, for each assumed value of distance and extinction, considering the magnitude and colour intervals populated by the observed CSP.

The weights a_j of the linear combination (and eventually distance, reddening and unresolved binary fraction) are varied until the merit function is minimized; the resulting values of the coefficients a_j provide the best-fit model for the SFR and AMR. In Figures 10.4 and 10.5 we display the results of this procedure for the LMC

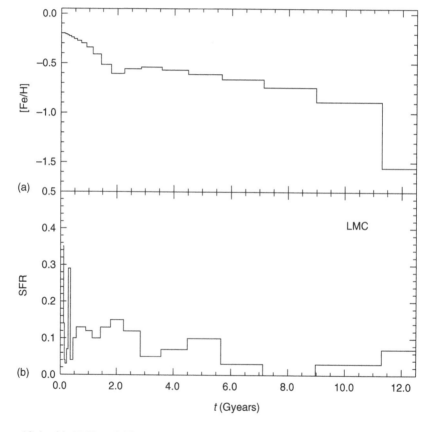

Figure 10.4 (a) AMR and (b) SFR estimated for the LMC, following the methods discussed in the text. An age equal to zero corresponds to the stars currently forming. The vertical axis in (b) provides the relative weights of the elementary stellar populations used in the derivation of the SFH. A scaled solar heavy element mixture has been employed in the adopted theoretical models

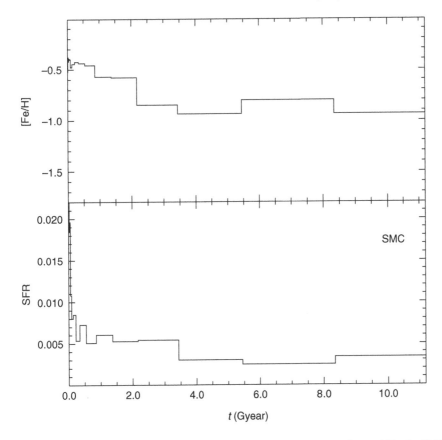

Figure 10.5 As in Figure 10.4 but obtained from an analysis of 351 regions within the SMC

and SMC (from [92] and [99]). Once the best-fit model has been found, one can estimate empirically the confidence interval of each coefficient a_j by exploring the χ^2 parameter space surrounding the location of the minimum value $\chi^2_{\min}$. The 1σ confidence interval on each parameter is then defined by the appropriate value of $\Delta\chi^2 = \chi^2 - \chi^2_{\min}$. It is important to notice that the denominator in Equation (10.10) should correspond to a Gaussian error $\sigma(i)^2$ for the χ^2 statistics to be meaningful. This is only approximately true when $N_o(i)$ is large (since we are comparing star counts and the error is essentially Poissonian). This problem can be avoided by using more complicated merit functions ([65]). One possibility is to minimize what is defined in [65] as the equivalent of χ^2 in the presence of Poisson errors, e.g.

$$-2\ln(\text{PLR}) = 2\sum_i N_s(i) - N_o(i) + N_o(i)\ln\left(\frac{N_o(i)}{N_s(i)}\right)$$

In general, one will not be able to estimate the age and metallicity of individual stars in the observed CSP, but it is possible to study the statistical distribution of the t

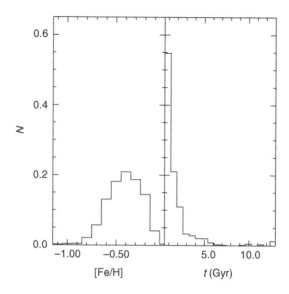

Figure 10.6 Age and [Fe/H] distribution for LMC stars populating the He-burning phase. In each panel, the total number of stars is normalized to unity

and Z along the CMD. In Figure 10.6 we show, as an example, age and metallicities of LMC stars along the HB phase obtained from the best-fit synthetic CMD. If some high resolution spectroscopic abundance determinations are available (this is the case of only the Milky Way and a few nearby galaxies) one can compare – as a test of the SFH – the metallicity distributions inferred from the derived SFH with observations. These measurements can also be used a priori to constrain the range of metallicity to use for the synthetic elementary populations.

There are a number of considerations to be made regarding the procedure outlined above. The first obvious point is that in principle the photometry should reach the faintest possible stars in the observed CSP, also to constrain the SFH for the first phases of galaxy evolution. To give some quantitative estimates, the typical TO magnitude in the V band for ages comparable to the age of the universe is $M_V \approx 4$, and this is the level to be reached (for a given distance modulus) in order to be sure to detect the SGB of the oldest stellar population in the CSP under scrutiny. Also, the method described before rests entirely on the accuracy of the colours (and magnitudes) predicted by the theoretical isochrones. As discussed in the previous chapter, there are non-negligible uncertainties affecting the predicted stellar colours that can be to some extent bypassed when studying SSPs, but cannot be avoided when inferring the SFH of a CSP from synthetic CMDs. The effect of these uncertainties on the estimated SFHs of galaxies is up to now largely unexplored. It is in principle possible to ignore in the sum of Equation (10.10) CMD cells for which theoretical models have the largest uncertainties, as in the case of stars with convective envelopes, but of course one cannot ignore all CMD regions populated by stars with surface convection, or else no information about the oldest CSP components (if present) can be obtained.

Another possibility to minimize the impact of badly modelled evolutionary stages, is to use variable-sized grid cells, by choosing the relative sizes in a way that gives less weight to the more uncertain phases.

It is useful to notice the following point. If bright MS stars exist in the observed CSP, a young population must be present; if these stars are vastly outnumbered by older stars, when using equal sized grid cells the fitting procedure will do its best to fit the older stars, even if it means sacrificing a good match to the younger objects. A way to circumvent this kind of problem is, again, to use grid cells with variable sizes, although this may introduce another degree of arbitrariness in the fitting procedure.

An additional warning concerns the fact that usually one employs scaled solar isochrones to determine the SFH of external galaxies, although in principle the metal mixture in the observed stellar population could be different. Different metal mixtures can alter the stellar colours at a given total metallicity, and lead to erroneous estimates of the SFH.

10.3 Distance indicators

In most cases the resolved CMD of a CSP is not deep enough to allow a meaningful determination of the SFH. However, it may still be very important to estimate the distance to the CSP, e.g. for cosmological distance scale calibrations, study of local deviations from the Hubble flow, relative distance analysis for groups of galaxies or comparisons of the intrinsic properties of some classes of stars in different environments.

If at least the bright part of a well-populated RGB is detected, one can employ the TRGB method to estimate the CSP distance. A note of warning is, however, necessary. Whenever a well-developed RGB is observed, it is natural to assume that it represents the counterpart of the Galactic globular cluster RGBs, with globular cluster-like ages. Therefore the TRGB method discussed in the previous chapter is applied, using calibrations for globular cluster stars, and the mean metallicity and distance of the observed RGB stars are estimated. Unfortunately, as shown in [180], appearances can be deceptive. An SFH like the LMC (or SMC) produces a CMD with an RGB that closely resembles the RGB of galactic globular clusters (see, for example, Figure 10.7). However, RGB stars in the LMC have a mean age of only ~4 Gyr, much younger than the typical globular cluster age. This is essentially due to the fact that in low-mass stars the ratio between RGB and MS timescale increases for increasing stellar mass until the mass is $\approx 1.6 M_\odot$. In the case of a constant SFR (or of an SFR increasing for decreasing age) the statistical weight of the RGB stellar sample related to the younger population is therefore larger than that of older RGB stars.

When applying the TRGB method to the LMC red giants one obtains a mean [Fe/H] much lower than the real one because, although the RGB dependence on age is weak, 4 Gyr old RGB stars are appreciably bluer than their ~12 Gyr counterpart at a given metallicity. If the real – unknown, in case of CSPs lacking estimates of

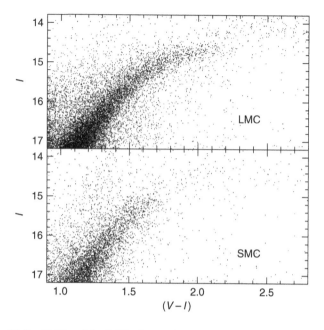

Figure 10.7   CMDs of the RGB observed in the LMC and SMC (from [223])

their SFH – metallicity is larger than ~ -0.7, the dependence of M_I^{TRGB} on the metal content is high, and an erroneous metallicity determination causes a large error on the inferred distance. When the a priori unknown RGB age gets closer to $\sim$2 Gyr, the TRGB brightness starts to decrease at a given metallicity, because of the approach to the transition from electron degenerate to non-degenerate He core ([7]) thus inducing erroneous distance determinations when the globular cluster TRGB calibration is applied.

Detection of a sizable sample of RR Lyrae stars could in principle be used to determine the average value of their mean apparent magnitudes, and estimate the CSP distance by applying the HB absolute magnitude calibrations described in the previous chapter. Given that the HB level is strongly affected by the metallicity, without [Fe/H] estimates this kind of distance determination technique is inapplicable. The same holds in the case of non-variable HB stars when detected in the observed CSP. The HB region of a CSP can be populated by objects with a mixture of metallicities and ages, as shown in Figure 10.6 for the LMC. Not only the metallicity but also the age of the He-burning objects affects the HB level, when t is lower than a few Gyr. The only way to estimate the expected absolute magnitude of He-burning stars in a CSP is to build a synthetic CMD with the population SFH. This approach has shown, as an example, that the difference between the mean V magnitude of the He-burning phase in the solar neighbourhood (for which accurate parallax-based determinations provide $M_V = 0.73 \pm 0.03$ – see [3]) and in the LMC, as predicted by the current SFH estimates, is equal to 0.26 mag (the LMC stars being brighter). For a fixed globular cluster-like age, the same magnitude difference predicted by considering only the difference of the mean [Fe/H] of the two HB populations (i.e. neglecting age effects) is of the order of $\sim$0.10 mag in V (again LMC stars being brighter). The effect of the SFH on the HB brightness is expected to be generally smaller if one employs the I, and especially the K (when metallicities are around solar) filters, but some knowledge of the SFH is absolutely necessary for a meaningful distance determination using HB stars ([83], [179]).

Cepheid stars are probably the most used objects to determine the distance to CSPs harbouring a young stellar population, from the fit of the observed periods and apparent magnitudes to a reference P–L relationship, in the same way as described in the previous chapter. In this case, there are potential uncertainties related to the not well-established metallicity (and probably also helium) dependence of the P–L relationship. If the P–L relationships are not universal, an estimate of the chemical composition of the observed Cepheids is necessary.

An important point to notice is that given the relatively narrow age range covered by Cepheid stars, one does not expect a substantial chemical evolution of a generic CSP during this time interval. This means that Cepheids in a CSP are expected to be characterized by the same mean metallicity (apart from an intrinsic spread around this constant mean value). This situation is simpler than the case of HB and TRGB stars discussed above.

10.3.1 The planetary nebula luminosity function (PNLF)

Planetary nebulae are bright objects, with bolometric luminosities typically higher than the tip of the RGB, and are sometimes used as distance indicators for stellar populations harbouring low- and/or intermediate-mass stars, e.g. with ages larger than $\approx 10^8$ yr. Empirical determinations of the number of planetary nebulae as a function of the flux they emit in the OIII line at 5007 Å (F_{5007}) has disclosed a remarkable homogeneity among galaxies of various types (see, for example, [109]). When the flux F_{5007} is transformed into a magnitude according to

$$m_{5007} = -2.5 \log(F_{5007}) - 13.74$$

the number of objects $N(m_{5007})$ with a given value of m_{5007} is well reproduced by the following empirical relationship

$$N(m_{5007}) = e^{0.307 m_{5007}} [1 - e^{3(m^* - m_{5007})}] \tag{10.11}$$

the so-called Planetary Nebula Luminosity Function (PNLF). The key quantity is the cut-off magnitude m^*. When comparing the PNLF of objects with known distances, a remarkably constant value of the absolute cut-off magnitude $M^* = -4.48 \pm 0.036$ is found.

Assuming M^* is universal, a distance estimate using the PNLF is straightforward. It is necessary to determine empirically the PNLF of the observed stellar population, fit the empirical star counts with Equation (10.11), and compare the apparent magnitude of the observed cut-off with its absolute value M^* given above. The difference between these two values gives an estimate of the distance modulus.

The reason for the existence of the PNLF cut-off is related to the combination of different effects. First, the brightness of the planetary nebulae increases with the increasing mass of the central star, essentially a CO electron degenerate core; since observations are made in the OIII emission line at 5007 Å, details of the emission processes also play a non-negligible role ([135]). Second, the evolutionary timescales along the PN phase sharply decrease with increasing stellar mass. Third, in CSPs with multiple stellar generations one has to take into account the interplay between the progenitor MS mass–final PN mass relationship (essentially defined by the mass-loss processes along RGB and AGB) and the IMF.

A modelling of the PNLF is extremely complicated; the most recent and sophisticated theoretical simulations ([135]) show that M^* depends strongly on the age of the latest episode of star formation, and on the oxygen abundance in the envelope of the observed PNe (determined not only by the initial metallicity of the progenitor, but also by the effect of the dredge-up episodes). In galaxies with a continuous SFR until the present time, or with a burst of star formation 1 Gyr ago, the cut-off is expected at M^* between -4 and -5, as observed. In galaxies without recent star formation the cut-off is expected to be about five magnitudes fainter. This is due to the lack of PNe with MS progenitor masses above $2M_\odot$, that produce the brightest PNe. The empirical agreement between M^* found in spiral, irregular galaxies and in some elliptical galaxy (of known distances) also seems to suggest a recent burst of star formation in this latter type of objects – traditionally expected to be SSPs – unless there is some additional effect not accounted for in theoretical models, or the cut-off absolute magnitude M^* is not really universal.

11 Unresolved Stellar Populations

11.1 Simple stellar populations

Stellar populations in distant galaxies cannot in general be resolved into individual stars. In this case photometric and spectroscopic observations can provide only integrated magnitudes, colours and spectra that include the contribution of all the stars belonging to the population. To study the evolutionary status of these unresolved stellar populations we cannot apply the techniques discussed in the last two chapters, given that they were based on the properties of individual stars in specific evolutionary phases. Alternative methods for dealing with unresolved populations have been devised in the last decades, starting with the works by [23], [29], [63], [90], [203], [218], until more recent investigations by [21], [133], [228], [235], [236]; in this section we first discuss techniques applied to unresolved SSPs.

The monochromatic integrated flux F_λ^I received from an unresolved SSP of age t and metallicity Z can be written as

$$F_\lambda^I(t, Z) = \int_{M_1}^{M_u} f_\lambda(M, t, Z)\Phi(M)dM \tag{11.1}$$

where $f_\lambda(M, t, Z)$ is the monochromatic flux emitted by a star of mass M, metallicity Z and age t, $\Phi(M)dM$ is the IMF (in the following we will always use the Salpeter IMF) M_1 is the mass of the lowest-mass star in the SSP, M_u is the mass of the highest-mass star still alive in the SSP. The value of M_u is typically the initial mass of the object evolving towards the WD sequence at the SSP age, and can be approximated to the TO mass at that age. The contribution of WDs to the integrated flux, especially that of the faintest more massive ones, produced by low- and intermediate-mass stars that started to evolve along the cooling sequence in the earlier evolution of the SSP, is usually negligible.

Equation (11.1) simply says that the integrated flux is the sum of the individual fluxes of the stars belonging to the SSP, represented by the term $f_\lambda(M, t, Z)$; the

Evolution of Stars and Stellar Populations Maurizio Salaris and Santi Cassisi
© 2005 John Wiley & Sons, Ltd

IMF gives the number of stars formed with a given mass M. In the hypothesis of a universal IMF the effect of age and chemical composition is included in $f_\lambda(M, t, Z)$ – since the energy output of a star of mass M and its wavelength distribution depend on both t and Z – and also in M_u. In a similar fashion the integrated magnitude in a generic photometric band A received from an unresolved SSP can be written as the sum of the energy fluxes within the appropriate wavelength range, i.e.

$$M_A(t, Z) = -2.5 \log \left(\int_{M_l}^{M_u} 10^{-0.4 M_A(M, t, Z)} \Phi(M) dM \right) \qquad (11.2)$$

Integrated colours follow directly from Equation (11.2) applied to two different photometric bands.

Before discussing how to use integrated colours and spectra to estimate the age and metallicity of unresolved SSPs, one has to learn more about the contribution of individual evolutionary phases to the integrated properties of an SSP. We start discussing the case of the bolometric luminosity L_T, the integral of F_λ^I over the whole wavelength range.

Figure 11.1 displays the fractional contribution of various evolutionary phases to the integrated bolometric luminosity L_T of solar metallicity SSPs spanning a large age range. In general the MS is the most populated evolutionary phase, due to its very long timescale compared with later phases; however, its contribution to L_T also depends on the maximum luminosity reached at the TO, compared with post-MS phases.

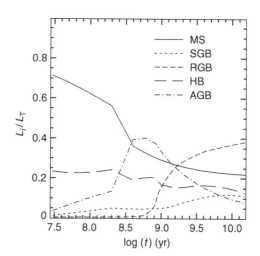

Figure 11.1 Contribution of different evolutionary phases to the total bolometric luminosity of an SSP (L_i/L_T is the ratio of the integrated bolometric luminosity produced by stars in the evolutionary phase i to the total integrated bolometric luminosity L_T of the population) with solar initial chemical composition and varying ages (data from [133]). Here the acronym HB denotes the phase of core He-burning, regardless of the value of stellar evolving mass. The contribution to L_T of phases not displayed in this diagram is negligible

For t below $\sim$300Myr the MS contributes the largest fraction of L_T, approximately 70 per cent at $t = 30$Myr and $\sim$40 per cent at $t = 300$Myr. This happens because at young ages the MS reaches extremely bright luminosities. Between $\sim$300Myr and $\sim$2Gyr the AGB becomes the largest contributor to L_T, due to the appearance of AGB stars with progenitor masses between $\sim$7 and $\sim$2$M_\odot$, that reach high luminosities due to their large final CO core mass, and are much brighter than the TO. Above 2–3 Gyr the RGB takes over as major contributor to L_T, since it now harbours low-mass stars with relatively long lifetimes (hence it is extremely well populated) and reaches much brighter luminosities than the MS. The fractional contribution of the He-burning phase is approximately constant, at about 20 per cent, and the SGB is even smaller and almost negligible at young ages, when it is severely depopulated.

Let us repeat the previous analysis by considering the luminosity in various wave-length ranges (see Figure 11.2). We choose four representative photometric bands covering a wide wavelength range, from the near ultraviolet to the near infrared, namely U, B, V and K. The contribution of the individual evolutionary phases depends clearly on the filter considered. The U and B bands are always dominated by the contribution of MS stars, mainly the hotter objects close to the TO, even at ages of the order of 10 Gyr. The second most relevant phase in U and B is the central He-burning at low ages, while SGB stars become the second largest contributor to the integrated luminosity when the age is above $\sim$1 Gyr. At these ages the HB and RGB contribute about 15–20 per cent each to the total integrated magnitude in B. We notice that the actual morphology, i.e. the colour extension of the HB, strongly affects its contribution to the blue and ultraviolet portion of the spectrum. In this example we are considering an HB populated only in the red part of the CMD, an assumption generally true for the solar metallicity stellar populations in the Galactic disk. Bluer HBs would provide a larger contribution to the flux in the U and B filters.

In V the situation is similar to the B filter, but in this case the RGB takes over as the second more relevant phase when the age is above $\sim$1 Gyr. In the K filter things are very different. For ages below $\sim$200 Myr the He-burning phase dominates the integrated luminosity due to masses below $\sim$12$M_\odot$ that experience the onset of central He-burning at the red side of the CMD (notice that below $\sim$10^7 yr He-burning starts at the blue side of the CMD). Between $\sim$200 Myr and $\sim$3 Gyr it is the TPAGB phase along the AGB that mainly determines L_T^K and at higher ages it is the RGB.[1]

The change of behaviour when considering different filters is easily understood when we recall that bluer filters are more sensitive to the hottest evolutionary phases, like MS and He-burning phases for SSPs below $\sim$10^9 yr, because higher temperatures shift the peak of the energy spectrum towards shorter wavelengths. Redder filters are necessarily more sensitive to cooler objects like RGB and AGB stars, that also happen to reach extremely high luminosities. It is therefore very important to bear in mind that the integrated flux in different wavelength bands may be dominated by different evolutionary phases. This has relevant consequences for the case of unresolved CSPs.

---

[1] It should be pointed out that post-AGB stars and hot, bright WDs contribute to the flux in the ultraviolet, at wavelengths much shorter than the U and B bands discussed in this section.

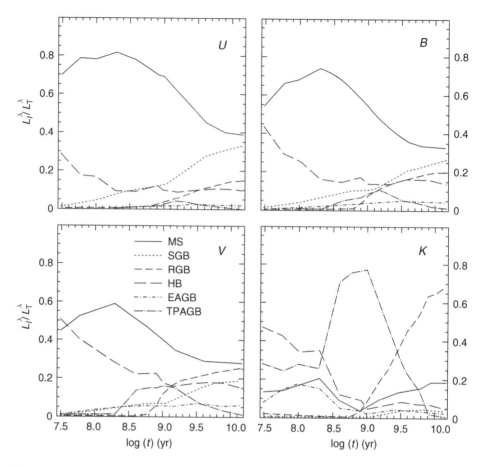

Figure 11.2 As in Figure 11.1 but for the luminosity in various wavelength bands (data from [133]). The AGB has been split into the phase up to the onset of thermal pulses (EAGB) and the thermal pulse phase (TPAGB)

The initial chemical composition can affect the fractional contribution of the various phases to integrated luminosities L_T^λ, although this is in general a second-order effect. What is important to consider is that in old SSPs with Z below ~ 0.001 the HB is also populated at its blue side; the contribution of He-burning stars to the total luminosity mainly in U and B will then be larger than the case of a solar metallicity SSP, which has a red HB.

11.1.1 Integrated colours

From an observational point of view it is much more convenient to measure broadband magnitudes and colours of distant unresolved SSPs, than performing spectroscopic observations to determine F_λ^I to some chosen resolution. It is therefore important to study the predicted behaviour of integrated magnitudes and colours to devise, if possible, theoretical tools able to infer the age and chemical composition of the observed SSP.

Figure 11.3 displays the behaviour of the integrated magnitudes in B, V, I and K for SSPs with solar metallicity (normalized assuming the constant factor in the Salpeter IMF equal to unity) and varying ages. The overall property is a general fading of the magnitudes for increasing age when t is larger than $\sim 10^7$ yr. The fading in B, V and I is approximately linear with $\log(t)$, with a slope of ≈ 2 mag dex^{-1}. For the K band there is a sudden brightening of the SSP luminosity at $t \sim 200$ Myr due to the onset of the AGB, followed again by a steady decrease of the luminosity. At ages of the order of 10^7 yr the K magnitude and, to a minor extent, also B, V and I show a brightening due to the appearance of He-burning stars located at the red side of the CMD.

The general fading of the integrated magnitudes in B, V and I arises mainly from the decrease of the MS extension (fainter TO) with increasing age, as well as the overall general decrease of the He-burning luminosity for increasing ages, at least until reaching old ages where the HB level is approximately constant. In the case of K the situation is complicated by the interplay between AGB and RGB, that give the most relevant contribution to the integrated magnitude, for ages above $\sim 10^8$ yr.

In practice, the absolute values of the integrated magnitudes of an unresolved SSP are not very useful, because of the unknown total stellar mass of the observed SSP

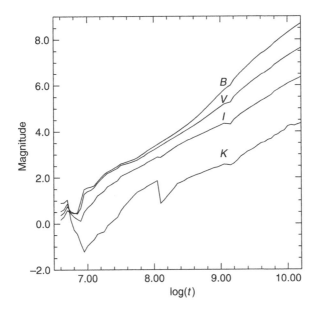

Figure 11.3 Time evolution of the integrated magnitudes in selected filters for solar metallicity SSPs (data from [84])

and also its possibly unknown distance. Integrated colours are more helpful, given
that they are unaffected by the distance and also, in principle (but see below), by the
mass of the SSP.

The run of three selected SSP integrated colours with age is displayed in
Figure 11.4, where the effect of metallicity is also considered. Their behaviour is
generally complex, non-monotonic at young ages, mirroring the trends shown by
the relevant integrated magnitudes. Also the dependence on the initial metallicity is
affected by the considered age range. At ages above ~1 Gyr, however, when SGB
and RGB become important contributors to the integrated U, B, V and I magnitudes,
all three colours show the same monotonic increase with age, and the metallicity
dependence is overall the strongest. At these ages, increasing the metal content always
produces redder colours.

For ages above ~1 Gyr, all displayed colours become redder in cases of both
increasing age at a fixed metallicity, and increasing metallicity at a fixed age. This
is the so-called age–metallicity degeneracy. For many colours the degeneracy follows the approximate rule $(\Delta t/t) \sim 1.5 - 2.0(\Delta Z/Z)$ for ages around 12 Gyr and
metallicities around solar. The meaning of this relationship is the following. Suppose
a generic colour index $(A - B)$ varies by an amount $\Delta(A - B)$. One can explain
$\Delta(A - B)$ by either a ΔZ variation, or a Δt variation related to ΔZ through the previous rule. To give a practical example of the implications for the age and metallicity
estimate, consider the case of a 12 Gyr old stellar population with solar metallicity

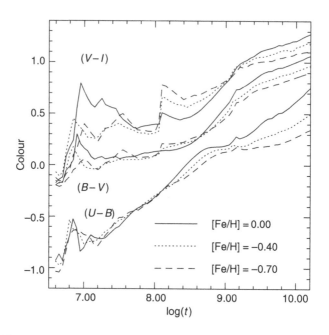

Figure 11.4 Time evolution of selected integrated colours of SSPs with various metallicities (data
from [84])

$(Z \sim 0.02)$. A decrease of Z by, for example, 20 per cent, will cause a change in colour that can be compensated by a 30–40 per cent increase of the age. This means that the same observed colour $(A - B)$ can be reproduced with $Z = 0.016$ and age $t = 15.6$–16.8 Gyr.

It is obvious that observing only one integrated colour of an unresolved SSP is not sufficient to determine univocally its metallicity and age. An additional colour is in principle needed in order to disentangle the effects of these two important evolutionary parameters.

Figure 11.5 shows two colour–colour diagrams that make use of the colours displayed in Figure 11.4. Some reference ages are marked. The paths described in these diagrams are very complicated when the age is young, a mirror of the non-monotonic metallicity and age dependence of the individual colours at low ages. When the age is older than $\sim$100–300 Myr the behaviour is smoother. Notice how the lines of different metallicity tend to overlap, and the points of constant age tend to be displaced along this same line. This means that the observed pair of colours for a given SSP can be reproduced, taking into account observational errors of even only a few hundredths of a magnitude, by different metallicites for different ages. For example, the $(U - B)$ and $(B - V)$ pair corresponding to 10 Gyr and $[Fe/H] = -0.7$ is also virtually equivalent to the colours of an SSP with solar metallicity and lower age.

Searching for pairs of colours able to break the age–metallicity degeneracy is one of the main goals of modern research in this field. One generally tries to devise

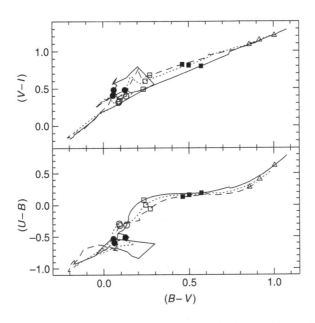

Figure 11.5 Colour–colour diagrams for the data displayed in Figure 11.4. Symbols denote some reference age. More specifically, filled circles correspond to an age of 30 Myr, open circles to 100 Myr, open squares to 350 Myr, filled squares to 1 Gyr, open triangles to 10 Gyr

colour–colour diagrams where the lines of constant age and constant metallicity are approximately orthogonal to each other, not parallel as in Figure 11.5.

Figure 11.6 displays the theoretical $(B - K) - (J - K)$ diagram for unresolved SSPs in the labelled age and metallicity range. The difference from Figure 11.5 is striking; this combination of colours, at least in the displayed age range, is able to largely disentangle age and metallicity effects. In fact, the colour $(B - K)$ appears to be mainly sensitive to age, whereas $(J - K)$ is weakly affected by age but very sensitive to the metal content. Studying the contribution of the individual phases to the B, J and K integrated magnitudes one finds that the J and K integrated fluxes are dominated by AGB stars when the SSP age is below ~ 1 Gyr, and by upper RGB objects for higher ages. As for the integrated B flux, the main contribution always comes from the upper MS and TO stars. This means that the integrated $(J - K)$ is mainly determined by the colour of AGB and/or bright RGB stars, whose location is strongly affected by the initial metallicity, whereas $(B - K)$ is sensitive to the magnitude and colour of the TO, hence to the SSP age. Although the TO colour is also affected by the metallicity, the dominant effect is the age one. Concerning RGB and AGB stars, their colours are also affected by the age, when age is below a few Gyr, but the strongest sensitivity is to metals.

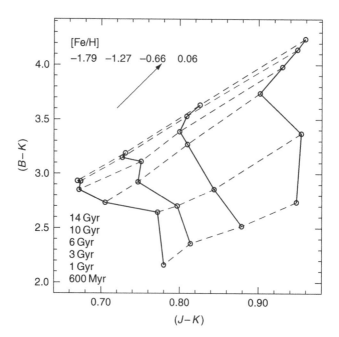

Figure 11.6 Theoretical $(B - K) - (J - K)$ diagram for the labelled selected ages (marked by open circles) and metallicities. Lines of constant metallicities are displayed with solid lines; lines of constant age are shown as dashed lines. Metallicity increases towards increasing $(J - K)$, age increases towards increasing $(B - K)$. The direction of the reddening vector is also displayed

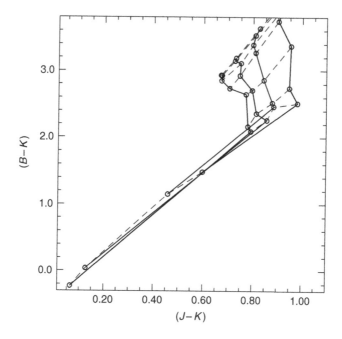

Figure 11.7 As in Figure 11.6 but including younger ages (i.e. two additional ages of, respectively, 300 and 40 Myr, the latter being the bluest point on the constant metallicity lines) that display the age–metallicity degeneracy

At younger ages (see Figure 11.7) below ~ 300 Myr the full age–metallicity degeneracy is again present, since the lines of different metallicities tend to overlap, as in the case of the diagrams in Figure 11.5.

As a practical application of this colour–colour diagram to real unresolved SSPs, we compare in Figure 11.8 the theoretical calibration shown in Figure 11.6 with the integrated colours of a sample of elliptical galaxies (generally considered to be SSPs) in the Coma cluster. The mean values of the observed $(B - K)$ and $(J - K)$ correspond, without ambiguity, to an age of ~ 10 Gyr and $[Fe/H] = +0.06$.

Statistical fluctuation

Given that the fast evolving upper RGB and AGB stars are the main contributors to the J and K integrated magnitudes, it is very important to assess the effect of small number statistics on the integrated $(B - K)$ and $(J - K)$ colours. When the mass of the SSP under scrutiny is small – by 'small' we mean total masses up to $\sim 10^5$–$10^6 M_{\odot}$, which correspond approximately to the upper end of the mass spectrum of star clusters – the number of stars in these fast evolutionary phases (populated by objects within an extremely narrow mass range) is subject to sizeable fluctuations from one SSP to another with the same age and metallicity. This can cause large

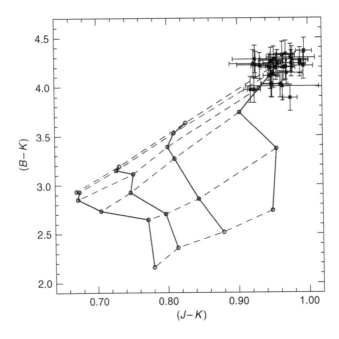

Figure 11.8 Theoretical $(B - K) - (J - K)$ diagram displayed in Figure 11.6 compared to the observed colours (corrected for extinction and K-correction) of a sample of elliptical galaxies in the Coma cluster

fluctuations of the integrated magnitudes and colours. To quantify this effect we show in Figure 11.9 the results of a series of MC simulations. For a metallicity [Fe/H] = −1.27 and ages of 600 Myr and 10 Gyr, respectively, we have drawn stars randomly from a Salpeter IMF and placed them in their evolutionary phases along the isochrone at that age, until a cluster mass of $10^5 M_\odot$ is reached. In general we found that the fluctuations are negligible in B since this wavelength range is dominated by the much more populous MS phase, whereas they are large in both J and K.

The open squares in Figure 11.9 represent the integrated colours obtained from 100 realizations each for the two adopted ages. Each realization corresponds to a different SSP with the same mass, age and metallicity. The $(J - K)$ colours show a very large spread due to statistical fluctuations of the number of stars along the upper RGB and AGB phases. This spread increases for decreasing age because of the shorter timescales − hence larger number fluctuations − along the AGB phase that dominates the $(J - K)$ colours at 600 Myr. This colour spread causes a 3σ uncertainty of ≈2 dex in the inferred metallicity at 600 Myr, and of ≈1 dex at 10 Gyr. The fluctuation of $(B - K)$ is entirely due to the fluctuation of the K magnitudes and it is interesting to notice that the path in the colour–colour plane described by the 100 realizations follows a vector that is almost parallel to lines of constant age. At 600 Myr the fluctuations cause a small uncertainty in the age, while the effect is more important at 10 Gyr, due to the lower sensitivity of $(B - K)$ to age in this regime.

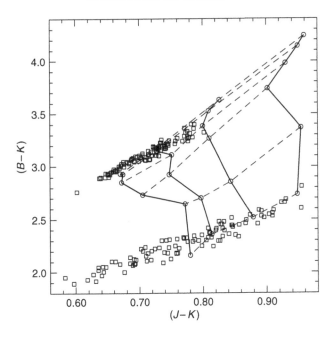

Figure 11.9 Distribution of the integrated $(B - K)$ and $(J - K)$ colours for two SSPs with a total actual mass of $10^6 M_\odot$, [Fe/H] $= -1.27$ and ages of 10 Gyr and 600 Myr, respectively

HB colour

When dealing with integrated colours (and spectra) of SSPs, one has to take into account that colours and spectral features involving the blue part of the spectrum are seriously affected by the possible presence of blue HB stars not accounted for in the theoretical calibration (this issue was discussed in Chapter 9). Such stars have high temperatures and can add an appreciable contribution to the flux, i.e. in the B filter. As an example we consider a synthetic 10 Gyr old SSP with [Fe/H] $=$ -1.27, whose integrated $(B - K)$ and $(J - K)$ colours are given by the reference calibration displayed in Figure 11.7. If we now compute the colours for the same population but considering a much bluer HB than the one previously used, the SSP will appear to be $\sim$6 Gyr old when the age is evaluated using the diagram of Figure 11.7, i.e. the reference calibration. The metallicity estimate is only minimally affected – slightly reduced – because of a small contribution of the HB to the J and K colours.

11.1.2 Absorption-feature indices

Another way to gain information about unresolved SSPs makes use of observations of their integrated spectra with resolution of the order of a few Angstroms.

From these spectra one tries to identify individual absorption features that are sensitive to the age and metallicity of the underlying unresolved stellar component. The observed spectrum of a galaxy is the convolution of the integrated spectrum of its stellar population (Equation (11.1)) by the instrumental broadening and the distribution of line-of-sight velocities of the stars. The instrumental and velocity-dispersion effects broaden the spectral features, which causes absorption features to appear weaker than they are intrinsically. Correction for these effects is therefore necessary before using the integrated spectrum of an SSP to study the properties of its parent stars.

In general, absorption-feature indices are composed of measurements of relative flux in a central wavelength interval corresponding to the absorption feature considered, and two flanking intervals (called sidebands) that provide a reference level (called pseudocontinuum) from which the strength of the absorption feature is evaluated. The average fluxes in the pseudocontinuum ranges are found and a line is drawn between the mid-points to represent the reference pseudocontinuum level. The difference in flux between this line and the observed spectrum within the feature wavelength interval determines the index. It must be noted that the sidebands themselves contain absorption features and consequently the value of an index is dependent on the strength of the lines present in the sidebands as well as that of the central feature. A graphical sketch of the index definition is displayed in Figure 11.10.

For narrow features, the indices are usually expressed in Angstroms; for broad molecular bands, in magnitudes. Specifically, if $F_{C,\lambda}$ represents the straight line connecting the mid-points of the flanking pseudocontinuum levels, and $F_{I,\lambda}$ the

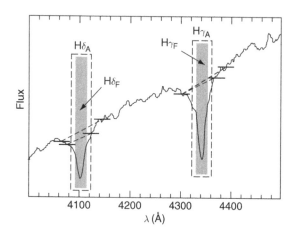

Figure 11.10 Sketch of the definition of four Balmer line indices and a representative stellar spectrum. The central passbands are shown as boxes (filled in for the case of the narrower F definition); pseudocontinuum sidebands are displayed as horizontal strokes at the average flux level. The pseudocontinuum used for the index measurement is drawn as a dashed line between the flanking sidebands. The blue sideband is the same for the A and F definitions of the H_γ index

observed flux per unit wavelength in the central wavelength range $\lambda_1 - \lambda_2$, the numerical value for a narrow absorption-feature ($I_{\text{Å}}$) is defined as

$$I_{\text{Å}} = \int_{\lambda_1}^{\lambda_2} \left(1 - \frac{F_{1,\lambda}}{F_{C,\lambda}} \right) d\lambda \tag{11.3}$$

The value of an index measured in magnitudes (I_{mag}) is

$$I_{\text{mag}} = -2.5 \log \left[\left(\frac{1}{\lambda_2 - \lambda_1} \right) \int_{\lambda_1}^{\lambda_2} \frac{F_{1,\lambda}}{F_{C,\lambda}} d\lambda \right] \tag{11.4}$$

A widely used set of absorption-feature indices is the so-called Lick system (e.g. [235]), that we will employ in the rest of this chapter. Alternative sets of indices can be found in the literature, see [124] and [169].

Definitions of the Lick indices are given in Table 11.1, together with the main chemical elements that contribute to their strength. It is perhaps interesting to notice that many of the indices do not in fact measure the abundances of the elements for which they were named. An important bonus is that, due to their narrowness, these indices are largely unaffected by interstellar extinction.

The theoretical computation of the values of the indices for an unresolved SSP is based on the following procedure. Consider an isochrone representing an SSP of a given age and initial chemical composition; the isochrone is populated according to a prescribed IMF, and if a spectral library of the adequate resolution is available (i.e. [124], [140]) one can determine the SSP integrated spectrum by applying Equation (11.1) to all relevant wavelengths. It is then straightforward to determine the strength of a given index from the integrated spectrum, by following the appropriate index definition.

To date, it has been traditional to use semi-empirical methods to determine the strenght of the absorption-line indices in an SSP. In brief, the values of the Lick indices have been measured from observations of samples of local stars to which a surface gravity, metallicity and effective temperature have been assigned. These measured values of the indices have been then parametrized as functions of T_{eff}, surface gravity g and [Fe/H], to produce a series of fitting functions. As an example we give here fitting functions for the $H\gamma_A$ and $H\delta_A$ Balmer lines, applicable to stars with $\Theta = 5040/T_{\text{eff}}$ between 0.75 and 1.0 ([237])

$$H\gamma_A = 99.846 - 180.61\Theta + 74.564\Theta^2 - 0.02066 \log(g)^3 - 2.56\Theta^2[\text{Fe/H}]$$

$$H\delta_A = 35.982 - 39.599\Theta - 0.4963[\text{Fe/H}]^2 - 0.01241 \log(g)^3$$
$$- 2.8349\Theta^2[\text{Fe/H}]$$

These fitting functions can be used to compute the expected index strengths of an SSP, starting from a theoretical isochrone populated according to a prescribed IMF.

Table 11.1 Various spectral indices (Lick indices) used in age and metallicity determinations of unresolved stellar populations. The chemical elements contributing to the strength of each index are given in column five. Elements in brackets increase the line strength when their abundance decreases

Name	Index band	Blue continuum	Red continuum	Units	Measures
$H\delta_A$	4083.500-4122.250	4041.600-4079.750	4128.500-4161.000	Å	
$H\delta_F$	4091.000-4112.250	4057.250-4088.500	4114.750-4137.250	Å	
CN_1	4142.125-4177.125	4080.125-4117.625	4244.125-4284.125	mag	C, N, (O)
CN_2	4142.125-4177.125	4083.875-4096.375	4244.125-4284.125	mag	C, N, (O)
Ca4227	4222.250-4234.750	4211.000-4219.750	4241.000-4251.000	Å	Ca, (C)
G4300	4281.375-4316.375	4266.375-4282.625	4318.875-4335.125	Å	C, (O)
$H_{\gamma A}$	4319.750-4363.500	4283.500-4319.750	4367.250-4419.750	Å	
$H_{\gamma F}$	4331.250-4352.250	4283.500-4319.750	4354.750-4384.750	Å	
Fe4383	4369.125-4420.375	4359.125-4370.375	4442.875-4455.375	Å	Fe, C, (Mg)
Ca4455	4452.125-4474.625	4445.875-4454.625	4477.125-4492.125	Å	(Fe), (C), Cr
Fe4531	4514.250-4559.250	4504.250-4514.250	4560.500-4579.250	Å	Ti, (Si)
$C_2$4668	4634.000-4720.250	4611.500-4630.250	4742.750-4756.500	Å	C, (O), (Si)
H_β	4847.875-4876.625	4827.875-4847.875	4876.625-4891.625	Å	
Fe5015	4977.750-5054.000	4946.500-4977.750	5054.000-5065.250	Å	(Mg), Ti, Fe
Mg_1	5069.125-5134.125	4895.125-4957.625	5301.125-5366.125	mag	C, Mg, (O), (Fe)
Mg_2	5154.125-5196.625	4895.125-4957.625	5301.125-5366.125	mag	Mg, C, (Fe), (O)
Mg_b	5160.125-5192.625	5142.625-5161.375	5191.375-5206.375	Å	Mg, (C), (Cr)
Fe5270	5245.650-5285.650	5233.150-5248.150	5285.650-5318.150	Å	Fe, C, (Mg)
Fe5335	5312.125-5352.125	5304.625-5315.875	5353.375-5363.375	Å	Fe, (C), (Mg), Cr
Fe5406	5387.500-5415.000	5376.250-5387.500	5415.000-5425.000	Å	Fe
Fe5709	5696.625-5720.375	5672.875-5696.625	5722.875-5736.625	Å	(C), Fe
Fe5782	5776.625-5796.625	5765.375-5775.375	5797.875-5811.625	Å	Cr
Na_D	5876.875-5909.375	5860.625-5875.625	5922.125-5948.125	Å	Na, C, (Mg)
TiO_1	5936.625-5994.125	5816.625-5849.125	6038.625-6103.625	mag	C
TiO_2	6189.625-6272.125	6066.625-6141.625	6372.625-6415.125	mag	C, V, Sc

The value of a generic index W_i for an SSP can be well approximated as follows. First, for each star with a given metallicity, temperature and gravity, one obtains the value of $F_{C,\lambda}(M, t, Z)$ from an appropriate low resolution spectrum (theoretical or empirical) as the flux at the central wavelength of the absorption feature (e.g. $F_{C,\lambda}(M, t, Z)$ is treated as a constant over the wavelength range spanned by the index). Second, one calculates the value $W_i(M, t, Z)$ of the index W_i for the same star, as given by the appropriate fitting function. Once $F_{C,\lambda}(M, t, Z)$ and $W_i(M, t, Z)$ are known, it is possible to invert Equations (11.3) and (11.4) to determine an effective value (assumed constant in the wavelength range of the index) of $F_{I,\lambda}(M, t, Z)$ that reproduces the observed $W_i(M, t, Z)$. This is repeated for all the stars in the SSP, and all the individual $F_{C,\lambda}(M, t, Z)$ and $F_{I,\lambda}(M, t, Z)$ values are summed up, to provide the integrated $F_{C,\lambda}$ and $F_{I,\lambda}$. These integrated fluxes are then inserted in Equations (11.3) and (11.4) to compute the line strengths for the SSP.

Breaking the age–metallicity degeneracy

Table 11.1 shows how to break the age–metallicity degeneracy with the Lick indices. Many of the indices are affected by the abundances of certain heavy elements, hence they are sensitive to the SSP metal content; however, the Hβ, Hγ and Hδ indices that measure the strength of the Balmer lines are largely unaffected by the chemical abundances. The strength of these lines is mostly sensitive to the temperature of hot TO stars. Although decreasing metallicity at a given age in principle produces stronger Balmer lines because of the hotter TO, the effect is somewhat muted compared with the effect of age. On the other hand, the indices affected by metal abundances are mainly sensitive to the temperature of cool stars, like RGB, AGB and in some cases the lower MS (like the Mg$_2$ index) which suffer from comparably minor changes due to age, at least in the regime of low-mass stars (i.e. ages above $\sim$1 Gyr).

A graphical sketch of these properties is given in Figure 11.11. This figure compares the observed values of the indices Hβ and Fe5270 for a sample of Galactic and extragalactic globular clusters, with a theoretical calibration for different ages and initial [Fe/H] values. The H$_\beta$ index appears to be largely insensitive to [Fe/H] for ages above $\sim$1 Gyr, whereas the Fe5270 index is clearly unaffected by age. This kind of diagram clearly allows one to estimate independently age and [Fe/H] of the observed SSP.

There is a very important point to discuss in connection with the metallicity estimated from the absorption-feature indices. Table 11.1 shows how different metal

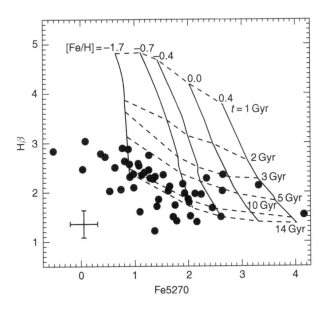

Figure 11.11 Theoretical calibration of the Hβ–Fe5270 diagram for the labelled values of age and metallicity, compared with observed data for a sample of Galactic and extragalactic globular clusters (data from [221]). The typical observational error bars are also shown

indices are actually affected by different elements. For example, Fe5270 is sensitive to Fe, C and Mg whereas Fe4531 measures the abundances of Ti and Si. Isochrones and spectra with scaled solar metal distributions will provide a calibration of the strength of the various indices parametrized as a function of [Fe/H] (or of the total metal content [M/H]) that is strictly accurate only if the observed SSP has the same metal mixture. In other words, let us imagine observing an SSP with a non-scaled solar metal distribution, and try to determine its [Fe/H] from metal indices calibrated with our scaled solar models. Due to the non-solar metal ratios, different metal indices will provide different values of [Fe/H]. For example, if the metal distribution is α-enhanced one expects to find a higher [Fe/H] estimate from the observed value of the Mg_2 feature, than from the Fe5270 one.

A consistent [Fe/H] from all indices can only be obtained using both spectra and/or fitting functions for the appropriate metal distribution, together with theoretical isochrones also computed with the appropriate metal mixture. In fact, different metal mixtures change the stellar spectra at a given gravity and effective temperature, but also the values of gravity and T_{eff} at a given evolutionary stage along an isochrone of fixed age. The combination of these two effects change the theoretical calibration of the index strengths. Figure 11.12 shows, as an example, the theoretical calibration of the Hβ–Fe5270 diagram for a scaled solar and α-enhanced metal mixture, respectively. The first noticeable effect, as expected, is a decrease of the Fe5270 strength at fixed age and [M/H] in the case of the α-enhanced mixture, because of its lower

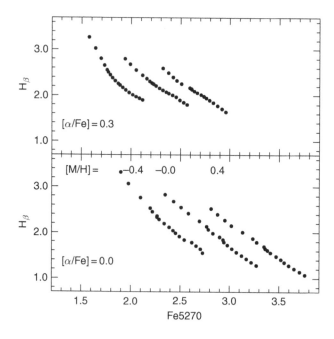

Figure 11.12 Effect of an α-element enhancement on the theoretical calibration of the Hβ–Fe5270 diagram (data from [214]). Ages range from 2 Gyr to 14 Gyr at steps of 0.05 in $\log(t(\text{yrs}))$

iron content. The second result is that that the [α/Fe] > 0 calibration leads to higher age estimates for old stellar populations.

Hot HB stars

As for the case of integrated colours, the presence of a number of hot HB stars, unaccounted for in the theoretical index calibration, can boost the strength of Balmer lines mimicking the signature of young stars in the integrated spectra of old SSPs. As shown by [195] the strength of the index $H\delta_F$ is more affected than H_β by the presence of blue HB stars. As a result, the spectroscopic ages look younger according to $H\delta_F$ than H_β. The presence of hot HB stars not included in the index calibration can therefore be inferred whenever the ages using these two different Balmer line indices differ.

11.2 Composite stellar populations

In the case of composite stellar populations of unknown SFH, one can only estimate some sort of mean age and mean metallicity when comparing the observed integrated colours or line indices with the corresponding calibrations for SSPs.

The mean age will, however, be strongly weighted towards any young population that might be present. The reason is that young MS stars have larger luminosity in the blue part of the spectrum, and stronger Balmer line indices than old populations. To give a quantitative estimate, for a uniform metallicity CSP just $\sim$1–2 per cent of a young (i.e. $\sim$300 Myr old) stellar population are sufficient to dominate the $(B - K)$ colour of a composite SSP. Similar results hold for the Balmer line strengths. In the case of a CSP with mixed age and metallicity, the dominant population will be the younger or metal-poorer one, since lower metallicities produce hotter stars at a given age.

An additional problem is that whereas blue colours and Balmer lines are sensitive to the hotter subpopulations, the metal line indices are dominated by the cooler and/or more metal-rich stars, since they produce stronger metal lines. The average metallicity estimated for an unresolved CSP will then be biased towards the older or more metal-rich population, i.e. a different component from the one controlling the strength of the age-sensitive colours and indices.

11.3 Distance to unresolved stellar populations

If a stellar population is sufficiently close for its individual stars to be resolved, we can apply a host of techniques to determine its distance, as discussed in the previous two chapters. A commonly used standard candle for unresolved stellar populations is the Type Ia supernova luminosity at the maximum of the light curve (see Section 7.6.3). These Type Ia SNe are so bright that they allow their detection at cosmological distances, when all the other stellar standard candles are too faint to be resolved. In addition, the discrete nature of the stars allows us to devise another distance-determination method based on photometric observations, the so-called Surface Brightness Fluctuation (SBF) technique ([220]), that with present capabilities is applicable out to distances in the range 10–100 Mpc.

The underlying idea is very simple. If we take an image of the target population with an angular resolution $\delta\phi$, each resolution element will contain an average number of unresolved stars $\bar{N} = n(D\delta\phi)^2$, where n is the number of stars per unit area across the observed face of the target population and D its distance. The average total flux $\bar{F}$ received from the stars in each resolution element will be

$$\bar{F} = \bar{N}f = \frac{nL\delta\phi^2}{4\pi}$$

where f is the flux at the Earth of a single star, and we have assumed that all stars have the same intrinsic luminosity L. Because of Poisson fluctuations in the number of objects populating a given resolution element, $\bar{F}$ will not be constant, but will fluctuate from one element to another, with variance

$$\sigma_{\bar{F}} = \bar{N}^{1/2}f = \frac{n^{1/2}L\delta\phi}{4\pi D}$$

This fluctuation of the surface brightness scales as the inverse of the distance. Combining $\sigma_{\bar{F}}$ with $\bar{F}$ we obtain

$$\frac{\sigma_{\bar{F}}^2}{\bar{F}} = \frac{L}{4\pi D^2} \tag{11.5}$$

This means that one can use the theoretical value for the representative stellar luminosity L together with the observed $\sigma_{\bar{F}}^2/\bar{F}$ in order to estimate the distance D.

The approximation of constant stellar luminosity is, however, not realistic because, as we have seen in detail in the last two chapters, stars in real SSPs and CSPs display a large range of luminosities. For an average number $\bar{N}_i$ of stars with luminosity L_i per resolution element, Equation (11.5) can be generalized as

$$\frac{\sigma_{\bar{F}}^2}{\bar{F}} = \frac{\bar{L}}{4\pi D^2} \quad \text{with} \quad \bar{L} = \frac{\sum_i \bar{N}_i L_i^2}{\sum_i \bar{N}_i L_i} \tag{11.6}$$

From theory one can determine $\bar{L}$ (the so called 'SBF luminosity') for an SSP or CSP, provided its age and metallicity or the SFH are known, and an IMF is adopted. Once $\bar{L}$ is determined, the observed $\sigma_{\bar{F}}^2/\bar{F}$ immediately provides the distance.

The method clearly works best when observing the external parts of galaxies, where the number of stars is low and Poisson fluctuations are larger. However, one cannot go too far away from the centre, because the brightness of the night sky places a limit on how far out it is possible to observe the galaxy stellar populations. To give a rough estimate of the photometric accuracy needed to apply this technique, measurements accurate to 1 per cent are needed to detect fluctuations in galaxies at a distance of ~ 20 Mpc.

In general $\bar{L}$ will depend on the SFH of the population under scrutiny; moreover, it is heavily weighted towards the brightest stars, since L_i^2 appears in the numerator of Equation (11.6). Given that observations are performed using photometric filters, it is the value of $\bar{L}$ in the chosen photometric band (the SBF magnitude) that has to be provided by the theoretical models. For a given stellar population the value of the SBF magnitude will depend on the adopted filter because, the brightest stars, e.g. in B, are not the same as in e.g. K. As an example, in populations containing intermediate- and low-mass stars, the RGB and AGB are much brighter in the infrared, and dominate the value of $\bar{L}$ in the near-infrared filters (e.g. I, J, H, K). On the other hand, for the same populations, $\bar{L}$ in the B filter is controlled by hotter (and fainter) stars and the contribution of the AGB is of merely $\approx 10\text{--}20$ per cent. After transforming $\bar{L}$ into magnitudes, in case of SSPs of solar metallicity and ages older than 1 Gyr, these typical SBF magnitudes are derived: $\bar{M}_K \sim -6$, $\bar{M}_J \sim -4$, $\bar{M}_I \sim -1.5$, $\bar{M}_V \sim 1.0$ and $\bar{M}_B \sim 2.5$.

It is important to notice that metallicity(ies) and age(s) of the observed SSP or CSP in general affect the predicted value of the SBF magnitudes. According to recent theoretical SBF computations ([131]) for metallicities around solar and ages larger than ~ 1 Gyr, the SBF in the K filter is very weakly dependent on age, with variations of the order of ~ 0.2 mag for a change of age from 1 to 15 Gyr. On the other hand, the same age variation changes the SBF in B, V and I by, respectively, ~ 0.7, ~ 1.0 and ~ 1.2 mag (in the sense of increasing the SBF magnitude when moving to higher ages).

Appendix I: Constants

Physical Constants

Quantity	Symbol	Value	Units
Speed of light	c	2.998×10^{10}	cm s^{-1}
Electron charge	e	1.602×10^{-19}	C
Planck's constant	h	6.626×10^{-27}	erg s
Gravitational constant	G	6.6742×10^{-8}	dyn cm^2 g^{-2}
Atomic mass unit	m_{H}	1.6605×10^{-24}	g
Electron rest mass	m_{e}	9.109×10^{-28}	g
Proton rest mass	m_{p}	1.6726×10^{-24}	g
Neutron rest mass	m_{n}	1.6749×10^{-24}	g
Boltzmann's constant	K_{B}	1.3807×10^{-16}	erg K^{-1}
Radiation constant	a	7.5659×10^{-15}	erg cm^{-3} K^{-4}
Stefan–Boltzmann constant	σ	5.67051×10^{-5}	erg cm^{-2} K^{-4} s^{-1}

Astronomical Constants

Quantity	Symbol	Value	Units
Solar mass	$M_{\odot}$	1.989×10^{33}	g
Solar radius	$R_{\odot}$	6.960×10^{10}	cm
Solar luminosity	$L_{\odot}$	3.846×10^{33}	erg s^{-1}
Solar effective temperature	$T_{\mathrm{eff}\odot}$	5770	K
Astronomical unit	AU	1.496×10^{13}	cm
Light year	ly	9.4607×10^{17}	cm

Evolution of Stars and Stellar Populations Maurizio Salaris and Santi Cassisi
© 2005 John Wiley & Sons, Ltd

Conversions

Quantity	Symbol	Value	Units
Electron volt	eV	1.6022×10^{-12}	erg
		1.1604×10^{4}	K
Parsec	pc	3.0857×10^{18}	cm
		3.2616	ly

Appendix II: Selected Web Sites

List of selected web sites (updated to March 2005) containing complete stellar evolution codes, input physics data and routines for stellar modelling, databases of evolutionary tracks, isochrones, integrated colours, and on-line software for generating CMDs of synthetic Single and Composite Stellar Populations.

STELLAR EVOLUTION CODES

http://chandra.as.arizona.edu/ dave/tycho-manual.html
The stellar evolution code TYCHO and all necessary physics inputs. The code TYCHO has been developed by D. Arnett.

STELLAR PHYSICS

http://www.webelements.com/
Extensive information about chemical elements.

http://www-phys.llnl.gov/Research/RRSN/
Reaction rates for stellar nucleosynthesis from the Lawrence Livermore National Laboratory.

http://www-phys.llnl.gov/Research/OPAL/index.html
Tables of radiative opacities from the Lawrence Livermore National Laboratory.

http://www.osc.edu/hpc/opacities/
Tables of radiative opacities from the Opacity Project database.

http://kurucz.harvard.edu/
Model atmospheres, low-temperature opacities and colour transformations computed with the ATLAS code.

http://webs.wichita.edu/physics/opacity/
Low-temperature opacities computed at Wichita State University.

Evolution of Stars and Stellar Populations   Maurizio Salaris and Santi Cassisi
© 2005 John Wiley & Sons, Ltd

http://www.ioffe.rssi.ru/astro/conduct/
Tables of electron conduction opacities at the Department of Theoretical Astrophysics of the Ioffe Physico-Technical Institute, St. Petersburg.

http://freeeos.sourceforge.net/
EOS of the stellar matter by A.W. Irwin.

http://www.cococubed.com/code pages/codes.shtml
A collection of free computer codes to compute EOS, neutrino energy loss rates, simplified model atmospheres, nuclear reaction networks and supernova explosions.

BOLOMETRIC CORRECTIONS AND COLOUR TRANSFORMATIONS

http://www.astro.mat.uc.pt/BaSeL/
University of Basel interactive server for bolometric corrections and colour transformations.

EVOLUTIONARY TRACKS, ISOCHRONES, INTEGRATED COLOURS OF SSPs
http://pleiadi.pd.astro.it/

Padova database of stellar evolutionary tracks and isochrones.

http://dipastro.pd.astro.it/galadriel/
The Padova GALaxies AnD Single StellaR PopulatIon ModELs. Tables of integrated colours and absorption feature index strength for SSPs.

http://www.te.astro.it/BASTI/index.php
A Bag of Stellar Tracks and Isochrones (Teramo Astronomical Observatory). Tables of evolutionary tracks and isochrones.

http://astro.df.unipi.it/SAA/PEL/Z0.html
Database of Pisa stellar models, tracks and isochrones.

http://webast.ast.obs-mip.fr/stellar/
Database of Geneva stellar evolution tracks and isochrones.

http://www-astro.physics.ox.ac.uk/ yi/yyiso.html
Tables of evolutionary tracks and isochrones (collaboration between the universities of Yonsei and Yale).

http://www.bo.astro.it/ eps/models.html
Tables of integrated colours and absorption feature index strength from A. Buzzoni's stellar population synthesis code.

http://www.cida.ve/ bruzual/bc2003
Database of integrated colours and absorption feature index strength using the stellar population synthesis code by G. Bruzual and S. Charlot.

POPULATION SYNTHESIS SOFTWARE

http://iac-star.iac.es/iac-star/

IAC-STAR, Synthetic CMD computation algorithm (Instituto de Astrofisica de Canarias). On-line software to compute CMD and integrated magnitudes of stellar populations with an arbitrary SFH.

http://www.cida.ve/ bruzual/bcXXI.html

On-line software to compute CMD and integrated magnitudes of SSPs (Centro de Investigaciones de Astronomia, Mérida).

References

[1] Alexander, D.R. and Ferguson, J.W., 1994, 'Low-temperature Rosseland opacities', *Astrophysical Journal*, **437**, 879

[2] Alongi, M. *et al.*, 1991, 'Effects of envelope overshoot on stellar models', *Astronomy & Astrophysics*, **244**, 95

[3] Alves, D., 2000, 'K-band calibration of the red clump luminosity', *Astrophysical Journal*, **539**, 732

[4] Angulo, C. *et al.* (NACRE Collaboration), 1999, 'A compilation of charged-particle induced thermonuclear reaction rates', *Nuclear Phyics A*, **656**, 3

[5] Arnett, W.A., 1972, 'Advanced evolution of massive stars. I. Helium burning', *Astrophysical Journal*, **176**, 681

[6] Asplund, M. *et al.*, 2004, 'Line formation in solar granulation. IV. [O I], O I and OH lines and the photospheric O abundance', *Astronomy & Astrophysics*, **417**, 751

[7] Barker, M.K., Sarajedini, A. and Harris, J., 2004, 'Variations in star formation history and the Red Giant Branch Tip', *Astrophysical Journal*, **606**, 869

[8] Beaudet, G., Petrosian, V. and Salpeter, E.E., 1967, 'Energy losses due to neutrino processes', *Astrophysical Journal*, **150**, 979

[9] Bellazzini, M., Ferraro, F.R. and Pancino, E., 2001, 'A step toward the calibration of the Red Giant Branch Tip as a standard candle', *Astrophysical Journal*, **556**, 635

[10] Bennett, C.L. *et al.*, 2003, 'First-year Wilkinson Microwave Anisotropy Probe (WMAP) observations: Preliminary maps and basic results', *Astrophysical Journal Supplement Series*, **148**, 1

[11] Berry, M.V., 1989, *Principles of Cosmology and Gravitation*, Adam Hilger

[12] Bessell, M.S., Castelli, F. and Plez, B., 1998, 'Model atmospheres broad-band colors, bolometric corrections and temperature calibrations for O - M stars', *Astronomy & Astrophysics*, **333**, 231

[13] Bildsten, L. *et al.*, 1997, 'Lithium depletion in fully convective Pre-Main-Sequence stars', *Astrophysical Journal*, **482**, 442

[14] Böhm-Vitense, E., 1958, 'Über die Wasserstoffkonvektionszone in Sternen verschiedener Effektivtemperaturen und Leuchtkräfte', *Zeitschrift für Astrophysik*, **46**, 108

[15] Böhm-Vitense, E., 1981, 'The effective temperature scale', *Annual Review of Astronomy and Astrophysics*, **19**, 295

[16] Bono, G. *et al.*, 1997, 'Nonlinear investigation of the pulsational properties of RR Lyrae variables', *Astronomy & Astrophysics Supplement Series*, **121**, 327

[17] Bono, G. *et al.*, 1999, 'Theoretical models for classical Cepheids. II. Period-Luminosity, Period-Color, and Period-Luminosity-Color relations', *Astrophysical Journal*, **512**, 711

[18] Bono, G., Castellani, V. and Marconi, M., 2000, 'Classical Cepheid pulsation models. III. The predictable scenario', *Astrophysical Journal*, **529**, 293

[19] Bono, G. *et al.*, 2005, 'Classical Cepheid pulsation models. X. The Period-Age relation', *Astrophysical Journal*, **621**, 966

[20] Bragaglia, A. *et al.*, 1990, 'Double Degenerates among DA white dwarfs', *Astrophysical Journal Letters*, **365**, 13

[21] Bressan, A., Chiosi, C. and Fagotto, F., 1994, 'Spectrophotometric evolution of elliptical galaxies. I: Ultraviolet excess and color-magnitude-redshift relations', *Astrophysical Journal*, **94**, 63

[22] Brown, T.M. *et al.*, 2001, 'Flash Mixing on the White Dwarf cooling curve: understanding Hot Horizontal Branch anomalies in NGC 2808', *Astrophysical Journal*, **562**, 368

[23] Bruzual, G. and Charlot, S., 1993, 'Spectral evolution of stellar populations using isochrone synthesis', *Astrophysical Journal*, **405**, 538

[24] Buonanno, R. *et al.*, 1994, 'The stellar population of the globular cluster M 3. I. Photographic photometry of 10 000 stars', *Astronomy & Astrophysics*, **290**, 69

[25] Buonanno, R. *et al.*, 1998, 'On the relative ages of galactic globular clusters. A new observable, a semiempirical calibration and problems with the theoretical isochrones', *Astronomy & Astrophysics*, **333**, 505

[26] Burbidge, E.M. *et al.*, 1957, 'Synthesis of the elements in stars', *Review of Modern Physics*, **29**, 547

[27] Burgers, J.M., 1969, *Flow Equations for Composite Gase*, Academic Press

[28] Burke, C.J., Pinsonneault, M.H. and Sills, A., 2004, 'Theoretical examination of the Lithium depletion boundary', *Astrophysical Journal*, **604**, 272

[29] Buzzoni, A., 1989, 'Evolutionary population synthesis in stellar systems. I - A global approach', *Astrophysical Journal Supplement Series*, **71**, 817

[30] Cameron, A.G.W., 1955, 'Origin of anomalous abundances of the elements in giant stars', *Astrophysical Journal*, **121**, 144

[31] Cameron, A.G.W., 1959, 'Pycnonuclear reactions and Nova explosions', *Astrophysical Journal*, **130**, 916

[32] Cameron, A.G.W., 1960, 'New neutron sources of possible astrophysical importance', *Astronomical Journal*, **65**, 485

[33] Cameron, A.G.W. and Fowler, W.A., 1971, 'Lithium and the s-process in Red Giant Stars', *Astrophysical Journal*, **164**, 111

[34] Canuto, V.M., Goldman, I. And Mazzitelli, I., 1996, 'Stellar turbulent convection: A self-consistent model', *Astrophysical Journal*, **473**, 550

[35] Caputo, F., Cayrel, R. and Cayrel de Strobel, G., 1983, 'The galactic globular cluster system - Helium content versus metallicity', *Astronomy & Astrophysics*, **123**, 135

[36] Caputo, F. *et al.*, 1989, 'The 'Red Giant Clock' as an indicator for the efficiency of central mixing in horizontal-branch stars', *Astrophysical Journal*, **340**, 241

[37] Caputo, F, Marconi, M. and Musella, I., 2000, 'Theoretical models for classical Cepheids. V. Multiwavelength relations', *Astronomy & Astrophysics*, **354**, 610

[38] Cardelli, J.A., Clayton, G.C. and Mathis, J.S., 1989, 'The relationship between infrared, optical, and ultraviolet extinction', *Astrophysical Journal*, **345**, 245

[39] Carretta, E. *et al.*, 2000, 'Distances, ages, and epoch of formation of globular clusters', *Astrophysical Journal*, **533**, 215

[40] Cassisi, S. and Castellani, V., 1993, 'An evolutionary scenario for primeval stellar populations', *Astrophysical Journal Supplement*, **88**, 509

[41] Cassisi, S., Iben, I.Jr. and Tornambé, A., 1998, 'Hydrogen-accreting Carbon–Oxygen white dwarfs', *Astrophysical Journal*, **496**, 376

[42] Cassisi, S. *et al.*, 2004, 'Color transformations and bolometric corrections for Galactic halo stars: α-Enhanced versus scaled-solar results', *Astrophysical Journal*, **616**, 498

[43] Cassisi, S. and Salaris, M., 1997, 'A critical investigation on the discrepancy between the observational and the theoretical red giant luminosity function bump', *Monthly Notices of the Royal Astronomical Society*, **285**, 593

[44] Cassisi, S., Salaris, M. and Irwin, A.W., 2003, 'The initial helium content of Galactic globular cluster stars from the R-Parameter: comparison with the Cosmic Microwave Background constraint', *Astrophysical Journal*, **588**, 862

[45] Cassisi, S. *et al.*, 2003, 'First full evolutionary computation of the helium flash-induced mixing in Population II stars', *Astrophysical Journal Letter*, **582**, L43

[46] Castellani, V. *et al.*, 1997, 'Heavy element diffusion and globular cluster ages', *Astronomy & Astrophysics*, **322**, 801

[47] Castellani, V. *et al.*, 2003, 'Stellar evolutionary models for Magellanic Clouds', *Astronomy & Astrophysics*, **404**, 645

[48] Castellani, V., Giannone, P. and Renzini, A., 1969, 'An Investigation on HR diagrams of globular clusters', *Astrophysics and Space Science*, **3**, 518

[49] Castellani, V., Giannone, P. and Renzini, A., 1971, 'Overshooting of convective cores in helium-burning Horizontal-Branch Stars', *Astrophysics and Space Science*, **10**, 340

[50] Castellani, V., Giannone, P. and Renzini, A., 1971, 'Induced semi-convection in helium-burning Horizontal-Branch stars II', *Astrophysics and Space Science*, **10**, 355

[51] Caughlan, G.R. and Fowler, W.A., 1988, 'Thermonuclear reaction rates V', *Atomic Data and Nuclear Data Tables*, **40**, 283

[52] Chaboyer, B., Demarque, P. and Pinsonneault, M.H., 1995, 'Stellar models with microscopic diffusion and rotational mixing. 1: Application to the Sun', *Astrophysical Journal*, **441**, 865

[53] Chaboyer, B., Demarque, P. and Sarajedini, A., 1996, 'Globular cluster ages and the formation of the Galactic Halo', *Astrophysical Journal*, **459**, 558

[54] Chandrasekhar, S., 1939, *An introduction to the study of stellar structure*, University of Chicago Press, Chicago

[55] Chapman, S. and Cowling, T.G., 1970, *The mathematical theory of non-uniform gases*, Cambridge University Press

[56] Chiosi, C., Bertelli, G. and Bressan, A., 1992, 'New developments in understanding the HR diagram', *Annual Review of Astronomy and Astrophysics*, **30**, 235

[57] Coles, P. and Lucchin, F., 2002, *Cosmology: the origin and evolution of cosmic structure*, John Wiley

[58] Colgate, S.A. and McKee, C., 1969, 'Early supernova luminosity', *Astrophysical Journal*, **157**, 623

[59] Colgate, S.A. and White, R.H., 1966, 'The hydrodynamic behavior of supernovae explosions', *Astrophysical Journal*, **143**, 626

[60] Contardo, G., 2001, 'Bolometric light curves of Type Ia Supernovae', PhD Thesis, Technical University, Munich

[61] Cox, J.P., 1980, *Theory of stellar pulsation*, Princeton University Press

[62] Cox, J.P. and Giuli, R.T., 2004, *Principles of Stellar Structure*, 2nd expanded edition by Weiss, Hillebrandt, Thomas, Ritter, Cambridge Scientific Publishers

[63] Crampin, J. and Hoyle, F., 1961, 'On the change with time of the integrated colour and luminosity of an M67-type star group', *Monthly Notices of the Royal Astronomical Society*, **122**, 27

[64] Da Costa, G.S. and Armandroff, T.E., 1990, 'Standard globular cluster giant branches in the $(M_I - (V - I)_0)$ plane', *Astronomical Journal*, **100**, 162

[65] Dolphin, A., 2002, 'Numerical methods of star formation history measurement and applications to seven dwarf spheroidals', *Monthly Notices of the Royal Astronomical Society*, **332**, 91

[66] Domínguez, I. *et al.*, 1996, 'On the formation of massive C-O white dwarfs: The lifting effect of rotation', *Astrophysical Journal*, **472**, 783

[67] Durrell, P.R. and Harris, W.E., 1993, 'A color-magnitude study of the globular cluster M15', *Astronomical Journal*, **105**, 1420

[68] Eddington, A.S., 1917, 'The pulsation theory of Cepheid variables', *The Observatory*, **40**, 290

[69] Eddington, A.S., 1918, 'The problem of the Cepheid variables', *Monthly Notices of the Royal Astronomical Society*, **79**, 2

[70] Endal, A.S. and Sofia, S., 1976, 'The evolution of rotating stars. I - Method and exploratory calculations for a $7M_\odot$ star', *Astrophysical Journal*, **210**, 184

[71] Feast, M.W. and Catchpole, R.M., 1997, 'The Cepheid period-luminosity zero-point from HIPPARCOS trigonometrical parallaxes', *Monthly Notices of the Royal Astronomical Society*, **286**, L1

[72] Ferrarese, L. *et al.*, 2000, 'A database of Cepheid distance moduli and tip of the Red Giant Branch, globular cluster luminosity function, Planetary Nebula luminosity function, and Surface Brightness Fluctuation data useful for distance determinations', *Astrophysical Journal Supplement Series*, **128**, 431

[73] Ferraro, F.R. *et al.*, 2000, 'A new infrared array photometric survey of Galactic globular clusters: A detailed study of the Red Giant Branch sequence as a step toward the global testing of stellar models', *Astronomical Journal*, **119**, 1282

[74] Fontaine, G., Graboske Jr., H.G. and Van Horn, H.M., 1977, 'Equations of State for stellar partial ionization zones', *Astrophysical Journal Supplement Series*, **35**, 293

[75] Freedman, W.L. *et al.*, 2001, 'Final results from the Hubble Space Telescope Key Project to measure the Hubble constant', *Astrophysical Journal*, **553**, 47

[76] Gallart, C. *et al.*, 1996, 'The Local Group dwarf irregular galaxy NGC 6822.II. The old and intermediate-age Star Formation History', *Astronomical Journal*, **112**, 1950

[77] García-Berro, E. *et al.*, 1988, 'Theoretical white dwarf luminosity functions for two phase diagrams of the carbon–oxygen dense plasma', *Astronomy & Astrophysics*, **193**, 141

[78] García-Berro, E., Isern, J. and Hernanz, M., 1997, 'The cooling of oxygen–neon white dwarfs', *Monthly Notices of the Royal Astronomical Society*, **289**, 973

[79] Gautschy, A. and Saio, H., 1995, 'Stellar pulsations across the HR Diagram: Part 1', *Annual Review of Astronomy and Astrophysics*, **33**, 75

[80] Gautschy, A. and Saio, H., 1996, 'Stellar pulsations across the HR Diagram: Part 2', *Annual Review of Astronomy and Astrophysics*, **34**, 551

[81] Girardi, L. *et al.*, 1996, 'Evolutionary sequences of stellar models with new radiative opacities. VI. Z=0.0001', *Astronomy & Astrophysics Supplement*, **117**, 113

[82] Girardi, L. *et al.*, 2000, 'Evolutionary tracks and isochrones for low- and intermediate-mass stars: From 0.15 to $7M_\odot$, and from Z=0.0004 to 0.03', *Astronomy & Astrophysics Supplement*, **141**, 371

[83] Girardi, L. and Salaris, M., 2001, 'Population effects on the red giant clump absolute magnitude, and distance determinations to nearby galaxies', *Monthly Notices of the Royal Astronomical Society*, **323**, 109

[84] Girardi, L. *et al.*, 2002, 'Theoretical isochrones in several photometric systems. I. Johnson-Cousins-Glass, HST/WFPC2, HST/NICMOS, Washington, and ESO Imaging Survey filter sets', *Astronomy & Astrophysics*, **391**, 195

[85] Gordon, K.D. *et al.*, 2003, 'A quantitative comparison of the Small Magellanic Cloud, Large Magellanic Cloud, and Milky Way ultraviolet to near-infrared extinction curves', *Astrophysical Journal*, **594**, 279

[86] Graboske, H.C. *et al.*, 1973, 'Screening factors for nuclear reactions. II. Intermediate screening and astrophysical applications', *Astrophysical Journal*, **181**, 457

[87] Gratton, R.G. *et al.*, 1997, 'Ages of globular clusters from HIPPARCOS parallaxes of local subdwarfs', *Astrophysical Journal*, **491**, 749

[88] Grevesse, N. and Noels, A., 1993, 'Cosmic abundances of the elements', in *Cosmic abundances of the elements*, eds. Prantzos, N., Vangioni-Flam, E. and Casseé, M., Cambridge University Press, p. 15

[89] Groenewegen, M.A.T. and Salaris, M., 2003, 'The distance to the LMC cluster NGC 1866; clues from the cluster Cepheid population', *Astronomy & Astrophysics*, **410**, 887

[90] Guiderdoni, B. and Rocca-Volmerange, B., 1987, 'A model of spectrophotometric evolution for high-redshift galaxies', *Astronomy & Astrophysics*, **186**, 1

[91] Hansen, B.M.S., 1999, 'Cooling models for old white dwarfs', *Astrophysical Journal*, **520**, 680

[92] Harris, J. and Zaritsky, D., 2004, 'The Star Formation History of the Small Magellanic Cloud', *Astronomical Journal*, **127**, 1531

[93] Hayashi, E., Hoshi, R. and Sugimoto, D., 1962, 'Evolution of the stars', *Progress of Theoretical Physics Supplement*, **22**, 1

[94] Haywood, M., 2002, 'On the metallicity of the Milky Way thin disc and photometric abundance scales', *Monthly Notices of the Royal Astronomical Society*, **337**, 151

[95] Heger, A., Langer, N. and Woosley, S.E., 2000, 'Presupernova evolution of rotating massive stars. I. Numerical method and evolution of the internal stellar structure', *Astrophysical Journal*, **528**, 368

[96] Henyey, L.G., Forbes, G.E. and Gould, N.L., 1964, 'A new method of automatic computation of stellar evolution', *Astrophysical Journal*, **139**, 306

[97] Hernández, X., Valls-Gabaud, D. and Gilmore, G., 1999, 'Deriving star formation histories: inverting Hertzsprung-Russell diagrams through a variational calculus maximum likelihood method', *Monthly Notices of the Royal Astronomical Society*, **304**, 705

[98] Hilker, M., 2000, 'Revised Strömgren metallicity calibration for red giants', *Astronomy & Astrophysics*, **355**, 994

[99] Holtzman, J.A. *et al.*, 1999, 'Observations and implications of the star formation history of the Large Magellanic Cloud', *Astronomical Journal*, **118**, 2262

[100] Hummer, D.G. and Mihalas, D., 1988, 'The equation of state for stellar envelopes. I - an occupation probability formalism for the truncation of internal partition functions', *Astrophysical Journal*, **331**, 794

[101] Iben, I.Jr., 1968, 'Low-mass red giants', *Astrophysical Journal*, **154**, 581

[102] Iben, I.Jr., 1968, 'Age and initial helium abundance of stars in the globular cluster M15', *Nature*, **220**, 143

[103] Iben, I.Jr., 1975, 'Thermal pulses; p-capture, alpha-capture, s-process nucleosynthesis; and convective mixing in a star of intermediate mass', *Astrophysical Journal*, **196**, 525

[104] Iben, I.Jr. and Renzini, A., 1983, 'Asymptotic giant branch evolution and beyond', *Annual Review of Astronomy and Astrophysics*, **21**, 271

[105] Iben, I.Jr. and Tutukov, A.V., 1984, 'Supernovae of type I as end products of the evolution of binaries with components of moderate initial mass (M not greater than about 9 solar masses)', *Astrophysical Journal Supplement Series*, **54**, 335

[106] Iglesias, C.A. and Rogers, F.J., 1996, 'Updated Opal Opacities', *Astrophysical Journal*, **464**, 943

[107] Isern, J. *et al.*, 2000, 'The energetics of crystallizing white dwarfs revisited again', *Astrophysical Journal*, **528**, 397

[108] Itoh, N. *et al.*, 1996, 'Neutrino energy loss in stellar interiors. VII. Pair, Photo-, Plasma, Bremsstrahlung, and Re-combination neutrino processes', *Astrophysical Journal Supplement Series*, **102**, 411

[109] Jacoby, G.H. *et al.*, 1992, 'A critical review of selected techniques for measuring extragalactic distances', *Publications of the Astronomical Society of the Pacific*, **104**, 599

[110] Jenkins, A. *et al.*, 1998, 'Evolution of Structure in Cold Dark Matter Universes', *Astrophysical Journal*, **499**, 20

[111] Johnson, H.L., 1952, 'Praesepe: magnitudes and colors', *Astrophysical Journal*, **116**, 640

[112] Jungman, G. *et al.*, 1996, 'Cosmological-parameter determination with microwave background maps', *Physical Review D*, **54**, 1332

[113] Kato, S., 1966, 'Overstable convection in a medium stratified in mean molecular weight', *Publication of the Astronomical Society of Japan*, **18**, 374

[114] Kippenhahn, R. and Thomas, H.-C., 1970, 'A simple method for the solution of the stellar structure equations including rotation and tidal forces', in *Stellar Rotation*, ed. A. Slattebak, IAU Colloquium 4, Reidel, Dordrecht, p. 20

[115] Kippenhahn, R. and Weigert, A., 1990, *Stellar Structure and Evolution*, Springer-Verlag

[116] Kochanek, C.S., 1997, 'Rebuilding the Cepheid distance scale. I. A global analysis of Cepheid mean magnitudes', *Astrophysical Journal*, **491**, 13

[117] Koester, D. and Chanmugam, G., 1990, 'Physics of white dwarf stars', *Reports on Progress in Physics*, **53**, 837

[118] Kolb, E.W. and Turner, M.S., 1990, *The Early Universe*, Addison-Wesley

[119] Krisciunas, K. *et al.*, 2003, 'Optical and infrared photometry of the nearby Type Ia Supernova 2001el', *The Astronomical Journal*, **125**, 166

[120] Kroupa, P., 2001, 'On the variation of the initial mass function', *Monthly Notices of the Royal Astronomical Society*, **322**, 231

[121] Kunz, R. *et al.*, 2002, 'Astrophysical reaction rate of $^{12}C(\alpha, \gamma)^{16}O$', *Astrophysical Journal*, **567**, 643

[122] Kurucz, R.L., 1995, 'The Kurucz Smithsonian atomic and molecular database', in *Astrophysical Applications of Powerful New Databases, Joint Discussion No. 16 of the 22nd IAU General Assembly*, ed. S.J. Adelman and W. L Wiese, ASP Conference Series, **78**, 205

[123] Landau, L.D. and Lifshitz, E.M., 1981, *Mechanics*, Butterworth-Heinemann

[124] Le Borgne, D. *et al.*, 2004, 'Evolutionary synthesis of galaxies at high spectral resolution with the code PEGASE-HR: Metallicity and age tracers', *Astronomy & Astrophysics*, **425**, 881

[125] Lee, M., Freedman, W.L. and Madore, B.F., 1993, 'The tip of the Red Giant Branch as a distance indicator for resolved galaxies', *Astrophysical Journal*, **417**, 553

[126] Leibundgut, B., 2000, 'Type Ia Supernovae', *The Astronomy and Astrophysics Review*, **10**, 179

[127] Leibundgut, B., 2004, 'Are Type Ia Supernovae standard candles?', *Astrophysics and Space Science*, **290**, 29

[128] Leibundgut, B. and Tammann, G.A., 1990, 'Supernova studies. III - The calibration of the absolute magnitude of supernovae of Type Ia', *Astronomy & Astrophysics*, **230**, 81

[129] Limongi, M. and Chieffi, A., 2003, 'Evolution, explosion, and nucleosynthesis of core-collapse supernovae', *Astrophysical Journal*, **592**, 404

[130] Limongi, M., Straniero, O. and Chieffi, A., 2000, 'Massive stars in the range 13-25$M_\odot$: Evolution and nucleosynthesis. II. The solar metallicity models', *Astrophysical Journal Supplement Series*, **129**, 625

[131] Liu, M.C., Charlot, S. and Graham, J.R., 2000, 'Theoretical predictions for Surface Brightness Fluctuations and implications for stellar populations of elliptical galaxies', *Astrophysical Journal*, **543**, 644

[132] Magni, G. and Mazzitelli, I., 1979, 'Thermodynamic properties and equations of state for hydrogen and helium in stellar conditions', *Astronomy & Astrophysics*, **72**, 134

[133] Maraston, C., 1998, 'Evolutionary synthesis of stellar populations: a modular tool', *Monthly Notices of the Royal Astronomical Society*, **300**, 872

[134] Marigo, P., Bressan, A. and Chiosi, C., 1996, 'The TP-AGB phase: a new model', *Astronomy & Astrophysics*, **313**, 545

[135] Marigo, P. *et al.*, 2004, 'Evolution of planetary nebulae. II. Population effects on the bright cut-off of the PNLF', *Astronomy & Astrophysics*, **423**, 995

[136] Meynet, G. and Maeder, A., 1997, 'Stellar evolution with rotation. I. The computational method and the inhibiting effect of the μ-gradient', *Astronomy & Astrophysics*, **321**, 465

[137] Mestel, L., 1952, 'On the theory of white dwarf stars. I. The energy sources of white dwarfs', *Monthly Notices of the Royal Astronomical Society*, **112**, 583

[138] Mihalas, D., 1967, 'Model of basic stellar atmosphere', *Methods in Computational Physics*, **7**, 1

[139] Mihalas, D., 1978, *Stellar Atmospheres*, 2nd ed., W.H. Freeman and Company

[140] Murphy, T. and Meiksin, A., 2004, 'A library of high-resolution Kurucz spectra in the range λ 3000–10 000', *Monthly Notices of the Royal Astronomical Society*, **351**, 1430

[141] Nomoto, K. and Sugimoto, D., 1977, 'Rejuvenation of helium white dwarfs by mass accretion', *Publications of the Astronomical Society of Japan*, **29**, 765

[142] Peacock, J.A., 1999, *Cosmological Physics*, Cambridge University Press

[143] Percival, S. and Salaris, M., 2003, 'An empirical test of the theoretical population corrections to the red clump absolute magnitude', *Monthly Notices of the Royal Astronomical Society*, **343**, 539

[144] Percival, S., Salaris, M. and Groenewegen, M., 2005, 'The distance to the Pleiades. Main sequence fitting in the near infrared', *Astronomy & Astrophysics*, **429**, 887

[145] Percival, S., Salaris, M. and Kilkenny, D., 2003, 'The open cluster distance scale. A new empirical approach', *Astronomy & Astrophysics*, **400**, 541

[146] Perlmutter, S. *et al.*, 1999, 'Measurements of omega and lambda from 42 high-redshift supernovae', *Astrophysical Journal*, **517**, 565

[147] Perry, C.L., Olsen, E.H. and Crawford, D.L., 1987, 'A catalog of bright *uvbyβ* standard stars', *Publication of the Astronomical Society of the Pacific*, **99**, 1184

[148] Perryman, M.A.C., 1997, The Hipparcos and Tycho catalogues. ESA Publications Division Series: ESA SP Series, vol. no:1200

[149] Petersen, J.O., 1973, 'Masses of double mode cepheid variables determined by analysis of period ratios', *Astronomy & Astrophysics*, **27**, 89

[150] Phillips, M.M., 1993, 'The absolute magnitudes of Type IA supernovae', *Astrophysical Journal Letters*, **413**, 105

[151] Piersanti, L. *et al.*, 1999, 'On the very long term evolutionary behaviour of hydrogen-accreting low-Mass CO white dwarfs', *Astrophysical Journal Letters*, **521**, 59

[152] Pietrinferni, A. *et al.*, 2004, 'A large stellar evolution database for population synthesis studies. I. Scaled solar models and isochrones', *Astrophysical Journal*, **612**, 168

[153] Poggianti, B.M., 1997, 'K and evolutionary corrections from UV to IR', *Astronomy & Astrophysics Supplement*, **122**, 399

[154] Potekhin, A.Y., 1999, 'Electron conduction in magnetized neutron star envelopes', *Astronomy & Astrophysics*, **351**, 787

[155] Prialnik, D., 2000, *An introduction to the theory of stellar structure and evolution*, Cambridge University Press

[156] Provencal, J.L. *et al.*, 1998, 'Testing the white dwarf mass-radius relation with HIPPARCOS', *Astrophysical Journal*, **494**, 759

[157] Renzini, A., 1981, 'Red giants as precursors of planetary nebulae', *Physical processes in red giants*, Proceedings of the Second Workshop, Dordrecht, D. Reidel Publishing Co., 431

[158] Renzini, A. *et al.*, 1996, 'The white dwarf distance to the globular cluster NGC 6752 (and its age) with the Hubble Space Telescope', *Astrophysical Journal Letters*, **465**, 23

[159] Renzini, A. and Fusi Pecci, F. 1988, 'Tests of evolutionary sequences using color-magnitude diagrams of globular clusters', *Annual Review of Astronomy and Astrophysics*, **26**, 199

[160] Renzini, A. *et al.*, 1992, 'Why stars inflate to and deflate from red giant dimensions', *Astrophysical Journal*, **400**, 280

[161] Richard, O. *et al.*, 2002, 'Models of metal-poor stars with gravitational settling and radiative accelerations. I. Evolution and abundance anomalies', *Astrophysical Journal*, **568**, 979

[162] Richer, H.B. *et al.*, 2000, 'Isochrones and luminosity functions for old white dwarfs', *Astrophysical Journal*, **529**, 318

[163] Rieke, G.H. and Lebofsky, R.M., 1985, 'The interstellar extinction law from 1 to 13 microns', *Astrophysical Journal*, **288**, 618

[164] Riess, A.G. *et al.*, 1998, 'Observational evidence from supernovae for an accelerating universe and a cosmological constant', *Astronomical Journal*, **116**, 1009

[165] Rogers, F.J. and Nayfonov, A., 2002, 'Updated and expanded OPAL Equation-of-State tables: Implications for helioseismology', *Astrophysical Journal*, **576**, 1064

[166] Rolfs, C.E. and Rodney, W.S., 1988, *Cauldrons in the Cosmos – nuclear astrophysics*, University of Chicago Press

[167] Romaniello, M. *et al.*, 2005, 'The influence of chemical composition on the properties of Cepheid stars. I. Period-Luminosity relation vs. iron abundance', *Astronomy & Astrophysics Letters*, **429**, 37

[168] Rood, R.T. *et al.*, 1999, 'The luminosity function of M3', *Astrophysical Journal*, **523**, 752

[169] Rose, J.A, 1994, 'The integrated spectra of M32 and of 47 Tuc: A comparative study at high spectral resolution', *Astronomical Journal*, **107**, 206

[170] Rosenberg, A. *et al.*, 1999, 'Galactic globular cluster relative ages', *Astronomical Journal*, **118**, 2306

[171] Ryan, S.G., Norris, J.E. and Beers, T.C., 1999, 'The Spite lithium plateau: Ultrathin but postprimordial', *Astrophysical Journal*, **523**, 654

[172] Sakai, S. *et al.*, 2004, 'The effect of metallicity on Cepheid-based distances', *Astrophysical Journal*, **608**, 42

[173] Salaris, M. and Cassisi, S., 1998, 'A new analysis of the red giant branch 'tip' distance scale and the value of the Hubble constant', *Monthly Notices of the Royal Astronomical Society*, **298**, 166

[174] Salaris, M. *et al.*, 2001, 'On the white dwarf distances to galactic globular clusters', *Astronomy & Astrophysics*, **371**, 921

[175] Salaris, M., Chieffi, A. and Straniero, O., 1993, 'The α-enhanced isochrones and their impact on the fits to the Galactic globular cluster system', *Astrophysical Journal*, **414**, 580

[176] Salaris, M. *et al.*, 1997, 'The cooling of CO white dwarfs: Influence of the internal chemical distribution', *Astrophysical Journal*, **486**, 413

[177] Salaris, M. *et al.*, 2000, 'The ages of very cool hydrogen-rich white dwarfs', *Astrophysical Journal*, **544**, 1036

[178] Salaris, M. *et al.*, 2004, 'The initial helium abundance of the Galactic globular cluster system', *Astronomy & Astrophysics*, **420**, 911

[179] Salaris, M. and Girardi, L., 2002, 'Population effects on the red giant clump absolute magnitude: the K band', *Monthly Notices of the Royal Astronomical Society*, **337**, 332

[180] Salaris, M. and Girardi, L., 2005, 'Tip of the Red Giant Branch distances to galaxies with composite stellar populations', *Monthly Notices of the Royal Astronomical Society*, **357**, 669

[181] Salaris, M. and Weiss, A., 1998, 'Metal-rich globular clusters in the galactic disk: new age determinations and the relation to halo clusters', *Astronomy & Astrophysics*, **335**, 943

[182] Salaris, M. and Weiss, A., 2001, 'Atomic diffusion in metal-poor stars. II. Predictions for the Spite plateau', *Astronomy & Astrophysics*, **376**, 955

[183] Salaris, M. and Weiss, A., 2002, 'Homogeneous age dating of 55 Galactic globular clusters. Clues to the Galaxy formation mechanisms', *Astronomy & Astrophysics*, **388**, 492

[184] Salasnich, B. *et al.*, 2000, 'Evolutionary tracks and isochrones for alpha-enhanced stars', *Astronomy & Astrophysics*, **361**, 1023

[185] Salpeter, E.E., 1955, 'The luminosity function and stellar evolution', *Astrophysical Journal*, **121**, 161

[186] Salpeter, E.E. and Van Horn, H.M., 1969, 'Nuclear reaction rates at high densities', *Astrophysical Journal*, **155**, 183

[187] Sandage, A., 1988, 'Observational tests of world models', *Annual Review of Astronomy and Astrophysics*, **26**, 561

[188] Sandage, A., 1990, 'The vertical height of the horizontal branch – The range in the absolute magnitudes of RR Lyrae stars in a given globular cluster', *Astrophysical Journal*, **350**, 603

[189] Sandquist, E., 2000, 'A catalogue of helium abundance indicators from globular cluster photometry', *Monthly Notices of the Royal Astronomical Society*, **313**, 571

[190] Sandquist, E.L. *et al.*, 1996, 'CCD photometry of the globular cluster M5. I. The Color-Magnitude Diagram and luminosity functions', *Astrophysical Journal*, **470**, 910

[191] Sarajedini, A. and Demarque, P., 1990, 'A new age diagnostic applied to the globular clusters NGC 288 and NGC 362', *Astrophysical Journal*, **365**, 219

[192] Saumon, D., Chabrier, G. and Van Horn, H.M., 1995, 'An Equation of State for low-mass stars and giant planets', *Astrophysical Journal Supplement Series*, **99**, 713

[193] Saumon, D. and Jacobson, S.B., 1999, 'Pure hydrogen model atmospheres for very cool white dwarfs', *Astrophysical Journal Letters*, **511**, 107

[194] Saviane, I. *et al.*, 2000, 'The red giant branches of Galactic globular clusters in the $[(V - I)_0, M_V]$ plane: metallicity indices and morphology', *Astronomy & Astrophysics*, **355**, 966

[195] Schiavon, R.P. *et al.*, 2004, 'The identification of blue Horizontal-Branch stars in the integrated spectra of globular clusters', *Astrophysical Journal Letters*, **608**, 33

[196] Schlattl, H. *et al.*, 2001, 'On the helium flash in low-mass Population III Red Giant stars', *Astrophysical Journal*, **559**, 1082

[197] Schlattl, H. and Salaris, M., 2003, 'Quantum corrections to microscopic diffusion constants', *Astronomy & Astrophysics*, **402**, 29

[198] Schuster, W.J. and Nissen, P.E., 1989, '*uvbyβ* photometry of high-velocity and metal-poor stars. II-Intrinsic color and metallicity calibrations', *Astronomy & Astrophysics*, **221**, 65

[199] Shapiro, S.L. and Teukolsky, S.A., 1983, *Black holes, white dwarfs and neutron stars. The physics of compact object*, John Wiley

[200] Skillman, E.D. *et al.*, 2003, 'Deep Hubble Space Telescope imaging of IC 1613. II. The star formation history', *Astrophysical Journal*, **596**, 253

[201] Siess, L., Dufour, E. and Forestini, M., 2000, 'An internet server for pre-main sequence tracks of low- and intermediate-mass stars', *Astronomy & Astrophysics*, **358**, 593

[202] Slattery, W.L., Doolen, G.D. and De Witt, H.E., 1980, 'Improved equation of state for the classical one-component plasma', *Physical Review A*, **21**, 2087

[203] Spinrad, H. and Taylor, D.J., 1971, 'The stellar content of the nuclei of nearby galaxies. I. M31, M32, and M81', *Astrophysical Journal Supplement Series*, **22**, 445

[204] Stauffer, J.R. *et al.*, 1999, 'Keck Spectra of brown dwarf candidates and a precise determination of the lithium depletion boundary in the α Persei open cluster', *Astrophysical Journal*, **527**, 219

[205] Stetson, P.B., VandenBerg, D.A. and Bolte, M., 1996, 'The relative ages of Galactic globular clusters', *Publication of the Astronomical Society of Pacific*, **108**, 560

[206] Stothers, R.B. and Chin, C-.W., 1991, 'Blue loops during core helium burning as the consequence of moderate convective envelope overshooting in stars of intermediate to high mass', *Astrophysical Journal*, **374**, 288

[207] Straniero, O., 1988, 'A tabulation of thermodynamical properties of fully ionized matter in stellar interiors', *Astronomy & Astrophysics Supplement*, **76**, 157

[208] Straniero, O. *et al.*, 1997, 'Evolution and nucleosynthesis in low-mass Asymptotic Giant Branch stars. I. Formation of Population I carbon stars', *Astrophysical Journal*, **478**, 332

[209] Sugimoto, D., 1971, 'Mixing between stellar envelope and core in advanced phases of evolution. III-Stellar core of initial mass $1.5M_{\odot}$', *Progress of Theoretical Physics*, **45**, 761

[210] Sweigart, A.V. and Demarque, P., 1972, 'Effects of semiconvection on the Horizontal-Branch', *Astronomy & Astrophysics*, **20**, 445

[211] Sweigart, A.V., Greggio, L. and Renzini, A., 1990, 'The development of the red giant branch. II-Astrophysical properties', *Astrophysical Journal*, **364**, 527

[212] Talbot, R.J. and Arnett, W.D., 1971, 'The evolution of galaxies. I. formulation and mathematical behavior of the one-zone model', *Astrophysical Journal*, **170**, 409

[213] Tammann, G.A., Sandage, A. and Reindl, B., 2003, 'New Period-Luminosity and Period-Color relations of classical Cepheids: I. Cepheids in the Galaxy', *Astronomy & Astrophysics*, **404**, 423

[214] Tantalo, R. and Chiosi, C., 2004, 'Measuring age, metallicity and abundance ratios from absorption-line indices', *Monthly Notices of the Royal Astronomical Society*, **353**, 917

[215] Thielemann, F.-K., Nomoto, K. and Yokoi, K., 1986, 'Explosive nucleosynthesis in carbon deflagration models of Type I supernovae', *Astronomy & Astrophysics*, **158**, 17

[216] Thorburn, J.A., 1994, 'The primordial lithium abundance from extreme subdwarfs: New observations', *Astrophysical Journal*, **421**, 318

[217] Thoul, A.A., Bahcall, J.N. and Loeb, A., 1994, 'Element diffusion in the solar interior', *Astrophysical Journal*, **421**, 828

[218] Tinsley, B.M. and Gunn, J.E., 1976, 'Evolutionary synthesis of the stellar population in elliptical galaxies. I-Ingredients, broad-band colors, and infrared features', *Astrophysical Journal*, **203**, 52

[219] Tolstoy, E., 1996, 'The Resolved stellar population of Leo A', *Astrophysical Journal*, **462**, 684

[220] Tonry, J.L. and Schneider, D.P., 1988, 'A new technique for measuring extragalactic distances', *Astronomical Journal*, **96**, 807

[221] Trager, S.C. *et al.*, 1998, 'Old stellar populations. VI. Absorption-line spectra of galaxy nuclei and globular clusters', *Astrophysical Journal Supplement Series*, **116**, 1

[222] Udalski, A. *et al.*, 1999, 'The Optical Gravitational Lensing Experiment. Cepheids in the Magellanic Clouds. III. Period-Luminosity-Color and Period-Luminosity relations of classical Cepheids', *Acta Astronomica*, **49**, 201

[223] Udalski, A. *et al.*, 2000, 'The Optical Gravitational Lensing Experiment. BVI maps of dense stellar regions. II. The Large Magellanic Cloud', *Acta Astronomica*, **50**, 307

[224] Van Albada T.S. and Baker, N., 1971, 'On the masses, luminosities, and compositions of Horizontal-Branch stars', *Astrophysical Journal*, **169**, 311

[225] VandenBerg, D.A. *et al.*, 2000, 'Models for old, metal-poor stars with enhanced α-element abundances. I. Evolutionary tracks and ZAHB Loci; observational constraints', *Astrophysical Journal*, **532**, 430

[226] VandenBerg, D.A., 2000, 'Models for old, metal-poor stars with enhanced α-element abundances. II. Their implications for the ages of the Galaxy's globular clusters and field halo stars', *Astrophysical Journal Supplement Series*, **129**, 315

[227] VandenBerg, D.A., Bolte, M. and Stetson, P.B., 1990, 'Measuring age differences among globular clusters having similar metallicities – A new method and first results', *Astronomical Journal*, **100**, 445

[228] Vazdekis, A., 1999, 'Evolutionary stellar population synthesis at 2Å spectral resolution', *Astrophysical Journal*, **513**, 224

[229] Versteeg, H.K. and Malalasekera, W., 1995, *An introduction to computational fluid dynamics*, Longman

[230] Whelan, J. and Iben, I.Jr, 1973, 'Binaries and supernovae of Type I', *Astrophysical Journal*, **186**, 1007

[231] Wood, P.R., Bessell, M.S. and Fox, M.W., 1983, 'Long-period variables in the Magellanic Clouds-Supergiants, AGB stars, supernova precursors, planetary nebula precursors, and enrichment of the interstellar medium', *Astrophysical Journal*, **272**, 99

[232] Woosley, S.E. and Weaver, T.A., 1986, 'The physics of supernova explosions', *Annual Review of Astronomy and Astrophysics*, **24**, 205

[233] Woosley, S.E. and Weaver, T.A., 1994, 'Sub-Chandrasekhar mass models for Type IA supernovae', *Astrophysical Journal*, **423**, 371

[234] Woosley, S.E., Heger, A. and Weaver, T.A., 2002, 'The evolution and explosion of massive stars', *Reviews of Modern Physics*, **74**, 1015

[235] Worthey, G., 1994, 'Comprehensive stellar population models and the disentanglement of age and metallicity effects', *Astrophysical Journal Supplement Series*, **95**, 107

[236] Worthey, G. *et al.*, 1994, 'Old stellar populations. 5: Absorption feature indices for the complete LICK/IDS sample of stars', *Astrophysical Journal Supplement Series*, **94**, 687

[237] Worthey, G. and Ottaviani, D.L., 1997, 'H_γ and H_δ absorption features in stars and stellar populations', *Astrophysical Journal Supplement Series*, **111**, 377

[238] Yakovlev, D.G. and Pethick, C.J., 2004, 'Neutron star cooling', *Annual Review of Astronomy and Astrophysics*, **42**, 169

[239] Zahn, J.-P., 1992, 'Circulation and turbulence in rotating stars', *Astronomy & Astrophysics*, **265**, 115

Index

κ mechanism, 180, 181

A

absolute magnitude, 186, 232, 245, 250
absorption-feature indices, 341–347
accreting white dwarfs, 225–228
adiabatic temperature gradient, 59, 112
age determination, 270, 271, 274, 276
age–metallicity degeneracy, 336, 337, 339,
  345–347
age–metallicity relationship, 271
alpha–element enhancement, 240, 346
angular diameter, 249
angular power spectrum, 4
antimatter, 20
apparent magnitude, 241–246, 250–251, 254,
  291, 292, 299, 311, 328–330
Astronomical Unit, 289
astrophysical factor (S-factor), 74–75
asymptotic giant branch, 172, 187–198
atomic diffusion, 49, 100–102, 127, 128,
  195, 210

B

Bailey's diagram, 183
baryons, 18, 21, 23
beryllium, 114
Big Bang, 5, 18, 25, 27–29, 155, 264
binding energy, 69, 222, 223, 230
binding energy per nucleon, 70
black body, 5, 49, 93, 244
black hole, 2, 105, 224, 236, 237

blue loop, 174–177, 183
bolometric correction, 245, 246, 265, 268, 271,
  272, 295–297, 307
bolometric flux, 245, 249
bolometric magnitude, 231, 245
bolometric magnitude of the sun, 245
born-again AGB scenario, 197
boron, 114
boundary conditions, 91–95, 97–98, 112–114,
  203, 212, 279
boundary of the convective zone, 86
bound–bound process, 66
bound–free process, 67
breathing pulses, 172, 173
bremsstrahlung, 81
brown dwarfs, 116, 138
burning front, 229

C

Cameron–Fowler mechanism, 193
carbon burning, 105, 195, 198, 199, 213,
  219, 229
carbon burning reactions, 219
carbon deflagration, 229–230
carbon detonation, 227–230
carbon–oxygen white dwarfs, 199–213
central conditions, 92
Cepheid variables, 183–186, 311–314
Cepheid instability strip, 183–186, 312
Cepheid period–luminosity relation, 185–186,
  312–313

Evolution of Stars and Stellar Populations   Maurizio Salaris and Santi Cassisi
© 2005 John Wiley & Sons, Ltd

Chandrasekhar mass, 195, 198, 199, 202, 203, 225, 226, 228
closed-box model, 317–319
CNO cycle, 118–121, 127, 128, 134, 135, 143, 156–158, 162, 192, 304
collision induced absorption, 136, 247, 278
colour index, 186, 241, 246, 247
color-magnitude diagram, 241
completeness fraction, 322
composite stellar population, 347
Compton scattering, 67, 81
conductive opacity, 67–68, 138, 148
conductive transport of energy, 55–56
continuity of mass, 50, 111, 201
convection, 49, 56–66, 86, 98, 102, 111, 113, 115, 127, 129, 133, 135, 136, 142, 144–152, 155, 168, 172, 173, 181, 188, 192, 193, 212, 214, 218, 221, 223, 270, 272, 275, 289, 301, 302, 306, 326
convective mixing, 85–86, 99, 193, 209, 211
convective flux, 63, 65
convective velocity, 63, 65, 86, 102
core collapse, 215, 220, 222–224
core collapse supernovae, 224, 226
core mass–luminosity relation, 144, 191, 193
cosmic scale factor, 6
cosmological constant, 13, 15, 18, 24–26, 28
cosmological parameters, 24–26, 29
cosmological principle, 4–6, 8, 23, 27
Coulomb barrier, 73, 77, 219
Coulomb interactions, 43–46, 136, 206, 207
critical density, 13, 22
cross section, 17, 20, 67, 71–78, 83, 119, 120, 162, 163, 215, 224
crystallization, 45, 138, 206–210, 214, 278

D
DA white dwarfs, 210, 211
dark matter, 2–4, 21, 24, 25, 316
DB white dwarfs, 197
Debye radius, 43, 77
Debye temperature, 207, 213
deceleration parameter, 9, 11
decline rate, 231, 232
decoupling, 20, 23, 24, 26, 27
degeneracy parameter, 38, 40, 41, 43
degenerate cores, 148, 198, 199, 204–205, 207, 209, 212, 330
delayed-detonation scenario, 230
density parameter, 13, 22

diameter distance, 11
diffusion approximation, 52–55, 92, 93
distance modulus, 26, 243, 251, 254, 279, 291–296, 299, 300, 303, 307, 308, 311, 313, 323, 326, 330
double degenerate (DD) scenario, 226

E
Eddington approximation, 93
Eddington luminosity, 227
edge-detection algorithm, 298
effective temperature, 93, 130, 136, 142, 147, 150, 154, 165, 167, 176, 181, 182, 184, 185, 187, 191, 192, 198, 211, 217, 241, 244, 245, 249, 259, 260, 287, 288, 343, 346
effective wavelength, 242, 244
electron degeneracy, 40, 41, 67, 68, 89, 105, 116, 124, 136–138, 140, 142, 143, 147–149, 152, 158, 163, 167, 174, 189, 198, 199, 201, 229, 233, 296
electron mean molecular weight, 32, 204
electron scattering, 23, 67
electron screening, 77
elementary stellar population, 320–321, 324
energy conservation equation, 52
energy generation coefficient, 69, 71, 78, 207, 209
energy transport, 52, 55, 56, 92, 138, 142, 148, 151, 181, 218, 249
equation of state, 31–47, 50, 66, 88, 90, 92, 93, 131, 190, 206–210
equilibrium, local thermodynamic, 49, 53, 101
equilibrium, nuclear statistical, 222
evolutionary tracks, 110, 126, 134, 144, 157, 161, 167, 168, 172, 173, 176–179, 183, 200, 247, 259, 260, 267
evolution of chemical elements, 83–86
extinction law, 251, 252

F
Fermi–Dirac distribution, 38
Fermi energy, 39, 89, 143, 234, 235
first dredge-up, 145–147, 151, 152, 164, 178
fitting functions, 343, 346
flame front, 229
flame front velocity, 229
flatness problem, 27
fragmentation, 108, 110
free energy, 33, 34, 36, 37, 40, 42, 44, 45, 207

free-fall timescale, 91, 109
free-free process, 67
Friedmann equation, 13
Friedmann-Robertson-Walker metric, 6, 7,
  11–13
fully convective star, 110–114

G
Galaxy, 3, 20, 32, 200, 240, 246, 264,
  268, 312
Gamow energy, 77
Gamow peak, 74, 75
G-dwarf problem, 319
globular clusters, 25, 173, 264, 268, 271, 273,
  280–284, 286, 291, 296, 299, 301–303, 319,
  327, 345
gluons, 18, 19
Grand Unified Theories (GUT), 19
gravitational collapse, 107
gravitational energy, 78–81, 87–89, 106,
  108, 109, 123, 142, 147, 148, 189, 199,
  201, 234
gravitational potential, 10, 80, 81, 91, 102,
  107, 140, 201
grey atmosphere, 93

H
hadrons, 18
Hayashi track, 110–116, 150, 177
HB type, 280–281
He-ionization zone, 180
helioseismology, 125, 126
helium burning, 161–179, 210, 260, 305
helium flash, 148–149, 296
Henyey method, 93–97
Hertzsprung gap, 142
Hertzsprung–Russel diagram, 110
H-ionization zone, 180
horizon problem, 27, 28
horizontal branch, 47, 154, 163–173, 280–281,
  300, 332–334
horizontal method, 274–175
hot-bottom burning, 193
Hubble constant, 3, 7, 9, 24, 26
Hubble flow, 4, 5, 242, 327
Hubble law, 3–5, 7, 10, 11, 25
Hubble time, 16, 226
hydrogen burning, 84, 114–139

hydrogen burning shell, 141–148, 151,
  163–164, 167, 174, 176, 178, 187,
  189–196, 215, 227
hydrostatic equilibrium equation, 50–52, 92,
  138, 235

I
inflationary paradigm, 26, 28
initial helium abundance, 284
initial lithium abundance, 287, 291
initial mass function, 109, 117
instability strip, 179, 181–186, 265, 271, 273,
  280, 285, 287, 300, 312–314
instantaneous recycling approximation, 317
integrated colour, 332, 336, 337, 339–341, 347
integrated magnitude, 331–333, 335, 336,
  338, 339
intermediate mass stars, 305
interstellar extinction, 249, 250, 253, 282,
  313, 343
interstellar matter, 105, 106, 237
ionization, 23, 31, 34, 36, 38, 41–44, 46, 47,
  61, 66, 67, 78, 101, 109, 110, 130, 150,
  180, 181
iron core, 105, 199, 216, 222, 223
isochrones, 135, 259–270, 272, 276–281, 284,
  289, 291, 292, 294, 301, 302, 304–309, 319,
  321, 326, 327, 346
isothermal core, 140, 141, 205, 212

J
Jeans mass, 107
Johnson system, 245, 246, 251

K
K-correction, 253–254
Kelvin–Helmholtz timescale, 91, 140, 142,
  217, 222
kinetic energy, 35, 40, 45, 52, 55, 64, 72, 73,
  107, 108, 233
Kramers' opacity, 67, 205

L
Large Magellanic Cloud, 3
Ledoux criterion, 60, 133, 214
leptons, 18–20
light elements, 115, 116, 123, 138
lithium depletion, 289, 290
lithium depletion boundary, 306–309
Local Group, 3, 291, 320

Lower Main Sequence, 123–125, 141
luminosity, 10, 11, 50, 52, 87, 91, 92, 94, 98,
    105, 106, 110, 111, 113, 125, 126, 128, 130,
    131, 135, 138, 142, 144, 145, 147, 149,
    151–155, 158, 163, 165, 167, 169, 172, 174,
    180–185, 187, 191–193, 196, 197, 199,
    203–206, 208, 212, 217–219, 225, 227, 232,
    233, 235, 241, 242, 244–250, 259, 261, 266,
    272, 278, 279, 284, 285, 287, 293, 296,
    300–305, 307, 309–312, 329 332–335,
    347, 348
luminosity distance, 10, 26, 243
luminosity function, 147, 151, 263, 279,
    301–304, 329

M
magnitude zero point, 245, 246
main sequence fitting, 292–294
main sequence lifetime, 136, 280, 306
main sequence phase, 118–140, 156–159
mass–luminosity relation, 138–140, 303
mass loss, 28, 99, 100, 126, 145, 165, 167,
    193, 195, 196, 199, 211, 214, 225–227, 264,
    265, 277, 280, 316
mass–radius relation, 201–203
matter-dominated era, 17
Maxwell distribution, 35, 38, 39, 45, 72, 74
mean molecular weight, 31, 32, 59–62, 124,
    127, 131, 133, 141, 146, 152, 205, 209
mean free path, 24, 49, 52–55, 57, 64, 67,
    68, 223
mesons, 18
Mestel law, 206–208, 212, 213
Milky Way, 1, 3, 106, 326
mixing length, 57, 62, 128, 135, 150–152, 275
mixing length theory, 58, 62–66, 102, 127, 249
model atmosphere, 92, 93, 244, 246, 248, 249
molecular cloud, 106, 109

N
neon-burning, 219–220
neutrino production, 69, 81–82, 235
neutrino luminosity, 218, 219, 235
neutrino photosphere, 223
neutrino-trapping surface, 223
neutrinos, 18–21, 69, 71, 72, 81, 82, 189, 218,
    223, 224, 235
neutrons, 18, 20, 21, 36, 69, 70, 76, 194, 221,
    233–235
neutron stars, 40, 45, 233–235, 317

neutronization, 220–223, 234
non resonant cross section, 74–75
nuclear reactions, 43, 69, 76, 78, 79, 83–86,
    91, 92, 105–107, 115–123, 124, 161, 162,
    219–221
nuclear reaction rate, 71–76, 83, 163
nuclear reaction screening, 77–78

O
opacity coefficient, 53, 66
open clusters, 254–256, 259, 279, 290, 291,
    293, 304, 308
optical depth, 54, 93, 250
overshoot, 62, 131, 132, 135, 168–170, 177,
    178, 195, 305–307, 309
oxygen-burning, 220, 221
oxygen–neon white dwarfs, 213–214

P
pair-annihilation process, 81
pair-creation, 45–46
parallax, 289–292, 294, 312, 316, 328
parsec, 1, 26, 243, 290, 292
particle horizon, 12
partition function, 33, 36, 42, 43
perfect gas, 35–37, 40–44, 59, 65, 72, 88, 98,
    107, 112, 139, 203, 206
Petersen diagram, 183, 185
phase diagram, 208, 209
phase transitions, 206–209
photometric filters, 230, 244, 254, 273, 349
photometric systems, 230, 242, 243, 296
photoneutrino process, 81
photosphere, 126, 128, 136, 223, 241, 245, 306
Planck time, 18
planetary nebula, 198, 329–330
planetary nebula luminosity fuction, 329
plasma neutrino process, 81
Population I, 120, 155, 179
Population II, 134, 155, 156, 179
Population III, 134, 155–159, 224
p–p chain, 118, 119, 121–124, 134, 138, 156,
    157, 304
Pre-Main Sequence, 114–116, 123, 306–309
pressure ionization, 43, 44, 46, 136
pressure scale height, 57, 62
primary element, 84
primordial fluctuations, 4–5, 24–25, 27, 28
primordial nucleosynthesis, 17–22, 28, 32,
    284, 287

prompt explosion, 223, 224
proper time, 7
protons, 18, 20, 21, 23, 36, 69, 70, 73, 118, 119, 194, 195, 221, 234
protostar, 109, 110
pseudocontinuum, 342
pulsating delayed-detonation, 230

Q

quarks, 18, 19, 234

R

r-elements, 224
r-process, 76, 224
radiation dominated era, 17
radiation pressure, 37, 54, 65, 129, 130, 140
radiative energy transport, 52–55
radiative flux, 53–55, 133, 136, 232
radiative temperature gradient, 52–55, 168–171
Rayleigh–Taylor instability, 61
recombination, 23, 24, 137
recombination netrino process, 82
reddening, 250–253, 271, 276, 277, 282, 292–294, 307, 309, 310, 312, 314, 321, 324, 338
reddening estimates, 281–283, 309–310
red giant branch, 142–148
red giant branch bump, 147, 149, 151–152, 188, 285, 302
redshift, 3, 8–12, 15, 17, 23, 25, 237, 253, 254, 264
Reimers mass loss formula, 264, 280
resolved stellar population, 319, 327
resonant cross section, 75–76
response function, 241, 242
Rosseland mean, 54
rotation, 49, 51, 99, 100, 102–103, 107, 126, 167, 281
rotational mixing, 49, 102–103, 195
RR Lyrae instability strip, 181–183, 265, 269, 272, 280, 285, 300
R-parameter, 285–286
R2-parameter, 172–173
RR Lyrae stars, 172–173, 181–183, 287, 328

S

s-elements, 192, 194, 219
s-process, 76, 194
Saha equation, 43, 45

Salpeter initial mass function, 110, 279, 303, 331, 335, 340
Schönberg–Chandrasekhar limit, 140–142, 305, 322
Schwarzschild criterion, 60, 129, 133, 214
Schwarzschild metric, 236
Schwarzschild radius, 236, 237
second dredge-up, 188
second parameter, 274, 280–281
secondary element, 84, 121–123, 162
self-induced nova, 197
semiconvection, 167–173, 174, 178, 210
shooting method, 93, 96, 112
sidebands, 342
silicon-burning, 221
simple model with instantaneous recycling, 317
single degenerate scenario, 225
singularity, 15, 17–21, 236, 237
Small Magellanic Cloud, 3
solar standard model, 135
spiral galaxies, 2
Spite-plateau, 288, 289
stability, 88
standard candle, 12, 291, 297, 347, 348
star formation, 106–110, 257, 259, 316, 317, 319, 330
star formation history, 315, 320–327
star formation rate, 315
static universe, 16
statistical weight, 36
stellar atmosphere, 55, 92–93, 113, 136
stellar model, 55, 56, 99, 100, 176, 240, 241, 244, 246, 317
stellar pulsations, 179–186
stellar spectra, 241–256, 249, 276, 346
Strömgren filters, 249, 276, 282–283, 309, 310
strong interaction, 18
structure formation, 24, 25
subgiant branch, 142
super-wind, 195–198
surface brightness fluctuations, 348–349
symmetry breaking, 19
synthetic CMD, 319, 323, 326, 328

T

thermal runaway, 148
thermodynamical equilibrium, 17, 37, 53, 54, 67, 101
third dredge-up, 191–193, 195

tip of the red giant branch, 148, 152–155, 296–300, 327–329
tunnelling, 73, 74
turbulence, 57
turn off, 124, 126, 129, 143, 260, 261, 262, 263, 264, 266, 268–277, 284, 304–305, 338, 345
type I 1/2 supernova, 199
type Ia supernova, 12, 25, 26, 198, 224–233, 348
type II supernova, 187, 215, 222–224, 240, 317

U
unresolved binaries, 268, 323
upper main sequence, 128–133
URCA process, 82

V
Vega, 243, 245, 246, 250
vertical method, 271–274, 305–306

Virgo cluster, 3
virial theorem, 86–89, 91, 105, 107–109, 114, 116, 140, 203, 205, 212

W
weak interaction, 19, 81
white dwarf cooling, 205–210, 212–214
white dwarf crystallization, 206–210
white dwarf envelope, 210–212
white dwarf fitting, 294–296, 310
white dwarf isochrones, 277–279
white dwarf luminosity function, 279

Y
yield, 317

Z
zero age horizontal branch, 47, 154, 163–167
zero age main sequence, 47, 123–124, 155–157
zero point photometric system, 245

Lightning Source UK Ltd.
Milton Keynes UK
UKOW07f2023260416

272990UK00004B/163/P